# NON-CLASSICAL VIBRATIONS OF ARCHES AND BEAMS

# NON-CLASSICAL VIBRATIONS OF ARCHES AND BEAMS

## Eigenvalues and Eigenfunctions


**Igor A. Karnovsky,**

*Professor Emeritus*

**Olga I. Lebed,**

*British Columbia Institute of Technology,*
*Condor Rebar Consultants, Canada*



**McGraw-Hill**

New York  Chicago  San Francisco  Lisbon  London
Madrid  Mexico City  Milan  New Dehli  San Juan
Seoul  Singapore  Sydney  Toronto

**Library of Congress Cataloging-in-Publication Data**

Karnovsky, I. A. (Igor Alekseevich)
    Non-classical vibrations of arches and beams : eigenvalues and eigenfuctions /Igor A.
    Karnovsky, Olga I. Lebed
        p. cm.
    Includes index.
    ISBN 0-07-143188-8
    1. Girders–Vibration.  I. Lebed, Olga I.  II. Title

TA660.B4K33 2004
624.1'77723—dc22                                                    2003061570



*The sponsoring editor for this book was Larry Hager and the production supervisor was Sherri Souffrance. It was set in Times Roman by Techset Composition Limited. The art director for the cover was Margaret Webster-Shapiro.*

*Printed and bound by R. R. Donnelley & Sons Company.*

McGraw-Hill books are available at special quantity discounts to use as premiums and sales promotions, or for use in corporate training programs. For more information, please write to the Director of Special Sales, McGraw-Hill, Two Penn Plaza, New York, NY 10121-2298. Or contact your local bookstore.

 This book is printed on recycled, acid-free paper containing a minimum of 50% recycled, de-inked fiber.

# Contents

## Chapter 4.  Bress–Timoshenko Uniform Prismatic Beams                57



## Chapter 5.  Non-Uniform One-Span Beams                              83



## Chapter 6.  Optimal Designed Beams                                  125



## Chapter 7.  Nonlinear Transverse Vibrations                         139



## Chapter 8.  Arches                                                  165

# Appendices                                                                     201



# Index                                                                          237

# Preface

Deformable systems (DS) with distributed parameters such as beams and arches, are widely used in modern engineering. These structures find wide applications in civil and transport engineering, in mechanical, robotics and radio-engineering (load-bearing members, electric drives for robotics and mechanisms, boards of a radio-electronic apparatus), etc. The common fundamental feature of these structures is that their dynamical behavior is described by partial differential equations.

Analysis of any vibration problem of DS starts from definition of its fundamental characteristics, such as eigenvalues and eigenfunctions. For their calculation, there exist modern analytical and numerical methods, which are applicable for different mathematical models of DS. These models describe specific effects and operating conditions.

With the development of high technologies, the purpose of DS and their functional peculiarities as part of an engineering system in whole is changed. Also, the operating conditions of the structures are changed. For example, elastic elements become objects of active control; elastic beam elements are used as mechanical filters in electronics; elastic DS are used in control and measurement systems, which include elements of different natures, such as electrical, acoustical, optical, magnetic; beam systems are widely used as resonant strain gauges in micro-mechanical systems for the measurement of forces, accelerations, displacements and pressure. There also exist multi-purpose mechanical systems where elastic elements act simultaneously as load-bearing elements and functional devices, for a special purpose. Of course, the list of engineering fields where elastic DS are used is much wider than presented above.

The higher demands to a dynamical structure in whole leads to increasing demands of each part of a structure and in particular, to its DS. It means that the requirements to the accuracy of calculation of eigenvalues and eigenfunctions of DS are increased. However, on the basis of the simplified mathematical models it is impossible to obtain refined fundamental characteristics and take into account the specific effects.

This handbook presents solutions of eigenvalues and eigenfunctions for the advanced analysis of beams and arches. Analysis of beams on the basis of mathematical models, which take into account different additional effects, such as the effects of rotary inertia and shear force, are considered. Analysis of DS with specific conditions of a structure in service, such as elastic foundation, axial tensile or compressive load, is shown. Also, the handbook presents different types of nonlinear vibration problems of the beams, some results dealing with vibration of optimal designed beams. Special attention is paid to eigenvalue and eigenfunction problems for arches with different boundary conditions and with non-uniform cross-section.

During the last thirty years, a vast amount of information dealing with eigenvalues and eigenfunctions of DS has been accumulated. However, this information is spread out over numerous articles that are published in journals, conference proceedings, guidelines, departmental reports and theses. Existing handbooks do not reflect, in a reasonable manner, this important problem. For practicing engineers and researchers at universities and institutions, searching the vast literature, even with ready access to computerized databases and the Internet for a specific type of problem, this is a difficult and time consuming task. Solutions of many important problems remain unknown to specialists, who could greatly benefit from such knowledge.

The objective of this handbook is to provide the most comprehensive, up-to-date reference of known solutions to a large variety of vibration problems of beams and arches. The intent is to provide information that is not available in current handbooks and to provide solutions for the eigenvalues and eigenfunctions problems that engineers and researchers use for the advanced analysis of dynamical behavior of beams and arches.

The most distinctive feature of this handbook is that it is the most complete collection of eigenvalues and eigenfunctions for different types of beams and arches that has ever been published. It includes a large number of cases of beams and arches with concentrated and distributed parameters with different types of elastic supports and boundary conditions. All problems in this book may be considered as non-classical problems. Authors understand that division of vibration of deformable structures as classical and non-classical is conventional. However, this division is convenient in order to consider non-classical problems separately, i.e., such problems, which take into account additional important effects (shear, rotary inertia, etc), consider nonlinear problems by analytical methods (static, physical, geometrical non-linearity), problems of optimal design, as well as some special problems, for example vibration of beams in magnetic field. Of course, this list of non-classical problems is not complete. The authors have conducted a very extensive research of published materials in many countries and compiled solutions to different cases of vibration of deformable systems. The criteria for the selection of problems included in the handbook were mainly based on the importance and the frequency of appearance of the problem in practical engineering applications. Problems selection is based on the 35 years' combined experience of the authors in the field of structural dynamics.

To compile the information presented in this handbook, the authors carefully reviewed monographs, journals, handbooks, proceedings, preprints and theses, as well as results of the authors' own research. The handbook contains the fundamental and most up-to-date results concerning eigenvalues and eigenfunctions of beams and arches. The majority of the sources consulted have been published in the USA, Canada, England, Russia, Germany, Japan, Israel, and Netherlands over the past 40 years. Each case presented in the handbook is properly referenced. The majority of the results, which are presented in the original sources, have been independently verified by the authors.

Authors will appreciate comments and suggestions to improve the current edition. All constructive criticism will be accepted with gratitude.

# Acknowledgments

We would like to thank professors Patricio A. A. Laura and Diana V. Bambill (Argentina), whose support and constructive criticism were very important for us. We also would like to thank our friend Lev Bulkovshteyn (Canada) for discussions of many topics related to this book. Our thanks go to professors Isaak Chaikovsky (Israel) and Akbar Tamboli (USA) for their contributions to many aspects of the book.

Thanks a lot to Larry Hager of the McGraw-Hill Publishing company for his help and guidance in making this book a reality. We also would like to thank all people from McGraw-Hill, who worked with us on this book.

# Notation

| | |
|---|---|
| $x, y, z$ | Cartesian coordinates |
| $x$ | Spatial coordinate |
| $t$ | Time |
| $E, G$ | Young's and shear modulus of the beam material, $2G(1 + v) = E$ |
| $v, \rho$ | Poisson's ratio and density of the beam material |
| $\sigma, \varepsilon$ | Stress and strain of material of a beam |
| $\sigma_{xx}, \sigma_{xy}$ | Longitudinal and shear stresses |
| $u_x, u_y$ | Longitudinal and transversal displacements of a rod |
| $c_t, c_b$ | Velocity of shear and longitudinal waves, $c_t^2 \rho = G$; $c_b^2 \rho = E$ |
| $k_b k_t$ | Longitudinal and shear propagation constant, $k_b = \omega/c_b$, $k_t = \omega/c_t$ |
| $k_0$ | Bending wave number for Bernoulli-Euler rod, $k_0^4 D_0^4 = \omega^2$ |
| $l$ | Length of the beam |
| $h, d, b$ | Geometrical dimensional of cross sectional of the beam |
| $I_n$ | Cross sectional area moment of inertia of order $n$ |
| $I, I_2$ | Second moment of inertia of a cross-section area with respect to neutral line |
| $m$ | Mass per unit length of the beam, $m = \rho A$ |
| $M, J$ | Lumped mass and moment inertia of the mass |
| $k_{tr}, k_{rot}$ | Translational and rotational stiffness coefficients |
| $k_{tr}^*, k_{rot}^*$ | Dimensionless stiffness coefficients, $k_{tr}^* EI = k_{tr} l^3$, $k_{rot}^* EI = k_{rot} l$ |
| $A(x)$ | Cross-sectional area of a beam |
| $EI, i$ | Bending stiffness and bending stiffness per unit length, $i = EI/l$ |
| $y(x, t), \psi(x, t)$ | Lateral displacement and slope of the beam |
| $E_F, \rho_F$ | Young's modulus and density of the foundation material |
| $G_0$ | Foundation modulus of rigidity (Pasternak model) |
| $k_{tr}, k_0$ | Elastic transverse translatory stiffness of medium (Winkler foundation modulus) |
| $k_{\text{slope}}, D_0$ | Elastic sloping stiffness of medium |
| $k_{\text{tilt}}$ | Elastic tilting (transverse rotating) stiffness of medium, Nm/m |
| $T$ | Axial load |
| $U$ | Dimensionless parameter, $2UEI = Tl^2$ |
| $T_E$ | First Euler critical load |
| $M, Q$ | Bending moment and shear force |
| $U, T$ | Potential and kinetic energy |

| | |
|---|---|
| $X(x), \psi(x)$ | Mode shapes |
| $k$ | Shear factor |
| $\lambda, k$ | Frequency parameters, $\lambda^4 EI = ml^4\omega^2, k^4 EI = m\omega^2, \lambda = kl$ |
| $\omega_{0i}$ | Frequency of transverse vibration of a beam no axial force in $i$th mode vibration |
| $\Omega_{0i}$ | Dimensionless frequency parameter of beam no axial force in $i$th mode vibration; $\Omega_{0i}\alpha = \omega_{0i}l^2, \alpha^2 = EI, \Omega_{0i} = \lambda^2$ |
| $\omega$ | Frequency of transverse vibration compressed beam (relative natural frequency); |
| $\Omega$ | Dimensionless frequency parameter of compressed beam (relative natural frequency); $\Omega\alpha = \omega l^2$ |
| $\Omega^* = \Omega/\Omega_{0i}$ | Normalized natural frequency parameter |
| $H$ | Hamiltonian |
| $k_1, k_2$ | Lagrange multipliers |
| $V$ | Volume of a beam |
| $V_-, V_+$ | Lower and upper limit of the volume of a beam |
| $\omega^-, \omega^+$ | Lower and upper bounds of the frequency vibration |
| $G$ | Gauge factor |
| $v, v_{cr}$ | Velocity and critical velocity of the moving liquid |
| $\xi, r$ | Dimensionless coordinate and radius of gyration, $\xi = x/l, \ 0 \le \xi \le 1, \ r^2 Al^2 = I$ |
| $s, \zeta$ | Dimensionless parameters, $s^2 kAGl^2 = EI$ |
| $R$ | Correction factor |
| $k$ | Parameter of magnetic field |
| $\beta, \beta_F$ | Nonlinear parameter of the beam and foundation |
| $P$ | Internal pressure |
| $s$ | Stiffness coefficients |
| $s(M\psi)$ | Flexural stiffness coefficient of moment due to rotational deformation |
| $s(MX)$ | Flexural stiffness coefficient of moment due to transverse deformation |
| $s(V\psi)$ | Flexural stiffness coefficient of shear due to rotational deformation |
| $s(VX)$ | Flexural stiffness coefficient of shear due to transverse deformation |
| $V_i$ | Puzyrevsky functions |
| $w, v$ | Radial and tangential displacements of an arch |
| $R(\alpha), f, l$ | Radius of curvature, rise and span of an arch |
| $(\bullet), (')$ | Differentiation with respect to time and space coordinate |

# NON-CLASSICAL VIBRATIONS OF ARCHES AND BEAMS

# CHAPTER 1

# TRANSVERSE VIBRATION EQUATIONS

The different assumptions and corresponding theories of transverse vibrations of beams are presented. The dispersive equation, its corresponding curve 'propagation constant–frequency' and its comparison with the exact dispersive curve are presented for each theory and discussed.

The exact dispersive curve corresponds to the first and second antisymmetrical Lamb's wave.

## 1.1  AVERAGE VALUES AND RESOLVING EQUATIONS

The different theories of dynamic behaviours of beams may be obtained from the equations of the theory of elasticity, which are presented with respect to average values. The object under study is a thin plate with rectangular cross-section (Figure 1.1).

### 1.1.1  Average values for deflections and internal forces

**1.** Average displacement and slope are

$$w = \int\limits_{-H}^{+H} \frac{u_y}{2H}\,\mathrm{d}y \tag{1.1}$$

$$\psi = \int\limits_{-H}^{+H} \frac{y u_x}{I_z}\,\mathrm{d}y \tag{1.2}$$

where $u_x$ and $u_y$ are longitudinal and transverse displacements.

1

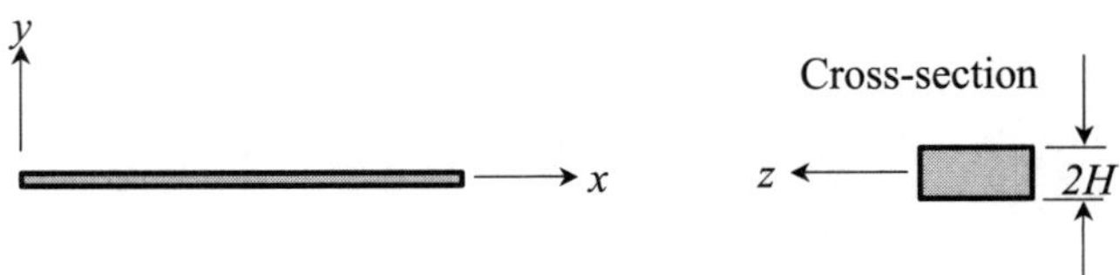

**FIGURE 1.1.**   Thin rectangular plate, the boundary conditions are not shown.

**2.** Shear force and bending moment are

$$Q = \int\limits_{-H}^{+H} \sigma_{xy}\, dy \tag{1.3}$$

$$M = \int\limits_{-H}^{+H} y\sigma_{xx}\, dy \tag{1.4}$$

where $\sigma_x$ and $\sigma_y$ are the normal and shear stresses that correspond to $u_x$ and $u_y$.

Resolving the equations may be presented in terms of average values as follows (Landau and Lifshitz, 1986)

**1.** Integrating the equilibrium equation of elasticity theory leads to

$$2\rho H\ddot{w} = Q' \tag{1.5}$$

$$\rho I_z\ddot{\psi} = M_z' - Q \tag{1.6}$$

**2.** Integrating Hooke's equation for the plane stress leads to

$$Q = 2HG\left[w' + \frac{u_x(H)}{H}\right] \tag{1.7}$$

$$M_z = E_1\{I_z\psi' + 2Hv[u_y(H) - w]\} = EI_z\psi' + v\int\limits_{-H}^{+H} y\sigma_{yy}\, dy \tag{1.7a}$$

Equations (1.5)–(1.7a) are complete systems of equations of the theory of elasticity with respect to average values $w$, $\psi$, $Q$ and $M$. These equations contain two redundant unknowns $u_x(H)$ and $u_y(H)$. Thus, to resolve the above system of equations, additional equations are required. These additional equations may be obtained from the assumptions accepted in approximate theories.

The solution of the governing differential equation is

$$w = \exp(ikx - i\omega t) \tag{1.8}$$

where $k$ is a propagation constant of the wave and $\omega$ is the frequency of vibration.

The degree of accuracy of the theory may be evaluated by a dispersive curve $k - \omega$ and its comparison with the exact dispersive curve. We assume that the exact dispersive curve is one that corresponds to the first and second antisymmetric Lamb's wave. The closer the dispersive curve for a specific theory to the exact dispersive curve, the better the theory describes the vibration process. Short analysis of equations for transversal vibrations on the basis of different theories are shown below. (Artobolevsky *et al.* 1979).

## 1.2  FUNDAMENTAL THEORIES AND APPROACHES

### 1.2.1  Bernoulli–Euler theory

The Bernoulli–Euler theory takes into account the inertia forces due to the transverse translation and neglects the effect of shear deflection and rotary inertia.

*Assumptions*

1. The cross-sections remain plane and orthogonal to the neutral axis ($\psi = -w'$).
2. The longitudinal fibres do not compress each other ($\sigma_{yy} = 0, \rightarrow M_z = EI_z\psi'$).
3. The rotational inertia is neglected ($\rho I_z\ddot{\psi} = 0$). This assumption leads to

$$Q = -M_z' = -EI_z w'''$$

Substitution of the previous expression in Equation (1.5) leads to the differential equation describing the transverse vibration of the beam

$$\frac{\partial^4 w}{\partial x^4} + \frac{1}{D_0^4}\frac{\partial^2 w}{\partial t^2} = 0, \quad D_0^4 = \frac{EI_z}{2\rho H} \tag{1.9}$$

Let us assume that displacement $w$ is changed according to Equation (1.8). The dispersive equation which establishes the relationship between $k$ and $\omega$ may be presented as

$$k^4 = \frac{\omega^2}{D_0^4} = k_0^4$$

This equation has two roots for a forward-moving wave in a beam and two roots for a backward-moving wave. Positive roots correspond to a forward-moving wave, while negative roots correspond to a backward-moving wave.

The results of the dispersive relationships are shown in Figure 1.2; dimensionless parameters are $\lambda = kH$, $\mu_t = k_t H$. Here, bold curves 1 and 2 represent the exact results. Curves 1 and 2 correspond to the first and second antisymmetric Lamb's wave, respectively. The second wave transfers from the imaginary zone into the real one at $k_t H = \pi/2$. Curves 3 and 4 are in accordance with the Bernoulli–Euler theory. Dispersion obtained from this theory and dispersion obtained from the exact theory give a close result when frequencies are close to zero. This elementary beam theory is valid only when the height of the beam is small compared with its length (Artobolevsky *et al.*, 1979).

### 1.2.2  Rayleigh theory

This theory takes into account the effect of rotary inertia (Rayleigh, 1877).

*Assumptions*

1. The cross-sections remain plane and orthogonal to the neutral axis ($\psi = -w'$).
2. The longitudinal fibres do not compress each other ($\sigma_{yy} = 0$, $M_z = EI_z\psi'$).

From Equation (1.6) the shear force $Q = M_z' - \rho I_z\ddot{\psi}$.

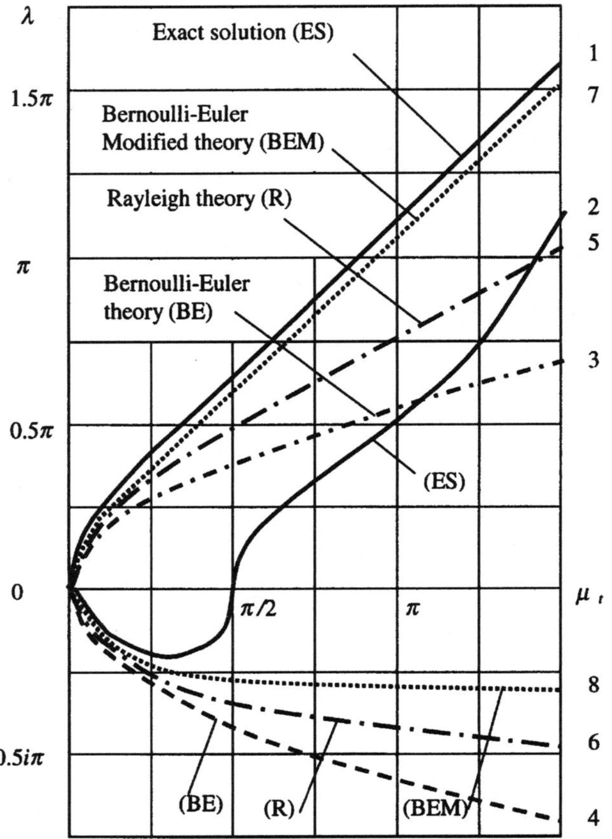



**FIGURE 1.2.**    Transverse vibration of beams. Dispersive curves for different theories. 1, 2–Exact solution; 3, 4–Bernoulli–Euler theory; 5, 6–Rayleigh theory, 7, 8–Bernoulli–Euler modified theory.

### *Differential equation of transverse vibration of the beam*

$$\frac{\partial^4 w}{\partial x^4} + \frac{1}{D_0^4}\frac{\partial^2 w}{\partial t^2} - \frac{1}{c_b^2}\frac{\partial^4 w}{\partial x^2 \partial t^2} = 0, \quad c_b^2 = \frac{E}{\rho} \tag{1.10}$$

where $c_b$ is the velocity of longitudinal waves in the thin rod.

The last term on the left-hand side of the differential equation describes the effect of the rotary inertia.

The dispersive equation may be presented as follows

$$2k_{1,2}^2 = k_b^2 \pm \sqrt{k_b^4 + 4k_0^4}, \quad k_b^2 = \frac{\omega^2}{c_b^2}$$

where $k_0$ is the wave number for the Bernoulli–Euler rod, and $k_b$ is the longitudinal wave number.

Curves 5 and 6 in Figure 1 reflect the effect of rotary inertia.

### 1.2.3 Bernoulli–Euler modified theory

This theory takes into account the effect of shear deformation; rotational inertia is negligible (Bernoulli, 1735, Euler, 1744). In this case, the cross-sections remain plane, but not orthogonal to the neutral axis, and the differential equation of the transverse vibration is

$$\frac{\partial^4 y}{\partial x^4} + \frac{1}{D_0^4}\frac{\partial^2 y}{\partial t^2} - \frac{1}{c_t^2}\frac{\partial^4 y}{\partial x^2 \partial t^2} = 0, \quad c_t^2 = \frac{G}{\rho} \tag{1.11}$$

where $c_t$ is the velocity of shear waves in the thin rod.

The dispersive equation may be presented as follows

$$2k_{1,2}^2 = k_t^2 \pm \sqrt{k_t^2 + 4k_0^4}, \quad k_t^2 = \frac{\omega^2}{c_t^2}$$

Curves 7 and 8 in Figure 1.2 reflect the effect of shear deformation.

The Bernoulli–Euler theory gives good results only for low frequencies; this dispersive curve for the Bernoulli–Euler *modified* theory is closer to the dispersive curve for exact theory than the dispersive curve for the Bernoulli–Euler theory; the Rayleigh theory gives a worse result than the modified Bernoulli–Euler theory.

### 1.2.4 Bress theory

This theory takes into account the rotational inertia, shear deformation and their combined effect (Bress, 1859).

*Assumptions*

**1.** The cross-sections remain plane.

**2.** The longitudinal fibres do not compress each other ($\sigma_{yy} = 0$).

*Differential equation of transverse vibration*

$$\frac{\partial^4 w}{\partial x^4} + \frac{1}{D_0^4}\frac{\partial^2 w}{\partial t^2} - \left(\frac{1}{c_b^2} + \frac{1}{c_t^2}\right)\frac{\partial^4 w}{\partial x^2 \partial t^2} + \frac{1}{c_b^2 c_t^2}\frac{\partial^4 w}{\partial t^4} = 0 \tag{1.12}$$

In this equation, the third and fourth terms reflect the rotational inertia and the shear deformation, respectively. The last term describes their combined effect; this term leads to the occurrence of a cut-off frequency of the model, which is a recently discovered fundamental property of the system.

### 1.2.5 Volterra theory

This theory, as with the Bress theory, takes into account the rotational inertia, shear deformation and their combined effect (Volterra, 1955).

*Assumption*

All displacements are linear functions of the transverse coordinates

$$u_x(x, y, t) = y\psi(x, t), \quad u_y(x, y, t) = w(x, t)$$

In this case the bending moment and shear force are

$$M_z = E_1 I_z \psi, \quad Q = 2HG(w' + \psi)$$

*Differential equation of transverse vibration*

$$\frac{\partial^4 w}{\partial x^4} + \frac{1 - v^2}{D_0^4}\frac{\partial^2 w}{\partial t^2} - \left(\frac{1}{c_s^2} + \frac{1}{c_t^2}\right)\frac{\partial^4 w}{\partial x^2 \partial t^2} + \frac{1}{c_s^2 c_t^2}\frac{\partial^4 w}{\partial t^4} = 0 \qquad (1.13)$$

where $c_s$ is the velocity of a longitudinal wave in the thin plate, $c_s^2 = (E_1/\rho)$, and $E_1$ is the longitudinal modulus of elasticity, $E_1 = E/(1 - v^2)$.

***Difference between Volterra and Bress theories.***   As is obvious from Equations (1.12) and (1.13), the bending stiffness of the beam according to the Volterra model is $(1 - v^2)^{-1}$ times greater than that given by the Bress theory (real rod). This is because transverse compressive and tensile stresses are not allowed in the Volterra model.


## 1.2.6   Ambartsumyan theory

The Ambartsumyan theory allows the distortion of the cross-section of the beam (Ambartsumyan, 1956).

*Assumptions*

**1.** The transverse displacements for all points in the cross-section are equal: $\partial u_y/\partial y = 0$.

**2.** The shear stress is distributed according to function $f(y)$:

$$\sigma_{xy}(x, y, t) = G\varphi(x, t)f(y)$$

In this case, longitudinal and transverse displacements may be given as

$$u_x(x, y, t) = -y\frac{\partial w(x, t)}{\partial x} + \varphi(x, t)g(y)$$

$$u_y(x, y, t) = w(x, t), \quad g(y) = \int_0^y f(\xi)d\xi$$

*Differential equation of transverse vibration*

$$\frac{\partial^4 w}{\partial x^4} + \frac{1 - v^2}{D_0^4}\frac{\partial^2 w}{\partial t^2} - \left(\frac{1}{c_s^2} + \frac{1}{ac_t^2}\right)\frac{\partial^4 w}{\partial x^2 \partial t^2} + \frac{1}{ac_s^2 c_t^2}\frac{\partial^4 w}{\partial t^4} = 0 \qquad (1.14)$$

where

$$a = \frac{I_z I_1}{2HI_0}, \quad I_1 = \int_{-H}^{H} f(\xi)d\xi, \quad I_0 = \int_{-H}^{H} yg(y)dy$$

***Difference between Ambartsumyan and Volterra theories.*** The Ambartsumyan's differential equation differs from the Volterra equation by coefficient $a$ at $c_t^2$. This coefficient depends on $f(y)$.

### Special cases

1. Ambartsumyan and Volterra differential equations coincide if $f(y) = 0.5$.
2. If shear stresses are distributed by the law $f(y) = 0.5(H^2 - y^2)$ then $a = 5/6$.
3. If shear stresses are distributed by the law $f(y) = 0.5(H^{2n} - y^{2n})$, then $a = (2n + 3)/(2n + 4)$.

## 1.2.7 Vlasov theory

The cross-sections have a distortion, but after deformation the cross-sections remain perpendicular to the surfaces $y = \pm H$ (Vlasov, 1957).

### Assumptions

1. The longitudinal and transversal displacements are

$$u_x(x, y, t) = -\varepsilon y - \frac{y^2}{2H^2 G} \sigma_{xy}^0$$

$$u_y(x, y, t) = w(x, y, t)$$

where $\varepsilon = -\left(\partial u_x / \partial y\right)_{y=0}$ and $\sigma_{xy}^0$ is the shear stress at $y = 0$.

This assumption means that the change in shear stress by the quadratic law is

$$\sigma_{xy} = \sigma_{xy}^0 \left(1 - \frac{y^2}{H^2}\right)$$

2. The cross-sections are curved but, after deflection, they remain perpendicular to the surfaces at

$$y = H \quad \text{and} \quad y = -H$$

This assumption corresponds to expression

$$\left(\frac{\partial u_x}{\partial y}\right)_{y=\pm H} = 0$$

These assumptions of the Vlasov and Ambartsumyan differential equations coincide at parameter $a = 5/6$. Coefficient $a$ is the improved dispersion properties on the higher bending frequencies.

## 1.2.8 Reissner, Goldenveizer and Ambartsumyan approaches

These approaches allow a transverse deformation, so differential equations may be developed from the Bress equation if additional coefficient $a$ is put before $c_t^2$.

***Assumptions***

**1.** $\sigma_{yy} = 0$.

**2.** $\sigma_{xy}(x, y, t) = G\varphi(x, t)f(y)$.

These assumptions lead to the Bress equation (1.12) with coefficient $a$ instead of $c_t^2$. The structure of this equation coincides with the Timoshenko equation (Reissner, 1945; Goldenveizer, 1961; Ambartsumyan, 1956).

### 1.2.9  Timoshenko theory

The Timoshenko theory takes into account the rotational inertia, shear deformation and their combined effects (Timoshenko, 1921, 1922, 1953).

***Assumptions***

**1.** Normal stresses $\sigma_{yy} = 0$; this assumption leads to the expression for the bending moment

$$M_z = EI_z \frac{\partial \psi}{\partial x}$$

**2.** The ratio $u_x(H)/H$ substitutes for angle $\psi$; this means that the cross-sections remain plane. This assumption leads to the expression for shear force

$$Q = 2qHG\left(\frac{\partial w}{\partial x} + \psi\right)$$

**3.** The fundamental assumption for the Timoshenko theory: arbitrary shear coefficient $q$ enters into the equation.

***Mechanical presentation of the Timoshenko beam.***  A beam can be substituted by the set of rigid non-deformable plates that are connected to each other by elastic massless pads.

***The complete set of the basic relationships***

$$2\rho H\ddot{w} = Q'$$

$$\rho I_z \ddot{\psi} = M_z' - Q$$

$$M_z = EI_z \frac{\partial \psi}{\partial x}$$

$$Q = 2qHG\left(\frac{\partial w}{\partial x} + \psi\right)$$

To obtain the differential equation of vibration eliminate from the basic relationships all variables except displacement.

***Timoshenko differential equation of the transverse vibration of a beam***

$$\frac{\partial^4 w}{\partial x^4} + \frac{1}{D_0^4}\frac{\partial^2 w}{\partial t^2} - \left(\frac{1}{c_b^2} + \frac{1}{qc_t^2}\right)\frac{\partial^4 w}{\partial x^2 \partial t^2} + \frac{1}{qc_b^2 c_t^2}\frac{\partial^4 w}{\partial t^4} = 0 \tag{1.15}$$

The fundamental difference between the Rayleigh and Bress theories, on one hand, and the Timoshenko theory, on the other, is that the correction factor in the Rayleigh and Bress theories appears as a result of shear and rotary effects, whereas in the Timoshenko theory, the correction factor is introduced in the initial equations. The arbitrary coefficient $q$ is the fundamental assumption in the Timoshenko theory.

Presenting the displacement in the form (1.8) leads to the dispersive equation

$$2k_{1,2}^2 = k_b^2 + \frac{k_t^2}{q} \pm \sqrt{\left(k_b^2 - \frac{k_t^2}{q}\right)^2 + 4k_0^4}$$

where $k_{1,2}$ are propagation constants;

$q$ is the shear coefficient;

$k_0$ is the wave number of the bending wave in the Bernoulli–Euler rod,

$$k_0^4 = \frac{\omega^2}{D_0^4}, \quad D_0^4 = \frac{EI_z}{2\rho H}$$

$k_b$ and $k_t$ are the longitudinal and shear propagation constants, respectively.

$$k_b^2 = \frac{\omega^2}{c_b^2}, \quad k_t^2 = \frac{\omega^2}{c_t^2}$$

$c_b$ and $c_t$ are the velocities of the longitudinal and shear waves

$$c_b^2 = \frac{E}{\rho}, \quad c_t^2 = \frac{G}{\rho}$$

***Practical advantages of the Timoshenko model.*** Figure 1.3 shows a good agreement between dispersive curves for both the Timoshenko model and the exact curve for high frequencies. This means that the two-wave Timoshenko model describes the vibration of short beams, or high modes of a thin beam, with high precision.

This type of problem is an important factor in choosing the shear coefficient (Mindlin, 1951; Mindlin and Deresiewicz, 1955).

Figure 1.3 shows the exact curves, 1 and 2, and the dispersive curves for different shear coefficients: curves 3 and 4 correspond to $q = 1$ (Bress theory), curves 5 and 6 to $q = \pi^2/12$, curves 7 and 8 to $q = 2/3$, curves 9 and 10 to $q = 1/2$.

### 1.2.10  Love theory

The equation of the Love (1927) theory may be obtained from the Timoshenko equation as a special case.

(a) Truncated Love equation

$$\frac{\partial^4 w}{\partial x^4} + \frac{1}{D_0^4}\frac{\partial^2 w}{\partial t^2} - \frac{1}{qc_t^2}\frac{\partial^4 w}{\partial x^2 \partial t^2} = 0, \quad c_t^2 = \frac{G}{\rho} \tag{1.16}$$

(b) Complete Love equation

$$\frac{\partial^4 w}{\partial x^4} + \frac{1}{D_0^4}\frac{\partial^2 w}{\partial t^2} - \left(\frac{1}{c_b^2} + \frac{1}{qc_t^2}\right)\frac{\partial^4 w}{\partial x^2 \partial t^2} = 0 \tag{1.17}$$

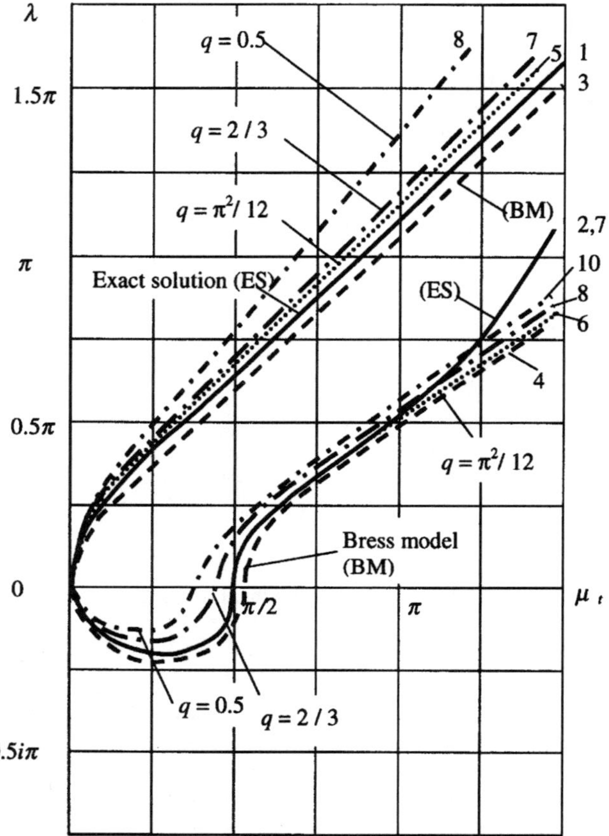


**FIGURE 1.3.**   Dispersive curves for the Timoshenko beam model. 1, 2–exact solution; 3, 4–Bress model; 5, 6–$q = \pi^2/12$; 7, 8–$q = 2/3$; 9, 10–$q = 1/2$.

## 1.2.11   Timoshenko modified theory

*Assumption*

More arbitrary coefficients are entered into the basic equations.

The bending moment and shear in the most general case are

$$M_z = pEI_z \frac{\partial \psi}{\partial x}, \quad Q = 2HG\left(q\frac{\partial w}{\partial x} + s\psi\right)$$

where $p$, $q$, and $s$ are arbitrary coefficients.

*Differential equation of transverse vibration*

$$\frac{\partial^4 w}{\partial x^4} + \left(\frac{s}{pq}\right)\frac{1}{D_0^4}\frac{\partial^2 w}{\partial t^2} - \left(\frac{1}{pc_b^2} + \frac{1}{qc_t^2}\right)\frac{\partial^4 w}{\partial x^2 \partial t^2} + \frac{1}{pqc_b^2 c_t^2}\frac{\partial^4 w}{\partial t^4} = 0 \qquad (1.18)$$

The dispersive equation may be written in the form

$$2k_{1,2}^2 = \frac{k_b^2}{p} + \frac{k_t^2}{q} \pm \sqrt{\left(\frac{k_b^2}{p} - \frac{k_t^2}{q}\right)^2 + 4k_0^4 \cdot \frac{s}{pq}}$$

The dispersive properties of the beam (and the corresponding dispersive curve) is sensitive to the change of parameters $p$, $q$ and $s$. Two additional relationships between parameters $p$, $q$, $s$, such that

$$s = pq \quad \text{and} \quad k_t^2 \frac{I_z}{A} = pq$$

define a differential equation with one optimal correct multiplier. The meaning of the above-mentioned relationships was discussed by Artobolevsky *et al.* (1979). The special case $p = q$ was studied by Aalami and Atzori (1974).

Figure 1.4 presents the exact curves 1 and 2 and dispersive curves for different values of coefficient $p$: $p = 0.62$, $p = 0.72$, $p = \pi^2/12$, $p = 0.94$ and $p = 1$ (Timoshenko model). The best approximation is $p = \pi^2/12$ for $k_t H$ in the interval from 0 to $\pi$.

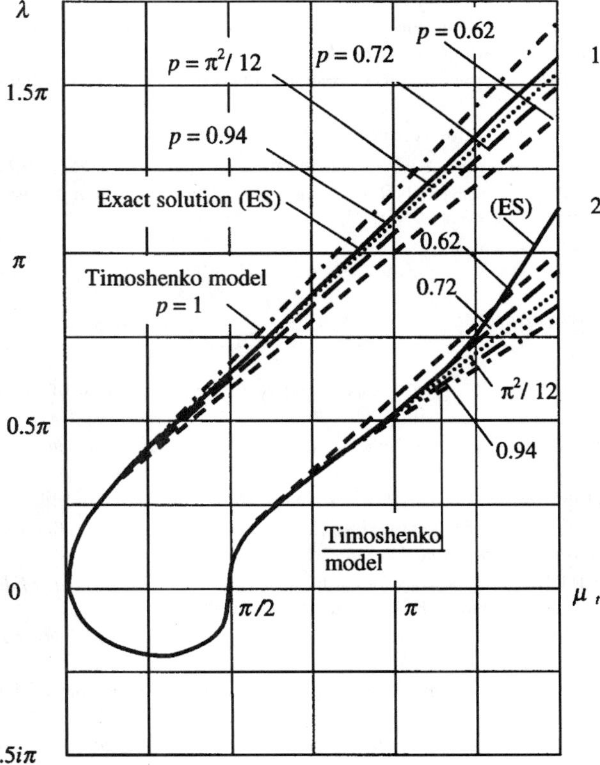


**FIGURE 1.4.** Dispersive curves for modified Timoshenko model. 1, 2–exact solution.

## REFERENCES


Aalami, B. and Atzori, B. (1974) Flexural vibrations and Timoshenko's beam theory, *Am. Inst. Aeronautics Astronautics J.*, **12**, 679–675.

Ambartsumyan, S.A. (1956) On the calculation of Shallow Shells, NACA TN 425, December 1956.

Artobolevsky, I.I., Bobrovnitsky, Yu.I. and Genkin, M.D. (1979) *An Introduction to Acoustical Dynamics of Machine* (Moscow: Nauka), (in Russian).

Bresse, M. (1859) *Cours de Mechanique Appliquee* (Paris: Mallet-Bachelier).

Bernoulli, D. (1735) Letters to Euler, Basel.

Euler, L. (1744) Methodus Inveniendi Lineas Curvas Maximi Minimive Proprietate Gaudenies, Berlin.

Goldenveizer, A.L. (1961) *Theory of Elastic Thin Shells* (New York: Pergamon Press).

Landau, L.D. and Lifshitz, E.M. (1986) *Theory of Elasticity* (Oxford: New York: Pergamon Press).

Love, E.A.H. (1927) *A Treatise on the Mathematical Theory of Elasticity* (New York: Dover).

Mindlin, R.D. (1951) Influence of rotary inertia and shear on flexural motion of isotopic elastic plates, *J. Appl. Mech. (Trans. ASME)*, **73**, 31–38.

Mindlin, R.D. and Deresiewicz, H. (1955) Timoshenko's shear coefficient for flexural vibrations of beams, *Proc. 2nd U.S. Nat. Cong. Applied Mechanics*, New York.

Rayleigh, J.W.S. (1877) *The Theory of Sound* (London: Macmillan) vol. 1, 326 pp.; vol. 2, 1878, 302 pp. 2nd edn (New York: Dover) 1945, vol. 1, 504 pp.

Reissner, E. (1945) The effect of transverse shear deformation on the bending of elastic plates, *J. Appl. Mech.* **12**.

Timoshenko, S.P. (1921) On the correction for shear of the differential equation for transverse vibrations of prismatic bars. *Philosophical Magazine*, Series 6, **41**, 744–746.

Timoshenko, S.P. (1922). On the transverse vibrations of bars of uniform cross sections. *Philosophical Magazine*, Series 6, **43**, 125–131.

Timoshenko, S.P. (1953) *Colected Papers* (New York: McGraw-Hill).

Vlasov, B.F. (1957) Equations of theory of bending plates. *Izvestiya AN USSR, OTN*, **12**.

Volterra, E. (1955) A one-dimensional theory of wave propagation in elastic rods based on the method of internal constraints. *Ingenieur-Archiv*, **23**, 6.


## FURTHER READING


Abbas, B.A.H. and Thomas, J. (1977) The secondary frequency spectrum of Timoshenko beams, *Journal of Sound and Vibration* **51**(1), 309–326.

Bickford, W.B. (1982) A consistent higher order beam theory. *Developments in Theoretical and Applied Mechanics*, **11**, 137–150.

Crawford, F.S. (1965) *Waves*. Berkeley Physics Course (New York: McGraw Hill).

Ewing, M.S. (1990) Another second order beam vibration theory: explicit bending warping flexibility and restraint. *Journal of Sound and Vibration*, **137**(1), 43–51.

Green, W.I. (1960) Dispersion relations for elastic waves in bars. In *Progress in Solid Mechanic*, Vol. 1, edited by I.N. Sneddon and R. Hill (Amsterdam: North-Holland).

Grigolyuk, E.I. and Selezov, I.T. (1973) *Nonclassical Vibration Theories of Rods, Plates and Shells*, Vol. 5. Mechanics of Solids Series (Moscow, VINITI).

Leung, A.Y. (1990) An improved third beam theory, *Journal of Sound and Vibration*, **142**(3) pp. 527–528.

Levinson, M. (1981) A new rectangular beam theory. *Journal of Sound and Vibration*, **74**, 81–87.

Pippard, A.B. (1989) *The Physics of Vibration* (Cambridge University Press).

Timoshenko, S.P. (1953) *History of Strength of Materials* (New York: McGraw Hill).

Todhunter, I. and Pearson, K. (1960) *A History of the Theory of Elasticity and of the Strength of Materials* (New York: Dover). Volume II. Saint-Venant to Lord Kelvin, part 1, 762 pp; part 2, 546 pp.

Wang, J.T.S. and Dickson, J.N. (1979) Elastic beams of various orders. *American Institute of Aeronautics and Astronautics Journal*, **17**, 535–537.

# BERNOULLI–EULER BEAMS ON ELASTIC LINEAR FOUNDATION

Chapter 2 describes the different mathematical models of an elastic foundation. A mechanical model of the Winkler model is discussed and natural frequencies of vibration of Bernoulli–Euler uniform and stepped one-span beams with different boundary conditions on the elastic foundation are presented.

## 2.1  MODELS OF FOUNDATION

The differential equation of the transverse vibration of a beam on an elastic foundation is

$$EI\frac{\partial^4 y}{\partial x^4} + N\frac{\partial^2 y}{\partial x^2} + \rho A\frac{\partial^2 y}{\partial t^2} + p(y, t) = 0 \qquad (2.1)$$

where $N$ is the axial force and $p(y, t)$ is the reaction of the foundation.

The models of the foundation describe the relation between the reaction of the foundation (or pressure) $p$, the deflection of the beam and the parameters of foundation.

### 2.1.1  Winkler foundation

The foundation may be presented as closely spaced independent linear springs. The foundation reaction equals $p = k_0 y$, where $y$ is the vertical deflection of the foundation surface (vertical deflection of the beam, plate), and $k_0$ is Winkler's foundation modulus (Winkler, 1867). Shear interactions between the foundation spring elements are neglected. This type of foundation is equivalent to a liquid base.

### 2.1.2 Viscoelastic Winkler foundation

The foundation reaction equals

$$p = k_0 y + d\frac{\partial y}{\partial t} \tag{2.2}$$

where second term takes into acount the viscoelastic properties of the Winkler foundation; $d$ is viscous damping coefficient of the foundation.

The governing equation is

$$EI\frac{\partial^4 y}{\partial x^4} + N\frac{\partial^2 y}{\partial x^2} + \rho A\frac{\partial^2 y}{\partial t^2} + k_0 y + d\frac{\partial y}{\partial t} = 0 \tag{2.3}$$

### 2.1.3 Hetenyi foundation

The relationship between load $p$ and deflection $y$ for the three-dimensional case is

$$p = k_0 y + D_0 \nabla^2 \nabla^2 y \tag{2.4}$$

where the parameter $D$ takes into acount the interaction of the spring elements (Hetenyi, 1946).

### 2.1.4 Viscoelastic Hetenyi foundation

The relation between lateral load $p$ and deflection $y$ for the 3-D case is presented as

$$p = k_0 y + d\frac{\partial y}{\partial t} + D_0 \nabla^2 \nabla^2 y \tag{2.5}$$

The governing equation is

$$(EI + D_0)\frac{\partial^4 y}{\partial x^4} + N\frac{\partial^2 y}{\partial x^2} + \rho A\frac{\partial^2 y}{\partial t^2} + k_0 y + d\frac{\partial y}{\partial t} = 0 \tag{2.6}$$

In this model, the overall bending stiffness beam ($EI$) has been increased by the 'bending stiffness' of the foundation (term $D_0$).

### 2.1.5 Pasternak foundation

The load–deflection relation is

$$p = k_0 y - G_0 \nabla^2 y \tag{2.7}$$

where the second term describes the effect of the shear interactions between the spring elements; $G_0$ is the shear foundation (Pasternak, 1954). For beam last term should be presented as

$$-G_0\frac{d^2 y}{dx^2}$$

Dynamic stability of an axially loaded beam on above mentioned elastic foundation with damping is presented by Engel (1991).

### 2.1.6  Viscoelastic Pasternak foundation

The load–deflection relation

$$p = k_0 y + d \frac{\partial y}{\partial t} - G_0 \nabla^2 y \tag{2.7a}$$

takes into acount the viscoelastic properties of the Pasternak foundation; $d$ is the viscous damping coefficient of the foundation.

For beam the governing equation is

$$EI \frac{\partial^4 y}{\partial x^4} + (N - G_0) \frac{\partial^2 y}{\partial x^2} + \rho A \frac{\partial^2 y}{\partial t^2} + k_0 y + d \frac{\partial y}{\partial t} = 0 \tag{2.8}$$

In this model, the effect of the compressive static load ($N$) has been reduced by the effective foundation shear (term $G_0$).

Some fundamental characteristics of the Pasternak foundation mathematical model are discussed by Kerr (1964).

### 2.1.7  Different model beams on a Pasternak foundation

The governing equation of the rectangular beam, with shear deformation and rotatory inertia being ignored, is

$$\frac{1+v}{6} Gh^3 \frac{\partial^4 y}{\partial x^4} + \rho h \frac{\partial^2 y}{\partial t^2} + k_0 y + d \frac{\partial y}{\partial t} - G_0 \frac{\partial^2 y}{\partial x^2} = 0 \tag{2.9}$$

where $h$ is height of the beam (Saito and Terasawa, 1980).

The governing equations of the rectangular beam, where shear deformation and rotatory inertia are incorporated, are

$$\frac{1+v}{6} Gh^3 \frac{\partial^2 \theta}{\partial x^2} - Gkh \left( \frac{\partial y}{\partial x} + \theta \right) - \frac{\rho h^3}{12} \frac{\partial^2 \theta}{\partial t^2} = 0$$

$$\rho h \frac{\partial^2 y}{\partial t^2} - Gkh \left( \frac{\partial^2 y}{\partial x^2} + \frac{\partial \theta}{\partial x} \right) + k_0 y + d \frac{\partial y}{\partial t} - G_0 \frac{\partial^2 y}{\partial x^2} = 0 \tag{2.10}$$

where $\theta$ is the bending slope, $k$ is the shear coefficient, and $k_0$ is the foundation modulus.

### 2.1.8  'Generalized' foundation

At the each point of the foundation the pressure $p$ is proportional to the deflection $y$ and the moment $m$ is proportional to the angle of rotation

$$p = k_0 y, \quad m = k_1 \frac{\mathrm{d}y}{\mathrm{d}n} \tag{2.11}$$

where $n$ is any direction at a point in the plane of the foundation surface; $k_0$ and $k_1$ are the corresponding moduli of elasticity (Pasternak, 1954).

### 2.1.9  Reissner foundation

*Assumptions*

1. The in-plane stresses throughout the foundation layer are negligibly small.
2. The horizontal displacements at the upper and lower surfaces of the foundation layer are zero.

The relationship between the reaction of the foundation $p$ and deflection $y$ is

$$c_1 y - c_2 \nabla^2 y = p - \frac{c_2}{4c_1}\nabla^2 p, \quad c_1 = \frac{E_0}{H}, \quad c_2 = \frac{HG_0}{3} \tag{2.12}$$

where $E_0$ and $G_0$ are the elastic constants of the foundation material, $H$ is the thickness of the foundation layer and $p$ is distributed lateral load acting on the foundation surface (Reissner, 1958).

The case when Reissner's and Pasternak's models of foundation coincide, as well as the Vlasov foundation model (Vlasov and Leontiev, 1966) have been discussed by Kerr (1964).

## 2.2  UNIFORM BERNOULLI–EULER BEAMS ON AN ELASTIC WINKLER FOUNDATION

The differential equation of the transverse vibration of the beam resting on an elastic Winkler foundation without damping is

$$EI\frac{\partial^4 y}{\partial x^4} + \rho A \frac{\partial^2 y}{\partial t^2} + k_0 y = 0 \tag{2.13}$$

**Solution.**   Method of the separation of variables $y(x, t) = X(x)T(t)$, where $X(x)$ is a space-dependent function and $T(t)$ is a time-dependent function. A shape function $X(x)$ depends on the boundary conditions.

The space-dependent function $X(x)$ can be obtained from

$$X^{IV}(x) - k^4 X(x) = 0, \quad k^4 = \frac{m\omega^2 - k_0}{EI} = -4\alpha^4 \tag{2.14}$$

The natural frequencies are defined by the formula (Weaver, Timoshenko and Yaung, 1990; Hetenyi, 1958; Blevins, 1979)

$$\omega = \frac{\lambda^2}{l^2}\sqrt{\frac{EI}{m}}\sqrt{1 + \frac{k_0 l^4}{EI\lambda^4}} \tag{2.14a}$$

Parameter $\lambda$ corresponds to beams with the same boundary conditions but without an elastic foundation. The Winkler elastic foundation increases the frequency vibration.

***Eigenfunction.*** The solutions of equation (2.14) may be presented in the following forms:

*Case 1.* The frequency parameter $k^4 > 0$.

The solutions of (2.14) are the same for $k_0 = 0$ and $k_0 \neq 0$. So, the elastic Winkler foundation has no effect on the mode shape vibration.

*Case of long beams* (Boitsov *et al.*, 1982)

$$X(kx) = e^{-\alpha x}(C_0 \cos \alpha x + C_1 \sin \alpha x) + e^{\alpha x}(C_2 \cos \alpha x + C_3 \sin \alpha x) \qquad (2.15)$$

*Case of short beams* (*especially for symmetric and antisymmetric forms*)

$$X(kx) = C_0 \cosh \alpha x \cos \alpha x + C_1 \cosh \alpha x \sin \alpha x + C_2 \sinh \alpha x \sin \alpha x + C_3 \sinh \alpha x \cos \alpha x \qquad (2.16)$$

Eigenfunction $X(x)$ may be presented in the form of Puzyrevsky functions

$$X(kx) = C_0 V_0(\alpha x) + C_1 V_1(\alpha x) + C_2 V_2(\alpha x) + C_3 V_3(\alpha x) \qquad (2.17)$$

$$V_0 = \cosh \alpha x \cos \alpha x \quad V_1 = \frac{1}{\sqrt{2}}(\cosh \alpha x \sin \alpha x + \sinh \alpha x \cos \alpha x)$$

$$V_2 = \sinh \alpha x \sin \alpha x \quad V_3 = \frac{1}{\sqrt{2}}(\cosh \alpha x \sin \alpha x - \sinh \alpha x \cos \alpha x) \qquad (2.18)$$

*Case 2.* The frequency parameter $k^4 < 0$.

The solution of (2.14) is

$$X = A \sin \frac{kx}{\sqrt{2}} \sinh \frac{kx}{\sqrt{2}} + B \sin \frac{kx}{\sqrt{2}} \cosh \frac{kx}{\sqrt{2}} + C \cos \frac{kx}{\sqrt{2}} \sinh \frac{kx}{\sqrt{2}} + D \cos \frac{kx}{\sqrt{2}} \cosh \frac{kx}{\sqrt{2}} \qquad (2.19)$$

which is different from expressions (2.15) and (2.16) (Wang, 1991).

## 2.2.1 Properties of Puzyrevsky functions

Puzyrevsky functions and their derivatives result in the diagonal matrix at $x = 0$.

$$\begin{aligned}
V_0(0) &= 1 & V_0'(0) &= 0 & V_0''(0) &= 0 & V_0'''(0) &= 0 \\
V_1(0) &= 0 & V_1'(0) &= \sqrt{2}\alpha & V_1''(0) &= 0 & V_1'''(0) &= 0 \\
V_2(0) &= 0 & V_2'(0) &= 0 & V_2''(0) &= 2\alpha^2 & V_2'''(0) &= 0 \\
V_3(0) &= 0 & V_3'(0) &= 0 & V_3''(0) &= 0 & V_3'''(0) &= 2\sqrt{2}\alpha^3
\end{aligned} \qquad (2.20)$$

***Derivatives of Puzyrevsky Functions***

$$V_3'(\alpha x) = \sqrt{2}\alpha V_2(\alpha x); \quad V_2'(\alpha x) = \sqrt{2}\alpha V_1(\alpha x)$$

$$V_1'(\alpha x) = \sqrt{2}\alpha V_0(\alpha x); \quad V_0'(\alpha x) = -\sqrt{2}\alpha V_3(\alpha x)$$

### 2.2.2　Beams on linear inertial foundation

The beam length $l$ and mass per unit $m$ rest on an elastic foundation. A linear inertial foundation is a two-way communication one. The model of the foundation represents separate rods with parameters: modulus $E_F$, cross-sectional area $A_F = b \times 1$, and density $\rho_F$; the length of the rods is $l_0$ (Fig. 2.1) (Bondar', 1971).

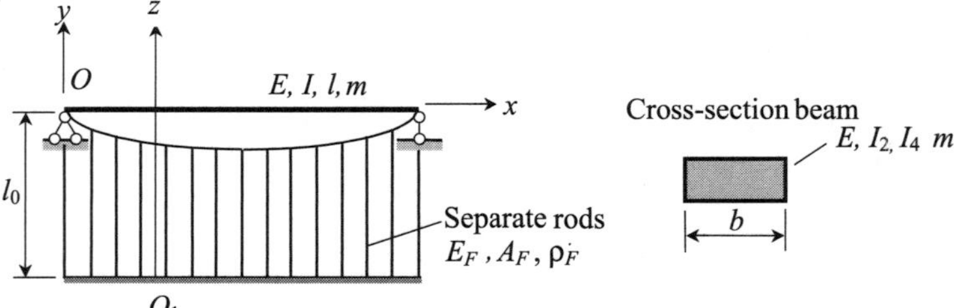


**FIGURE 2.1.**　Mechanical model of elastic foundation. System coordinates: for beam $xOy$; for rods $O_1z$.

Reaction of the rods

$$q_0 = -E_F A_F \left. \frac{\partial u}{\partial z} \right|_{z=l_0}$$

where $u$ is the longitudinal displacement of the rod.

### *Differential equations*

(a) Longitudinal vibration of the rods

$$\frac{\partial^2 u}{\partial t^2} = a^2 \frac{\partial^2 u}{\partial z^2} \tag{2.21}$$

where $a^2 = \dfrac{E_F A_F}{m_F}$,　$A_F = b \times 1$　$m_F = \rho_F A_F$

(b) Transverse vibration of the beam

$$EI_2 \frac{\partial^4 y}{\partial x^4} + m \frac{\partial^2 y}{\partial t^2} + E_F A_F \left. \frac{\partial u}{\partial z} \right|_{z=l_0} = 0 \tag{2.22}$$

The moment of inertia of the cross-sectional area of order $n$ is

$$I_n = \int\limits_{(A)} z^n \, dA$$

where $z$ is a distance from the neutral axis. For a rectangular cross-section, $b \times h$:

$$I_2 = bh^3/12, \quad I_4 = bh^5/80$$

The differential equation for the mode shape of vibration is

$$X^{IV} + \left(C\,\frac{\omega}{a}\cot\frac{\omega}{a}l_0 - b^2\omega^2\right)X = 0 \tag{2.23}$$

where

$$C = \frac{E_F A_F}{EI_2}, \quad b^2 = \frac{m}{EI_2}$$

The frequency equation may be presented in the form

$$\left(\frac{n\pi}{l}\right)^4 + C\,\frac{\omega}{a}\cot\frac{\omega}{a}l_0 - b^2\omega^2 = 0 \tag{2.24}$$

or

$$\left(\frac{n\pi}{l}\right)^4\tan\gamma = \left(\frac{ab\gamma}{l_0}\right)^2\tan\gamma - \frac{C\gamma}{l_0} \tag{2.24a}$$

where $\gamma = \dfrac{\omega}{a}l_0$, $\quad \omega = \gamma\dfrac{a}{l_0} = \dfrac{\gamma}{l_0}\sqrt{\dfrac{E_F}{\rho_F}}$.

This equation takes into account the bending stiffness of the beam and the elastic foundation. The fundamental natural frequency of vibration

$$\omega = \frac{\gamma}{l_0}\sqrt{\frac{E_F}{\rho_F}} \tag{2.25}$$

where $\gamma$ is the minimal root of the frequency equation (2.24). For soil of average density the length of the rods, $l_0$, approximately equals $5l$. For the condition $l_0 = 5l$, displacement of the bottom ends of the rods makes up 2.5% of the displacement of the beam. For more details see Section 7.4.

### Special cases

*1. No-foundation condition.* In this case, $E_F = 0$, then $C/a = 0$ and the frequency equation of the beam becomes

$$\left(\frac{n\pi}{l}\right)^4 - b^2\omega^2 = 0$$

The frequency of the transverse vibration of the beam becomes

$$\omega_n = \left(\frac{n\pi}{l}\right)^2\sqrt{\frac{EI_2}{m}}, \quad (n = 1, 2, 3, \ldots)$$

*2. No-beam condition.* In this case $EI_2 = 0$, then $b = \infty$, $C = \infty$ and the frequency equation of the clamped–free rod becomes

$$\tan\gamma = \infty, \quad \gamma = \frac{i\pi}{2}, \quad i = 1, 3, 5, \ldots$$

The frequency of the longitudinal vibration of the clamped–free rod is

$$\omega_{\text{cl–fr}} = \frac{i\pi}{2l_0}\sqrt{\frac{E_F}{m_F}}, \quad i = 1, 3, 5, \ldots$$

*3. Beam is absolutely rigid.* In this case $EI_2 = \infty$ then $b = 0$, $C = 0$ and the frequency equation becomes $\tan\gamma = 0$, $\gamma = i\pi$, $i = 1, 2, 3, \ldots$. The frequency of the longitudinal vibration of the clamped–clamped rod is

$$\omega_{\text{cl–cl}} = \frac{i\pi}{l_0}\sqrt{\frac{E_F}{m_F}}, \quad i = 1, 2, 3, \ldots$$

Thus, all limiting cases lead to well-known classical results.

For the fundamental mode ($i = 1$) the frequency of vibration $\omega$ of the system's 'beam-inertial foundation' satisfies condition

$$\omega_{\text{cl–fr}} \leq \omega \leq \omega_{\text{cl–cl}}$$

## 2.3  PINNED–PINNED BEAM UNDER COMPRESSIVE LOAD

The design diagram of a pinned–pinned uniform beam on an elastic foundation with compressive load $N$ is presented in Fig. 2.2. The parameters of the elastic foundation are $k_{\text{slope}} = D_0$, $k_{\text{tr}} = k_0$ and $k_{\text{tilt}}$ (Nielsen, 1991).

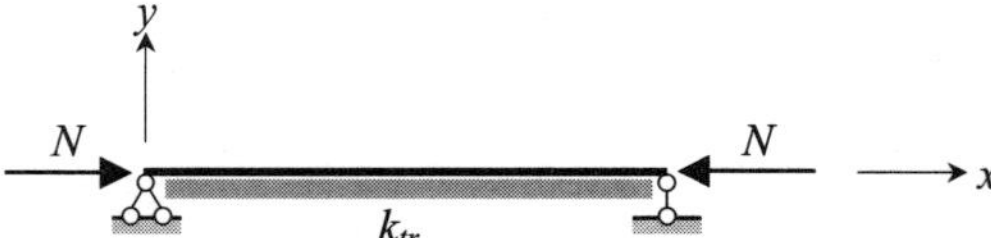


**FIGURE 2.2.**   Beam on an elastic foundation.

### 2.3.1  Bernoulli–Euler beam theory

***Winkler foundation.***   The differential equation of the transverse vibration is

$$EIy^{IV} + Ny'' + k_{\text{tr}}y + m\ddot{y} = 0 \tag{2.26}$$

where $y$ denotes the transverse displacement of the beam axis at position $x$ and time $t$. The elastic foundation does not change the boundary condition.

Frequency of vibration

$$\omega_n = \frac{n^2\pi^2}{l^2}\sqrt{\frac{EI}{m}}\sqrt{1 - \frac{Nl^2}{EIn^2\pi^2} + \frac{k_{\text{tr}}l^4}{EIn^4\pi^4}}, \quad n = 1, 2, \ldots \tag{2.27}$$

***Pasternak foundation.*** The differential equation of the transverse vibration is

$$EIy^{IV} + (N - k_{\text{slope}})y'' + k_{\text{tr}}y + m\ddot{y} = 0 \tag{2.28}$$

The natural frequency of vibration is

$$\omega_n = \frac{n^2\pi^2}{l^2}\sqrt{\frac{EI}{m}}\sqrt{1 - \frac{(N - k_{\text{slope}})l^2}{EIn^2\pi^2} + \frac{k_{\text{tr}}l^4}{EIn^4\pi^4}} \tag{2.29}$$

The elastic foundation leads to the increment of the eigenfrequencies, whereas the compressive force leads to the decrement of the eigenfrequencies.

### 2.3.2  Rayleigh–Timoshenko beam theory

The differential equations of the transverse vibration for an undamped beam are

$$\begin{aligned}
kGA(y' + \phi)' &= m\ddot{y} + k_{\text{tr}}y + Ny'' \\
EI\phi'' - kGA(y' + \phi) + k_{\text{slope}}\,y' &= mr^2\ddot{\phi} + k_{\text{tilt}}\phi
\end{aligned} \tag{2.30}$$

where $y(x, t)$ and $\phi(x, t)$ denote the transverse displacement of the beam axis and the transverse rotation (tilting) of the beam cross-section at position $x$ and time $t$; $I$ and $r$ are the moment of inertia and the radius of gyration of the cross-section with respect to the $z$-axis (Nielsen, 1991).

***Solution***

$$y = A_1 \sin\frac{n\pi x}{l}\exp(i\omega t), \quad \phi = A_2 \cos\frac{n\pi x}{l}\exp(i\omega t)$$

The eigenvalues of the pinned–pinned beam are

$$\begin{bmatrix} (N - kGA)\left(\dfrac{n\pi}{l}\right)^2 + m\omega^2 - k_{\text{tr}} & -kGA\dfrac{n\pi}{l} \\[2ex] -(kGA - k_{\text{slope}})\dfrac{n\pi}{l} & mr\omega^2 - kGA - k_{\text{tilt}} - EI\left(\dfrac{n\pi}{l}\right)^2 \end{bmatrix} = 0 \tag{2.31}$$

The frequency equation may be presented in the form

$$\omega_n^2 = \frac{kGA}{2mr^2}\left[B_1 \pm \sqrt{B_1^2 - \frac{4r^2B_2}{kGA}}\right] \tag{2.32}$$

$$B_1 = 1 + \frac{k_{\text{tilt}} + EI\mu_n^2 + r^2 k_{\text{tr}}}{kGA} + r^2\left(1 - \frac{N}{kGA}\right)\mu_n^2 \quad \mu_n = \frac{n\pi}{l}$$

$$B_2 = k_{\text{tr}}\left(1 + \frac{k_{\text{tilt}} + EI\mu_n^2}{kGA}\right) + \left(1 - \frac{N}{kGA}\right)(k_{\text{tilt}} + EI\mu_n^2)\mu_n^2 + (k_{\text{slope}} - N)\mu_n^2$$

It may occur that the minimum eigenfrequency does not correspond to the simplest mode of vibration ($n = 1$).

## 2.4   A STEPPED BERNOULLI–EULER BEAM SUBJECTED TO AN AXIAL FORCE AND EMBEDDED IN A NON-HOMOGENEOUS WINKLER FOUNDATION

A design diagram of a stepped beam is presented in Fig. 2.3. Boundary conditions are not shown. The elastic foundation is non-uniform with translational stiffness coefficients $k_0$ and $k_1$. The exact fundamental eigenfrequencies for a beam with different boundary

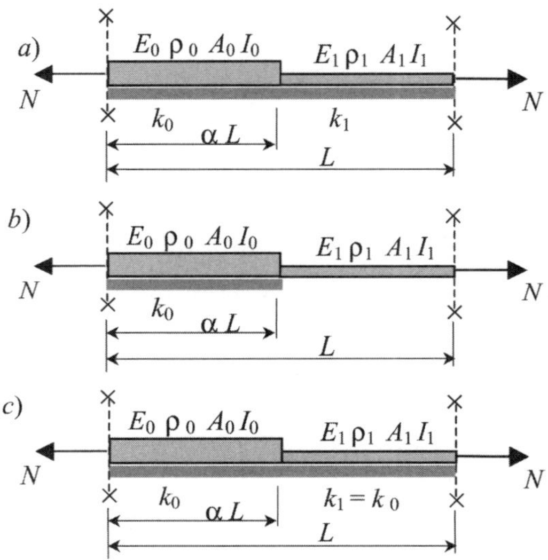



**FIGURE 2.3.**   Stepped beam embedded in a non-homogeneous foundation.

**TABLE 2.1.**   One-span stepped beam partially embedded in a Winkler foundation: Fundamental frequency vibration for beams with different boundary conditions and axial force

| Force | $p^2$ | Pinned–pinned | Clamped–clamped | Pinned–clamped | Clamped–pinned |
|---|---|---|---|---|---|
| Tensile | 0 | 9.2874 | 20.0711 | 14.0381 | 14.0926 |
|  | 5 | 11.9450 | 21.7646 | 16.0357 | 16.4150 |
|  | 10 | 14.0977 | 23.3174 | 17.8034 | 18.4058 |
| Compressive | 2 | 7.9717 | 19.3454 | 13.1513 | 13.0287 |
|  | 3 | 7.2219 | 18.9705 | 12.6840 | 12.4577 |
|  | 5 | 5.4119 | 18.1941 | 11.6917 | 11.2168 |
|  | 10 | — | 16.0647 | 8.7171 | 7.1072 |
|  | 20 | — | 10.3917 | — | — |
|  | 25 | — | 5.5051 | — | — |

**TABLE 2.2.** One-span stepped beam completely embedded in a Winkler foundation: Fundamental frequency vibration for beams with different boundary conditions and axial force

| Force | $p^2$ | Pinned–pinned | Clamped–clamped | Pinned–clamped | Clamped–pinned |
|---|---|---|---|---|---|
| Tensile | 0 | 10.1041 | 20.4739 | 14.4310 | 14.8147 |
| | 5 | 12.5774 | 22.1324 | 16.3829 | 17.0245 |
| | 10 | 14.6286 | 23.6578 | 18.1185 | 18.9401 |
| Compressive | 2 | 8.8187 | 19.7650 | 13.5689 | 13.8146 |
| | 3 | 8.2607 | 19.3993 | 13.1159 | 13.2818 |
| | 5 | 6.7499 | 18.6430 | 12.1576 | 12.1355 |
| | 10 | — | 16.5784 | 9.3279 | 8.5241 |
| | 20 | — | 11.1956 | — | — |
| | 25 | — | 6.9393 | — | — |

conditions, beam parameters, load and foundation are presented in Tables 2.1 and 2.2. The method of separation of variables is applied (Filipich *et al.*, 1988).

The dimensionless parameters of the system are

$$w_0^2 = \frac{k_0 L^4}{E_0 I_0}, \quad w_1^2 = \frac{k_1 L^4}{E_0 I_0}, \quad \beta = \frac{E_1 I_1}{E_0 I_0}, \quad \gamma = \frac{\rho_1 A_1}{\rho_0 A_0}$$

$$p^2 = \frac{NL^2}{E_0 I_0}(\geq 0), \quad \varepsilon = \left(\frac{w_1}{w_0}\right)^2 = \frac{k_1}{k_0}, \quad \Omega^2 = \omega^2 L^4 \frac{\rho_0 A_0}{E_0 I_0}$$

### 2.4.1 The stepped beam is partially embedded in a Winkler foundation (Fig. 2.3(b))

In this case $k_1 = 0$ and $w_0 = \varepsilon = 0$. The fundamental eigenvalues, $\Omega_1$, for $\alpha = 0.5$; $\beta = 0.512$; $\gamma = 0.8$ and $w_0^2 = 25$ are presented in Table 2.1.

### 2.4.2 The stepped beam is completely embedded in a homogeneous Winkler foundation (Fig. 2.3(c))

In this case $k_0 = k_1$ and $\varepsilon = 1$. The fundamental eigenvalues, $\Omega_1$ for $\alpha = 0.5$; $\beta = 0.512$; $\gamma = 0.8$ and $w_0^2 = 25$ are presented in Table 2.2.

**FIGURE 2.4.** (a) Infinite beam with lumped mass on elastic foundation (b) Corresponding frequency spectrum.

## 2.5  INFINITE UNIFORM BERNOULLI–EULER BEAM WITH A LUMPED MASS ON AN ELASTIC WINKLER FOUNDATION

An infinite uniform Bernoulli–Euler beam with a distributed mass $m$ and lumped mass $M$ on an elastic Winkler foundation with modulus elasticity $k$ is presented in Fig. 2.4(a).

The spectrum of this system is mixed (discrete and continuous) and is presented in Fig. 2.4(b). The discrete frequency $\omega_M$ is a real root of the characteristics equation (Bolotin, 1978)

$$\frac{\omega^2 M}{8EI} = \left(\frac{k - m\omega^2}{4E}\right)^{3/4} \tag{2.33}$$

The distributed spectrum begins in the frequency $\omega_0 = \sqrt{\dfrac{k}{m}}$.

## REFERENCES


Bojtsov, G. V., Paliy O. M., Postnov, V. A. and Chuvikovsky, V. S. (1982) Dynamics and Stability of Construction, Vol. 3, 317 p., In Handbook: *Structural Mechanics of a Ship*, Vol. 1–3, Leningrad, Sudostroenie, 1982 (In Russian).

Engel, R.S. (1991) Dynamic stability of an axially loaded beam on an elastic foundation with damping. *Journal of Sound and Vibration*, **146**(3), 463–477.

Filipich, C.P., Laura, P.A.A., Sonenblum, M. and Gil, E. (1988) Transverse vibrations of a stepped beam subject to an axial force and embedded in a non-homogeneous Winkler foundation. *Journal of Sound and Vibration*, **126**(1), 1–8.

Issa, M.S. (1988) Natural frequencies of continuous curved beams on Winkler-type foundation. *Journal of Sound and Vibration*, **127**(2), 291–301.

Hetenyi, M. (1958) *Beams on Elastic Foundation* (Ann Arbor: The University of Michigan Press).

Kerr, A.D. (1964) Elastic and viscoelastic foundation models. *ASME Journal of Applied Mechanics*, **31**, 491–498.

Lunden, R. and Akesson, B. (1983) Damped second-order Rayleigh–Timoshenko beam vibration in space – an exact complex dynamic member stiffness matrix. *International Journal for Numerical Methods in Engineering*, **19**, 431–449.

Mathews, P.M. (1958, 1959) Vibrations of a beam on elastic foundation. *Zeitschrift fur Angewandte Mathematik und Mechanik*, **38**, 105–115; **39**, 13–19.

Nielsen, J.C.O. (1991) Eigenfrequencies and eigenmodes of beam stuctures on an elastic foundation. *Journal of Sound and Vibration*, **145**(3), 479–487.

Pasternak, P.L. (1954) *On a New Method of Analysis of an Elastic Foundation by Means of Two Foundation Constants* (Moscow, USSR: Gosizdat).

Reissner, E. (1958) A note on deflections of plates on a viscoelastic foundation. *Journal of Applied Mechanics, Trans. ASME*, **25**, 144–145.

Saito, H. and Terasawa, T. (1980) Steady-State Vibrations of a Beam on a Pasternak Foundation for Moving Loads. *ASME Journal of Applied Mechanics*, **47**, 879–883.

Vlasov, V.Z. and Leontev, U.N. (1966) *Beams, Plates and Shells on Elastic Foundations*, NASA TTF-357.

Wang, T.M. and Gagnon, L.W. (1978) Vibrations of continuous Timoshenko beams on Winkler–Pasternak foundations. *Journal of Sound and Vibration*, **59**, 211–220.

Wang, T.M. and Stephens, J.E. (1977) Natural frequencies of Timoshenko beams on Pasternak foundations. *Journal of Sound and Vibration*, **51**, 149–155.

Weaver, W., Timoshenko, S.P. and Young, D.H. (1990) *Vibration Problems in Engineering*, Fifth edition (New York: Wiley).

Winkler, E. (1867) *Die Lehre von der Elasticitaet und Festigkeit* (Prag: Dominicus).

Yokoyama, T. (1991) Vibrations of Timoshenko beam-columns on two-parameter elastic foundations. *Earthquake Engineering and Structural Dynamics*, **20**, 355–370.

Yokoyama, T. (1987) Vibrations and transient responses of Timoshenko beams resting on elastic foundations. *Ingenieur-Archiv*, **57**, 81–90.


## FURTHER READING

Bert, C.W. (1987) Application of a version of the Rayleigh technique to problems of bars, beams, columns, membranes and plates. *Journal of Sound and Vibration*, **119**, 317–327.

Blevins, R.D. (1979) *Formulas for Natural Frequency and Mode Shape* (New York: Van Nostrand Reinhold).

Bolotin, V.V. (Ed) (1978) *Vibration of Linear Systems*, vol. 1. In Handbook: *Vibration in Tecnnik*, vols 1–6 (Moscow: Mashinostroenie) (in Russian).

Bondar', N.G. (1971) *Non-Linear Problems of Elastic System* (Kiev: Budivel'nik) (in Russian).

Capron, M.D. and Williams, F.W. (1988) Exact dynamic stiffnesses for an axially loaded uniform Timoshenko member embedded in an elastic medium. *Journal of Sound and Vibration*, **124**(3), 453–466.

Cheng, F.Y. and Pantelides, C.P. (1988) Dynamic Timoshenko beam-columns on elastic media. *ASCE Journal of Structural Engineering*, **114**, 1524–1550.

Crandall, S.H. (1957) The Timoshenko beam on an elastic foundation. *Proceedings of the Third Midwestern Conference on Solid Mechanics*, Ann Arbor, Michigan, pp. 146–159.

De Rosa, M.A. (1989) Stability and dynamics of beams on Winkler elastic foundations. *Earthquake Engineering and Structural Dynamics*, **18**, 377–388.

Djodjo, B.A. (1969) Transfer matrices for beams loaded axially and laid on an elastic foundation. *The Aeronautical Quarterly*, **20**(3), 281–306.

Doyle, P.F. and Pavlovic, M.N. (1982) Vibration of beams on partial elastic foundations. *Earthquake Engineering and Structural Dynamics*, **10**, 663–674.

Eisenberg, M., Yankelevsky, D.Z. and Adin, M.A. (1985) Vibrations of beams fully or partially supported on elastic foundations. *Earthquake Engineering and Structural Dynamics*, **13**, 651–660.

Eisenberg, M. and Clastornik, J. (1987) Beams on variable two-parameter elastic foundation. *ASCE Journal of Engineering Mechanics*, **113**, 1454–1466.

Eisenberg, M. (1990) Exact static and dynamic stiffness matrices for general variable cross section members. *AIAA Journal*, **28**, 1105–1109.

Engel, R.S. (1991) Dynamic stability of an axially loaded beam on an elastic foundation with damping *Journal of Sound and Vibration* **146**(3), 463–477.

Filipich, C.P. and Rosales, M.B. (1988) A variant of Rayleigh's method applied to Timoshenko beams embedded in a Winkler–Pasternak medium. *Journal of Sound and Vibration*, **124**(3), 443–451.

Fletcher, D.Q. and Hermann, L.R. (1971) Elastic foundation representation of continuum. *ASCE Journal of Engineering Mechanics*, **97**, 95–107.

Jones, R. and Xenophontos, J. (1977) The Vlasov foundation model. *International Journal of Mechanical Science*, **19**, 317–323.

Karamanlidis, D. and Prakash, V. (1988) Buckling and vibration analysis of flexible beams resting on an elastic half-space. *Earthquake Engineering and Structural Dynamics*, **16**, 1103–1114.

Karamanlidis, D. and Prakash, V. (1989) Exact transfer and stiffness matrices for a beam/column resting on a two-parameter foundation. *Comput. Methods Appl. Mech Eng.* **72**, 77–89.

Kassem, S.A. (1986) Lateral vibration of cantilevers on viscoelastic foundations. *Armed Forces Science Research Journal*, **XVII**(39), 34–41.

Kerr, A.D. (1961) Viscoelastic Winkler foundation with shear interactions. *Proc ASCE*, **87**(EM3), 13–30.

Kukla, S. (1991) Free vibration of a beam supported on a stepped elastic foundation. *Journal of Sound and Vibration*, **149**(2), 259–265.

Laura, P.A.A. and Cortinez, V.H. (1987) Vibrating beam partially embedded in Winkler-type foundation. *ASCE Journal of Engineering Mechanics*, **113**, 143–147.

Pavlovic, M.N. and Wylie, G.B. (1983) Vibration of beams on non-homogeneous elastic foundations. *Earthquake Engineering and Structural Dynamics*, **11**, 797–808.

Richart, F.E. Jr., Hall, J.R. Jr. and Woods, R.D. (1970) *Vibrations of Soils and Foundations* (Englewood Cliffs, New Jersey: Prentice-Hall).

Selvadurai, A.P.S. (1979) *Elastic Analysis of Soil-Foundation Interaction* (Amsterdam: Elsevier).

Scott, R.F. (1981) *Foundation Analysis* (Englewood Cliffs, New Jersey: Prentice-Hall).

Sundara Raja Iyengar, K.T. and Anantharamu, S. (1963) Finite beam-columns on elastic foundations. *ASCE Journal of Engineering Mechanics*, **89**(6), 139–160.

Taleb, N.J. and Suppiger, E.W. (1962) Vibrations of stepped beams. *Journal of Aeronautical Science*, **28**, 295–298.

Valsangkar, A.J. and Pradhanang, R.B. (1988) Vibrations of beam-columns on two-parameter elastic foundations. *Earthquake Engineering and Structural Dynamics*, **16**, 217–225.

Wang, J. (1991) Vibration of stepped beams on elastic foundations. *Journal of Sound and Vibration*, **149**(2), 315–322.

# CHAPTER 3

# PRISMATIC BEAMS UNDER COMPRESSIVE AND TENSILE AXIAL LOADS

This chapter focuses on prismatic Bernoulli–Euler beams under compressive and tensile loading. Analytic results for frequency equations and mode shape functions for beams with classical boundary conditions are presented. Galef's formula is discussed in detail. Upper and lower values for the frequency of vibrations are evaluated.

## 3.1 BEAMS UNDER COMPRESSIVE LOAD

### 3.1.1 Principal equations

The notation for a beam without axial load and under compressive constant axial load $T$ is presented in Figs. 3.1(a) and (b), respectively; boundary conditions of the beam are not shown. Parameter $\alpha^2 = EI/\rho A$.

*Notation*

- $\omega_{0i}$ and $\Omega_{0i} = \omega_{0i}l^2/\alpha$ are the circular natural frequency and dimensionless natural frequency parameters of the beam with no axial force in the $i$th mode of vibration;

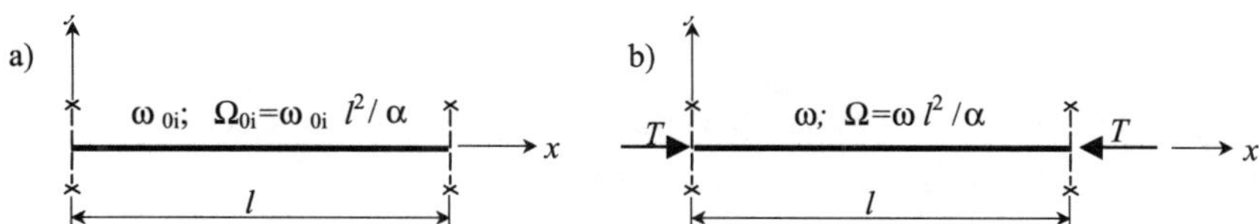

**FIGURE 3.1.** Notation of a beam  (a) Beam without axial load; (b) Beam under axial compressed load.

- $\omega$ and $\Omega = \omega l^2/\alpha$ are the circular natural frequency and dimensionless natural frequency parameters of the compressed beam (relative natural frequency);
- $\Omega^* = \Omega/\Omega_{0i}$ is the normalized natural frequency parameter.

Differential equation of transversal vibration of a beam, which is subjected to a compressed load $T$ is given by

$$EI\frac{\partial^4 y}{\partial x^4} + T\frac{\partial^2 y}{\partial x^2} + \rho A\frac{\partial^2 y}{\partial x^2} = 0 \tag{3.1}$$

### Solution
The solution of the equation (3.1) is assumed in the form

$$y(x,t) = X(x)\cos \omega t$$

which leads to the following differential equation for modal displacement

$$EI\frac{d^4 X}{dx^4} + T\frac{d^2 X}{dx^2} - \rho A\omega^2 X = 0 \tag{3.2}$$

### Modal displacement.
The solution of this equation may be presented in two forms:
Form 1

$$X(x) = X(l\xi) = C_1 \sinh M\xi + C_2 \cosh M\xi + C_3 \sin N\xi + C_4 \cos N\xi \tag{3.3}$$

where    $\xi = x/l$ is a dimensionless beam coordinate;
             $C_i(i = 1, 2, 3, 4)$ are constants to be determined from the boundary conditions;
             $M$ and $N$ are parameters, which may be written as

$$M = l \cdot \sqrt{-\left(\frac{T}{2EI}\right) + \sqrt{\left(\frac{T}{2EI}\right)^2 + \left(\frac{\rho A}{EI}\right)\omega^2}} = \sqrt{-U + \sqrt{U^2 + \Omega^2}}$$

$$\tag{3.3a}$$

$$N = l \cdot \sqrt{\left(\frac{T}{2EI}\right) + \sqrt{\left(\frac{T}{2EI}\right)^2 + \left(\frac{\rho A}{EI}\right)\omega^2}} = \sqrt{U + \sqrt{U^2 + \Omega^2}}$$

$U = Tl^2/2EI$ is a dimensionless compression parameter (the relative axial force);
$\Omega = \omega l^2/\alpha$,  $\alpha^2 = EI/m$ is a dimensionless natural frequency of vibration.

Form 2 (Initial parameter form) (Nowacki, 1963)

$$X(\xi) = X(0)[H(\xi) + \alpha^2 F(\xi)] + X'(0)[G(\xi) + \alpha^2 E(\xi)] + X''(0)F(\xi) + X'''(0)E(\xi) \tag{3.4}$$

where $X(0), X'(0), X''(0)$ and $X'''(0)$ are lateral displacement, slope, bending moment, and shear force at $x = 0$

$$E(\xi) = \frac{1}{N^2 + M^2}\left(\frac{1}{M}\sinh M\xi - \frac{1}{N}\sin N\xi\right)$$

$$F(\xi) = \frac{1}{N^2 + M^2}(\cosh M\xi - \cos N\xi)$$

$$G(\xi) = \frac{1}{N^2 + M^2}(M\sinh M\xi + N\sin N\xi)$$

$$H(\xi) = \frac{1}{N^2 + M^2}(M^2\cosh M\xi + N^2\cos N\xi)$$

$$(3.4a)$$

Galef's formula is a useful relationship between the frequency of vibration and the critical load of the compressed beam. The existence of this relationship is obviously because the frequency of vibration and the critical load are eigenvalues of the deformable system.

For bending vibration, it is worthwhile to cite Amba–Rao (1967) and Bokaian (1988) for Galef's formula:

*The fundamental natural frequency of a compressed beam/natural frequency of uncompressed beam = $(1 - compressive\ load/Euler\ buckling\ load)^{0.5}$.*

$$\Omega^* = \sqrt{1 - U^*}$$

$$\Omega^* = \frac{\Omega}{\Omega_{0i}} = \frac{\omega}{\omega_{0i}}, \quad i = 1 \quad \text{and} \quad U^* = \frac{T}{T_E}$$

$$(3.5)$$

where $T_E$ is the Euler critical buckling load in the first mode and $U^*$ is the normalized compression force parameter.

1. Galef's formula for the fundamental mode of vibration is
   (a) exact for pinned–pinned, sliding–pinned and sliding–sliding beams;
   (b) approximate for sliding–free, clamped–free, clamped–pinned, clamped–clamped and clamped–sliding beams;
   (c) not valid for pinned–free and free–free beams.
2. Galef's formula is valid for the third and higher modes of vibrations for all types of boundary conditions.

***Example.*** Find the fundamental frequency of vibration of the pinned–pinned uniform beam under compressive load, if $T/T_{\mathrm{cr}} = 0.2$ (Fig. 3.2).

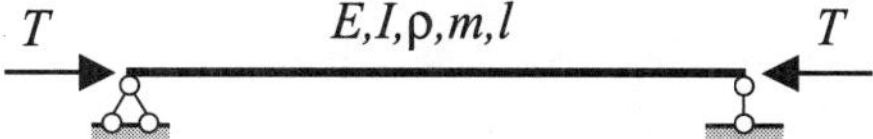


**FIGURE 3.2.** Pinned–pinned uniform beam under compressive axial load.

***Solution.***     According to Galef's formula

$$\Omega^* = \sqrt{1 - U^*} \quad \Omega^* = \frac{\Omega}{\Omega_{0i}} = \frac{\omega}{\omega_{0i}}, \quad i = 1, \quad U^* = \frac{T}{T_e}$$

so the fundamental frequency of vibration of a compressed beam equals

$$\omega = \omega_0 \sqrt{1 - \frac{T}{T_e}} = \frac{3.14159^2}{l^2} \sqrt{\frac{EI}{m}} \sqrt{1 - 0.2} = \frac{2.97113^2}{l^2} \sqrt{\frac{EI}{m}}$$

### 3.1.2   Frequency equations

Table 3.1 contains the frequency equation for compressed beams with classical boundary conditions (Bokaian, 1988). Parameters $M$ and $N$ are presented in Section 3.1.1. The relative natural frequency parameter and the frequency of vibration are

$$\Omega = \frac{\omega l^2}{a} = \lambda_i^2 \qquad \omega_i = \frac{\lambda_i^2}{l^2} \sqrt{\frac{EI}{\rho A}}$$

Table 3.2 predicts eigenvalues for axial compressed beams. They include the critical load and frequency of vibration for beams with different boundary conditions.

The critical buckling load parameter corresponding to the $i$th mode is $U_{\mathrm{mi}} = T_{\mathrm{mi}} l^2 / 2EI$.

### 3.1.3   Modal displacement and mode shape coefficients

The modal displacement may be written in the form

$$X(x) = X(l\xi) = C_1 \sinh M\xi + C_2 \cosh M\xi + C_3 \sin N\xi + C_4 \cos N\xi \qquad (3.6)$$

The mode shape coefficients $C_n$ ($n = 1, 2, 3, 4$) for a beam with different boundary conditions are presented in Table 3.3 (Bokaian, 1988). Parameters $M$ and $N$ are listed in Section 3.1.1, formulae (3.3a).

***Example.***     Find the frequencies of vibration for the simply-supported compressed beam shown in Fig. 3.2.

***Solution.***     The frequency equation is $\sin N = 0$, so $N = i\pi$, $i = 1, 2$, or

$$\sqrt{U + \sqrt{U^2 + \Omega^2}} = i\pi$$

Because $U = \dfrac{Tl^2}{2EI}$ and $\Omega = \omega l^2 \sqrt{\dfrac{m}{EI}}$, the expression for $N$ leads to the exact expression for the frequency of the system

$$\omega_i = \frac{i^2 \pi^2}{l^2} \sqrt{\frac{EI}{m}} \sqrt{1 - \frac{T}{\dfrac{EIi^2\pi^2}{l^2}}}$$

Let $i = 1$ (fundamental mode) and $T/T_E = 0.3$. In this case, the frequency of vibration is

$$\omega = \omega_0 \sqrt{1 - 0.3} = 0.8366\omega_0, \qquad \omega_0 = \frac{\pi^2}{l^2} \sqrt{\frac{EI}{m}}$$

**TABLE 3.1.** Uniform one-span beams with different boundary conditions under compressive axial load: frequency equations

| Beam type | Boundary condition | | Frequency equation |
|---|---|---|---|
| | Left end $(x = 0)$ | Right end $(x = l)$ | |
| Free–free | $X''(0) = 0$ <br> $X'''(0) +$ <br> $(T/EI)X'(0) = 0$ | $X''(l) = 0$ <br> $X'''(l) +$ <br> $(T/EI)X'(l) = 0$ | $\Omega^3[1 - \cosh M \cos N] +$ <br> $(4U^3 + 3U\Omega^2)\sinh M \sin N = 0$ |
| Sliding–free | $X'(0) = 0$ <br> $X'''(0) = 0$ | $X''(l) = 0$ <br> $X'''(l) + (T/EI)X'(l) = 0$ | $M^3 \cosh M \sin N + N^3 \cos N \sinh M = 0$ <br> or $M^3 \tan N + N^3 \tanh M = 0$ |
| Clamped–free | $X(0) = 0$ <br> $X'(0) = 0$ | $X''(l) = 0$ <br> $X'''(l) + (T/EI)X'(l) = 0$ | $\Omega^2 - \Omega U \sinh M \sin N +$ <br> $(2U^2 + \Omega^2)\cosh M \cos N = 0$ |
| Pinned–free | $X(0) = 0$ <br> $X''(0) = 0$ | $X''(l) = 0$ <br> $X'''(l) + (T/EI)X'(l) = 0$ | $N^3 \cosh M \sin N - M^3 \sinh M \cos N = 0$ <br> or $N^3 \tan N - M^3 \tanh M = 0$ |
| Pinned–pinned | $X(0) = 0$ <br> $X''(0) = 0$ | $X(l) = 0$ <br> $X''(l) = 0$ | $\sin N = 0$ |
| Clamped–pinned | $X(0) = 0$ <br> $X'(0) = 0$ | $X(l) = 0$ <br> $X''(l) = 0$ | $M \cosh M \sin N - N \sinh M \cos N = 0$ <br> or $M \tan N - N \tanh M = 0$ |
| Clamped–clamped | $X(0) = 0$ <br> $X'(0) = 0$ | $X(l) = 0$ <br> $X'(l) = 0$ | $\Omega - U \sinh M \sin N - \Omega \cosh M \cos N = 0$ |
| Clamped–sliding | $X(0) = 0$ <br> $X'(0) = 0$ | $X'(l) = 0$ <br> $X'''(l) = 0$ | $N \cosh M \sin N + M \sinh M \cos N = 0$ <br> or $N \tan N + M \tanh M = 0$ |
| Sliding–pinned | $X'(0) = 0$ <br> $X'''(0) = 0$ | $X(l) = 0$ <br> $X''(l) = 0$ | $\cos N = 0$ |
| Sliding–sliding | $X'(0) = 0$ <br> $X'''(0) = 0$ | $X'(l) = 0$ <br> $X'''(l) = 0$ | $\sin N = 0$ |

Special case: If compressed load $T = 0$, then $U = V = 0$ and $M = N = \lambda$.

**TABLE 3.2.** Uniform one-span beams with different boundary conditions under compressive axial load: frequency parameter and critical buckling load

| Beam type | Critical buckling load parameter $U_{\mathrm{m}i}$ | Euler critical buckling load $T_E$ | Parameter $\Omega_{0i}$ | Galef formula for first mode |
|---|---|---|---|---|
| Free–free | $i^2\pi^2/2$ | $\pi^2 EI/l^2$ | $(2i+1)^2\pi^2/4^\dagger$ | Not valid |
| Sliding–free | $(2i-1)^2\pi^2/8$ | $\pi^2 EI/4l^2$ | $(4i-1)^2\pi^2/16^\dagger$ | Approximate |
| Clamped–free | $(2i-1)^2\pi^2/8$ | $\pi^2 EI/4l^2$ | $(2i-1)^2\pi^2/4$ | Approximate |
| Pinned–free | $i^2\pi^2/2$ | $\pi^2 EI/l^2$ | $(4i+1)^2\pi^2/16^\dagger$ | Not valid |
| Pinned–pinned | $i^2\pi^2/2$ | $\pi^2 EI/l^2$ | $i^2\pi^2$ | $\Omega^* = \sqrt{1-U^*}$ |
| Clamped–pinned | $(2i+1)^2\pi^2/8$ | $2.05\pi^2 EI/l^2$ | $(4i+1)^2\pi^2/16$ | Approximate |
| Clamped–clamped | $(i+1)^2\pi^2/2$ | $4\pi^2 EI/l^2$ | $(2i+1)^2\pi^2/4$ | Approximate |
| Clamped–sliding | $i^2\pi^2/2$ | $\pi^2 EI/l^2$ | $(4i-1)^2\pi^2/16$ | Approximate |
| Sliding–pinned | $(2i-1)^2\pi^2/8$ | $\pi^2 EI/4l^2$ | $(2i-1)^2\pi^2/4$ | $\Omega^* = \sqrt{1-U^*}$ |
| Sliding–sliding | $i^2\pi^2/2$ | $\pi^2 EI/l^2$ | $i^2\pi^{2\dagger}$ | $\Omega^* = \sqrt{1-U^*}$ |

[†] The asymptotic formulas are also presented by Karnovsky and Lebed (2003). The numerical results concerning variation of $\Omega$ with $U$ for beams with classical boundary conditions are presented by Bokaian (1988).

**TABLE 3.3.**  Uniform one-span beams with different boundary conditions under compressive axial loads: mode shape coefficients

| Beam type | $C_1$ | $C_2$ | $C_3$ | $C_4$ |
|---|---|---|---|---|
| Free–free | 1 | $\dfrac{N^3(-\cosh M + \cos N)}{N^3 \sinh M + M^3 \sin N}$ | $\dfrac{N}{M}$ | $\dfrac{M^2 N(\cos N - \cosh M)}{N^3 \sinh M + M^3 \sin N}$ |
| Sliding–free | 0 | 1 | 0 | $-\dfrac{N}{M}\dfrac{\sinh M}{\sin N}$ |
| Clamped–free | 1 | $-\dfrac{M^2 \sinh M + MN \sin N}{M^2 \cosh M + N^2 \cos N}$ | $-\dfrac{M}{N}$ | $\dfrac{M^2 \sinh M + MN \sin N}{M^2 \cosh M + N^2 \cos N}$ |
| Pinned–free | 1 | 0 | $\dfrac{M^2}{N^2}\dfrac{\sinh M}{\sin N}$ | 0 |
| Pinned–pinned | 0 | 0 | 1 | 0 |
| Clamped–pinned | 1 | $-\tanh M$ | $-\dfrac{M}{N}$ | $\dfrac{M}{N}\tan N$ |
| Clamped–clamped | 1 | $\dfrac{M \sin N - N \sinh M}{N(\cosh M - \cos N)}$ | $-\dfrac{M}{N}$ | $-\dfrac{M \sin N - N \sinh M}{N(\cosh M - \cos N)}$ |
| Clamped–sliding | 1 | $-\dfrac{M(\cosh M - \cos N)}{M \sinh M + N \sin N}$ | $-\dfrac{M}{N}$ | $\dfrac{M(\cosh M - \cos N)}{M \sinh M + N \sin N}$ |
| Sliding–pinned | 0 | 0 | 0 | 1 |
| Sliding–sliding | 0 | 0 | 0 | 1 |

Calculate the following parameters

$$U = \frac{Tl^2}{2EI} = \frac{0.3 T_E l^2}{2EI} = \frac{0.3\pi^2 EI}{l^2}\frac{l^2}{2EI} = 1.4804$$

$$\Omega = \omega l^2 \sqrt{\frac{m}{EI}} = 0.8336\omega_0 l^2 \sqrt{\frac{m}{EI}} = 8.2569$$

$$\sqrt{U + \sqrt{U^2 + \Omega^2}} = \sqrt{1.4804 + \sqrt{1.4804^2 + 8.2569^2}} = 3.1415$$

The modal displacement is

$$X(x) = \sin\left[(U + \sqrt{U^2 + \Omega^2})^{1/2}\frac{x}{l}\right] = \sin 3.1415\frac{x}{l}$$

***Example.***   Find the frequencies of vibration for a clamped–pinned compressed beam, if $T/T_E = 0.3$.

***Solution.***   The frequency equation is

$$M \cosh M \sin N - N \sinh M \cos N = 0$$

The first Euler critical force, $T_E = \dfrac{\pi^2 EI}{(0.7l)^2}$, so parameter

$$U = \frac{Tl^2}{2EI} = 0.3T_E\frac{l^2}{2EI} = 3.0201$$

and the frequency equation becomes

$$\frac{\sqrt{-3.0201 + \sqrt{3.0201^2 + \Omega^2}}}{\sqrt{3.0201 + \sqrt{3.0201^2 + \Omega^2}}}\tan\left(\sqrt{3.0201 + \sqrt{3.201^2 + \Omega^2}}\right)$$
$$- \tanh\left(\sqrt{-3.0201 + \sqrt{3.0201^2 + \Omega^2}}\right) = 0$$

The root of this equation is $\Omega = \omega l^2\sqrt{\dfrac{m}{EI}} = 12.954$, so the fundamental frequency of vibration of a compressed clamped–pinned beam equals

$$\omega = \frac{12.954}{l^2}\sqrt{\frac{EI}{m}} = \frac{3.599^2}{l^2}\sqrt{\frac{EI}{m}}$$

If $T = 0$, then $\omega = \dfrac{3.9266^2}{l^2}\sqrt{\dfrac{EI}{m}}$.

Parameters

$$M = \sqrt{-3.0201 + \sqrt{3.0201^2 + \Omega^2}} = 3.2064$$
$$N = \sqrt{3.0201 + \sqrt{3.0201^2 + \Omega^2}} = 4.0399$$

The mode shape coefficients are

$$C_1 = 1, \quad C_2 = -\tanh M = -0.9967$$
$$C_3 = -\frac{M}{N} = -0.7937, \quad C_4 = \frac{M}{N}\tan N = 0.9968$$

The modal displacement and slope are

$$X(x) = X(l\xi) = \sinh 3.2064\xi - 0.9967 \cosh 3.2064\xi - 0.7937 \sin 4.0399\xi$$
$$+ 0.9968 \cos 4.0399\xi$$
$$X'(l\xi) = 3.2064 \cosh 3.2064\xi - 3.1958 \sinh 3.2064\xi - 3.2065 \cos 4.0399\xi$$
$$- 4.0273 \sin 4.0399\xi$$

## 3.2  SIMPLY SUPPORTED BEAM WITH CONSTRAINTS AT AN INTERMEDIATE POINT

The design diagram of the compressed simply supported uniform beam with translational and rotational spring supports at an intermediate point, is presented in Fig. 3.3.

The differential equation for eigenfunctions in the $i$th mode is

$$X_i^{IV} + k^2 X_i'' - \lambda^4 X_i = 0, \quad i = 1, 2 \tag{3.7}$$

where

$$k^2 = \frac{Tl^2}{EI} = 2U, \quad \omega = \frac{\lambda^2}{l^2}\sqrt{\frac{EI}{m}}$$

Boundary and compatibility conditions are

$$
\begin{aligned}
x = 0 \quad & X_1 = X_1'' = 0 \\
x = l - c \quad & X_1 = X_2; \;\; X_1' = X_2', \;\; k_1^* X_1' + X_1'' = X_2'', \;\; X_1''' - k_2^* X_1 = X_2''' \\
x = l \quad & X_2 = X_2'' = 0
\end{aligned}
\tag{3.8}
$$

where the dimensionless parameters of rotational and translational spring supports are, respectively

$$k_1^* = \frac{k_{\mathrm{rot}}l}{EI}, \quad k_2^* = \frac{k_{\mathrm{tr}}l^3}{EI}$$

The fundamental natural frequencies parameter $\lambda$ for a simply-supported beam with axial compressive force and various restraint parameters and their spacing are presented in Table 3.4 (Liu and Chen, 1989). The normalized compression force parameter and the

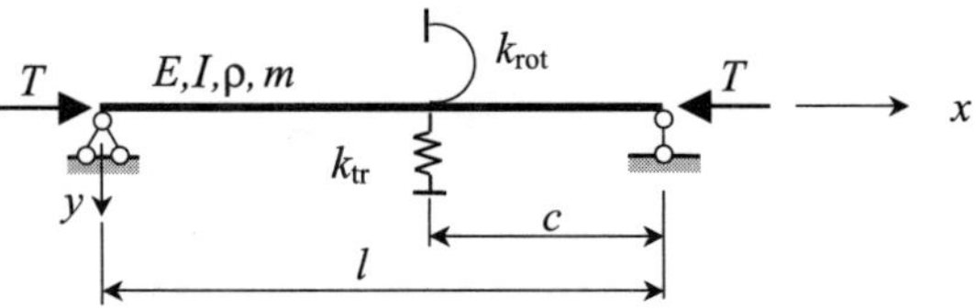


**FIGURE 3.3.**  Compressed beam with elasic restrictions at any point.

**TABLE 3.4.** Compressed uniform pinned–pinned beam with translational and rotational spring support at intermediate point: fundamental frequency parameter $\lambda$

| $c/l$ | $k_1^*$ | $k_2^*$ | $\alpha = T/T_{cr}$ | | | | $c/l$ | $k_1^*$ | $k_2^*$ | $\alpha = T/T_{cr}$ | | | |
|---|---|---|---|---|---|---|---|---|---|---|---|---|---|
| | | | $\alpha = 0.2$ | 0.4 | 0.6 | 0.8 | | | | 0.2 | 0.4 | 0.6 | 0.8 |
| | | 0 | 2.97113 | 2.76495 | 2.49842 | 2.10091 | | | 0 | 2.97113 | 2.76495 | 2.49842 | 2.10091 |
| | | 1 | 2.97872 | 2.77201 | 2.50479 | 2.10627 | | | 1 | 2.98627 | 2.77903 | 2.51114 | 2.11161 |
| | 0 | 10 | 3.04355 | 2.83235 | 2.55933 | 2.15215 | | 0 | 10 | 3.11273 | 2.89673 | 2.61749 | 2.20104 |
| | | 100 | 3.48250 | 3.24305 | 2.93274 | 2.46842 | | | 100 | 3.90783 | 3.63712 | 3.28701 | 2.76452 |
| | | 1000 | 4.33520 | 4.04904 | 3.67278 | 3.10098 | | | 1000 | 5.94227 | 5.52991 | 4.99683 | 4.20182 |
| | | 0 | 3.04069 | 2.82969 | 2.55691 | 2.15010 | | | 0 | 2.97113 | 2.76495 | 2.49842 | 2.10091 |
| | | 1 | 3.04782 | 2.83631 | 2.56290 | 2.15514 | | | 1 | 2.98627 | 2.77903 | 2.51114 | 2.11161 |
| | 1 | 10 | 3.10886 | 2.89314 | 2.61426 | 2.19834 | | 1 | 10 | 3.11273 | 2.89673 | 2.61749 | 2.20104 |
| | | 100 | 3.52883 | 3.28646 | 2.97232 | 2.50207 | | | 100 | 3.90783 | 3.63712 | 3.28701 | 2.76452 |
| | | 1000 | 4.36019 | 4.07293 | 3.69500 | 3.12025 | | | 1000 | 6.01379 | 5.59647 | 5.05700 | 4.25242 |
| 0.25 | | | | | | | 0.50 | | | | | | |
| | | 0 | 3.39257 | 3.15735 | 2.85319 | 2.39942 | | | 0 | 2.97113 | 2.76495 | 2.49842 | 2.10091 |
| | | 1 | 3.39790 | 3.16231 | 2.85768 | 2.40319 | | | 1 | 2.98627 | 2.77903 | 2.51114 | 2.11161 |
| | 10 | 10 | 3.44407 | 3.20532 | 2.89658 | 2.43595 | | 10 | 10 | 3.11273 | 2.89673 | 2.61749 | 2.20104 |
| | | 100 | 3.78345 | 3.52638 | 3.19261 | 2.69125 | | | 100 | 3.90783 | 3.63712 | 3.28701 | 2.76452 |
| | | 1000 | 4.50527 | 4.21180 | 3.82429 | 3.23246 | | | 1000 | 6.43922 | 5.99329 | 5.41648 | 4.55556 |
| | | 0 | 3.85833 | 3.59212 | 3.24734 | 2.73202 | | | 0 | 2.97113 | 2.76495 | 2.49842 | 2.10091 |
| | | 1 | 3.86221 | 3.59574 | 3.25061 | 2.73478 | | | 1 | 2.98627 | 2.77903 | 2.51114 | 2.11161 |
| | 100 | 10 | 3.89615 | 3.62752 | 3.27957 | 2.75940 | | 100 | 10 | 3.11273 | 2.89673 | 2.61749 | 2.20104 |
| | | 100 | 4.16402 | 3.89167 | 3.53679 | 2.99840 | | | 100 | 3.90783 | 3.63712 | 3.28701 | 2.76452 |
| | | 1000 | 4.73881 | 4.43407 | 4.02981 | 3.40937 | | | 1000 | 7.18649 | 6.69520 | 6.05723 | 5.10049 |
| | | 0 | 3.97417 | 3.70029 | 3.34544 | 2.81486 | | | 0 | 2.97113 | 2.76495 | 2.49842 | 2.10091 |
| | | 1 | 3.97779 | 3.70367 | 3.34851 | 2.81744 | | | 1 | 2.98627 | 2.77903 | 2.51114 | 2.11161 |
| | 1000 | 10 | 4.00957 | 3.73351 | 3.37580 | 2.84082 | | 1000 | 10 | 3.11273 | 2.89673 | 2.61749 | 2.20104 |
| | | 100 | 4.26481 | 3.98976 | 3.67121 | 3.08568 | | | 100 | 3.90783 | 3.63712 | 3.28701 | 2.76452 |
| | | 1000 | 4.80240 | 4.49381 | 4.08428 | 3.45554 | | | 1000 | 7.40653 | 6.90130 | 6.24476 | 5.25936 |

Euler critical buckling load in the first mode for a pinned–pinned beam without elastic constraints is

$$T_{\mathrm{cr}} = \frac{\pi^2 BEI}{l^2}, \; B = 1$$

The buckling coefficients $B$ for a pinned–pinned beam with various values of $k_1^*$, $k_2^*$, and spacing ratio $c/l$ are presented in Fig. 3.4.

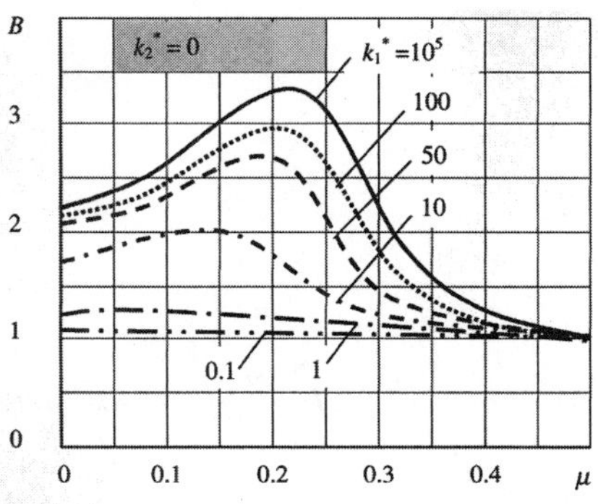


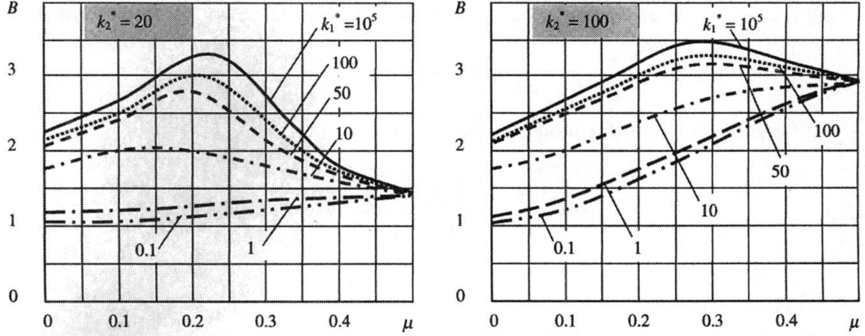


**FIGURE 3.4.** Buckling coefficient $B$ for a simply-supported compressed beam with elastic restrictions at any point for various parameters $k_1^* = k_{\mathrm{rot}}l/EI$, $k_2^* = k_{\mathrm{tr}}l^3/EI$ and spacing ratio $\mu = c/l$.

## 3.3 BEAMS ON ELASTIC SUPPORTS AT THE ENDS

A uniform one-span beam with ends elastically restrained against translation and rotation and initially loaded with an axial constant compressive force $T$ is presented in Fig. 3.5.

*Differential equation of vibration*

$$EI\frac{\partial^4 y}{\partial x^4} + T\frac{\partial^2 y}{\partial x^2} + \rho A\frac{\partial^2 y}{\partial t^2} = 0 \tag{3.9}$$

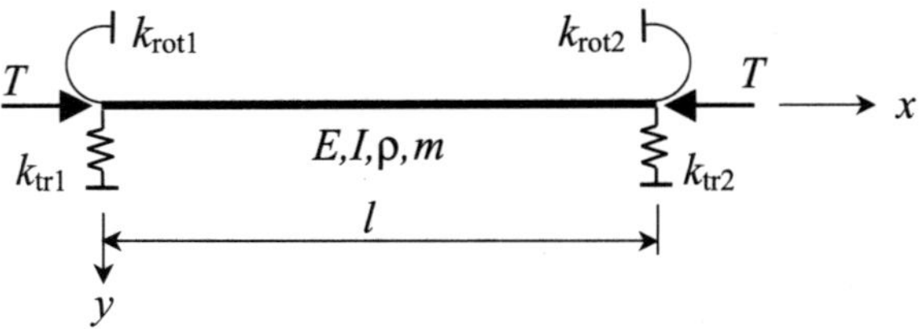

**FIGURE 3.5.**  Compressed beam with elastic restrictions at both ends.

### Boundary Conditions

The boundary conditions for given system are

$$\text{at } x = 0: \quad K_{\text{rot1}}\frac{\partial y(0, t)}{\partial x} = EI\frac{\partial^2 y(0, t)}{\partial x^2} \quad EI\frac{\partial^3 y(0, t)}{\partial x^3} = -K_{\text{tr1}}y(0, t) - T\frac{\partial y(0, t)}{\partial x}$$

$$\text{at } x = l: \quad K_{\text{rot2}}\frac{\partial y(l, t)}{\partial x} = -EI\frac{\partial^2 y(l, t)}{\partial x^2} \quad EI\frac{\partial^3 y(l, t)}{\partial x^3} = -K_{\text{tr2}}y(l, t) - T\frac{\partial y(l, t)}{\partial x}$$

$$(3.10)$$

### Solution

Using the technique of the separation of variables, the displacement $y$ can be written as

$$y(x, t) = X(x)\cos \omega t$$

The differential equation for modal displacement is

$$EI\frac{d^4 X}{dx^4} + T\frac{d^2 X}{dx^2} - \rho A \omega^2 X = 0$$

### Modal displacement

The solution of the ordinary differential equation is

$$X(x) = X(l\xi) = C_1 \sinh M\xi + C_2 \cosh M\xi + C_3 \sin N\xi + C_4 \cos N\xi$$

$$M = l\sqrt{-\left(\frac{T}{2EI}\right) + \sqrt{\left(\frac{T}{2EI}\right)^2 + \left(\frac{\rho A}{EI}\right)\omega^2}} = \sqrt{-U + \sqrt{U^2 + \Omega^2}}$$

$$(3.11)$$

$$N = l\sqrt{\left(\frac{T}{2EI}\right) + \sqrt{\left(\frac{T}{2EI}\right)^2 + \left(\frac{\rho A}{EI}\right)\omega^2}} = \sqrt{U + \sqrt{U^2 + \Omega^2}}$$

where $U = Tl^2/2EI$ is the dimensionless compression parameter and $\Omega = \omega l^2/\alpha$ is the dimensionless natural frequency parameter of the compressed beam, $\alpha^2 = EI/\rho A$.

The frequency equation may be written as follows (Maurizi and Belles, 1991)

$$
\begin{aligned}
T_1 T_2 R_1 R_2 &\{(\cos N \cosh M - 1)[2M^5N^5 + 4U(M^3N^5 - M^5N^3) - 8U^2M^3N^3] \\
&+ \sin N \sinh M[(M^4N^6 - M^6N^4) - 8UM^4N^4 + 4U^2(M^4N^2 - M^2N^4)]\} \\
&+ (R_1T_1T_2 + R_2T_1T_2)\{\sin N \cosh M[(M^5N^4 + M^3N^6) - 2U(M^5N^2 + M^3N^4)] \\
&+ \cos N \sinh M[(M^4N^5 + M^6N^3) + 2U(M^2N^5 + M^4N^3)]\} \\
&+ T_1T_2 \sin N \sinh M(M^6N^2 + M^2N^6 + 2M^4N^4) \\
&+ (R_1R_2T_1 + R_1R_2T_2)\{\sin N \cosh M[(M^5N^2 + M^3N^4) + 2U(M^3N^2 + MN^4)] \\
&- \cos N \sinh M[(M^2N^5 + M^4N^3) + 2U(M^2N^3 + M^4N)]\} \\
&- (R_1T_2 + R_2T_1)\cos N \cosh M[(M^5N + MN^5) + 8M^3N^3] \\
&- (R_1T_1 + R_2T_2)[(MN^5 + M^5N) + 2M^3N^3 \cos N \cosh M \\
&+ 2U(MN^3 - M^3N)(\cos N \cosh M - 1) \\
&- (M^4N^2 - M^2N^4 + 4U^2M^2N)\sin N \sinh M] \\
&- (T_1 + T_2)[(M^3N^2 + MN^4)\sin N \cosh M + (M^2N^3 + M^4N)\cos N \sinh M] \\
&- R_1R_2(M^4 + N^4 + 2M^2N^2)\sin N \sinh M - (M^2 - N^2)\sin N \sinh M \\
&- (R_1 + R_2)[(M^3 + MN^2)\sin N \cosh M - (N^3 + M^2N)\cos N \sinh M] \\
&+ 2M^3N^3(\cos N \cosh M - 1) = 0
\end{aligned}
$$

where the dimensionless stiffness parameters are

$$
R_1 = \frac{EI}{K_{r1}l}, \quad T_1 = \frac{EI}{K_{t1}l^3}, \quad R_2 = \frac{EI}{K_{r2}l}, \quad T_2 = \frac{EI}{K_{t2}l^3}
$$

To reduce the system presented in Fig. 3.5 to the system with classical boundary conditions, the stiffness coefficients in the above frequency equation must be changed accordingly to data presented in Table 3.5.

**TABLE 3.5.** Special cases: compressed uniform beam with elastic restrictions at both ends: stiffness parameters for limiting cases

| Beam type | $R_1$ | $T_1$ | $R_2$ | $T_2$ |
|---|---|---|---|---|
| Free–free | $R_1 \to \infty$ | $T_1 \to \infty$ | $R_2 \to \infty$ | $T_2 \to \infty$ |
| Sliding–free | $R_1 = 0$ | $T_1 \to \infty$ | $R_2 \to \infty$ | $T_2 \to \infty$ |
| Clamped–free | $R_1 = 0$ | $T_1 = 0$ | $R_2 \to \infty$ | $T_2 \to \infty$ |
| Pinned–free | $R_1 \to \infty$ | $T_1 = 0$ | $R_2 \to \infty$ | $T_2 \to \infty$ |
| Pinned–pinned | $R_1 \to \infty$ | $T_1 = 0$ | $R_2 \to \infty$ | $T_2 = 0$ |
| Clamped–pinned | $R_1 = 0$ | $T_1 = 0$ | $R_2 \to \infty$ | $T_2 = 0$ |
| Clamped–clamped | $R_1 = 0$ | $T_1 = 0$ | $R_2 = 0$ | $T_2 = 0$ |
| Clamped-sliding | $R_1 = 0$ | $T_1 = 0$ | $R_2 = 0$ | $T_2 \to \infty$ |
| Sliding–pinned | $R_1 = 0$ | $T_1 \to \infty$ | $R_2 \to \infty$ | $T_2 = 0$ |
| Sliding–sliding | $R_1 = 0$ | $T_1 \to \infty$ | $R_2 \to \infty$ | $T_2 \to \infty$ |

## 3.4  BEAMS UNDER TENSILE AXIAL LOAD

### 3.4.1  Principal equations

The notation for a beam without axial load and under tensile constant axial load $T$ is presented in Figs. 3.6(a) and (b), respectively; the boundary conditions of the beams are shown in Table 3.6.

Parameter $\alpha^2 = EI/\rho A$.

*Notation*

- $\omega_{0i}$ and $\Omega_{0i} = \omega_{0i}l^2/\alpha$ are the circular natural frequency and the dimensionless natural frequency parameters of a beam with no axial force in the $i$th mode of vibration;
- $\omega$ and $\Omega = \omega l^2/\alpha$ are the circular natural frequency and the dimensionless natural frequency parameter of a compressed beam (relative natural frequency);
- $\Omega^* = \Omega/\Omega_{0i}$ is the normalized natural frequency parameter.

The differential equation of transversal vibration of the uniform beam subjected to tensile axial load is given by

$$EI\frac{\partial^4 y}{\partial x^4} - T\frac{\partial^2 y}{\partial x^2} + \rho A\frac{\partial^2 y}{\partial t^2} = 0 \tag{3.12}$$

*Solution*
The solution of this partial differential equation can be obtained by using the technique of the separation of variables

$$y(x, t) = X(x)\cos \omega t$$

The differential equation for modal displacement is

$$EI\frac{d^4 X}{dx^4} - T\frac{d^2 X}{dx^2} - \rho A\omega^2 X(x) = 0 \tag{3.13}$$

*Modal displacement*
The solution of the last equation is given by

$$X(x) = X(l\xi) = C_1\sinh M\xi + C_2\cosh M\xi + C_3\sin N\xi + C_4\cos N\xi, \quad \xi = \frac{x}{l} \tag{3.14}$$

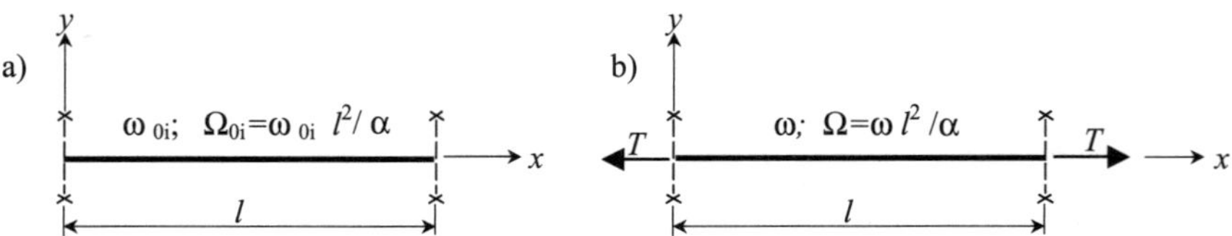


**FIGURE 3.6.**  Notation of a beam. (a) Beam without axial load; (b) Beam under axial tensile load.

As opposed to compressed beams (Section 3.1.1) the parameters $M$ and $N$ are

$$M = l\sqrt{\left(\frac{T}{2EI}\right) + \sqrt{\left(\frac{T}{2EI}\right)^2 + \left(\frac{\rho A}{EI}\right)\omega^2}} = \sqrt{U + \sqrt{U^2 + \Omega^2}}$$

$$N = l\sqrt{\left(\frac{T}{2EI}\right) + \sqrt{\left(\frac{T}{2EI}\right)^2 + \left(\frac{\rho A}{EI}\right)\omega^2}} = \sqrt{-U + \sqrt{U^2 + \Omega^2}}$$

(3.14a)

The modal shape coefficients $C_k$, $k = 1, 2, 3, 4$ are presented in Table 3.7.

### 3.4.2 Relationship of the normalized natural frequency parameter, $\Omega^*$, with the normalized tension parameter, $U^*$

The Rayleigh quotient is

$$\omega^2 = \frac{\int\limits_0^l EIX''^2\,dx}{\int\limits_0^l mX^2\,dx}\left[1 + \frac{\beta_1}{\beta_2}\right]$$

(3.15)

where parameters $\beta_1$ and $\beta_2$ are

$$\beta_1 = T\int\limits_0^l X'^2\,dx \qquad \beta_2 = \int\limits_0^l EIX''^2,\,dx$$

so the normalized natural frequency parameter $\Omega^*$ may be presented in the form

$$\Omega^* = \sqrt{1 + \gamma U^*}$$

(3.16)

This relationship $(\Omega^* - U^*)$ is exact if the exact $X(x)$ is employed, and is only true if the vibrating mode shape is identical with the buckling mode shape. So, for pinned–pinned, sliding–pinned and sliding–sliding beams, the coefficient $\gamma = 1$ (Table 3.7). The values of coefficient $\gamma$ for a beam with different boundary conditions are presented in Table 3.9.

For third and higher modes of vibration, the expression $\Omega^* = \sqrt{1 + U^*}$ is valid for any boundary conditions of one-span beams. The exact frequency equations for a beam with different boundary conditions are presented in Table 3.6.

***Frequency equations.*** The dimensionless parameters $M$ and $N$ for tensile beams are (Bokaian, 1990)

$$M = (U + \sqrt{U^2 + \Omega^2})^{1/2}, \quad N = (-U + \sqrt{U^2 + \Omega^2})^{1/2}$$

The Table 3.6 contains the frequency equation for uniform one-span beams with different boundary conditions under tensile axial load.

***Example.*** Find the fundamental frequency of vibration of the pinned–pinned uniform beam under tensile load (Fig. 3.7)

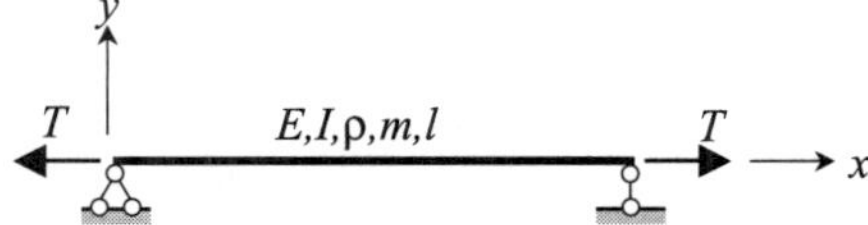


**FIGURE 3.7.** Pinned–pinned uniform beam under tensile axial load.

**TABLE 3.6.**   Uniform one-span beams with different boundary conditions under tensile axial load: frequency equation

| Beam type | Boundary condition | | Frequency equation |
|---|---|---|---|
| | Left end ($x = 0$) | Right end ($x = l$) | |
| Free–free | $X''(0) = 0$ | $X''(l) = 0$ | $\Omega^3[1 - \cosh M \cos N]$ |
| | $X''' + (T/EI)X' = 0$ | $X''' + (T/EI)X' = 0$ | $-(4U^3 + 3U\Omega^2)\sinh M \sin N = 0$ |
| Sliding–free | $X'(0) = 0$ | $X''(l) = 0$ | $M^3 \cosh M \sin N + N^3 \cos N \sinh M = 0$ |
| | $X'''(0) = 0$ | $X''' - (T/EI)X' = 0$ | or $(M^3 \tan N + N^3 \tanh M = 0)$ |
| Clamped–free | $X(0) = 0$ | $X''(l) = 0$ | $\Omega^2 + \Omega U \sinh M \sin N$ |
| | $X'(0) = 0$ | $X'''(l) - (T/EI)X'(l) = 0$ | $+ (2U^2 + \Omega^2)\cosh M \cos N = 0$ |
| Pinned–free | $X(0) = 0$ | $X''(l) = 0$ | $N^3 \cosh M \sin N - M^3 \sinh M \cos N = 0$ |
| | $X''(0) = 0$ | $X'''(l) - (T/EI)X'(l) = 0$ | or $(N^3 \tan N - M^3 \tanh M = 0)$ |
| Pinned–pinned | $X(0) = 0$ | $X(l) = 0$ | $\sin N = 0$ |
| | $X''(0) = 0$ | $X''(l) = 0$ | |
| Clamped–pinned | $X(0) = 0$ | $X(l) = 0$ | $M \cosh M \sin N - N \sinh M \cos N = 0$ |
| | $X'(0) = 0$ | $X''(l) = 0$ | or $(M \tan N - N \tanh M = 0)$ |
| Clamped–clamped | $X(0) = 0$ | $X(l) = 0$ | $\Omega + U \sinh M \sin N - \Omega \cosh M \cos N = 0$ |
| | $X'(0) = 0$ | $X'(l) = 0$ | |
| Clamped–sliding | $X(0) = 0$ | $X'(l) = 0$ | $N \cosh M \sin N + M \sinh M \cos N = 0$ |
| | $X'(0) = 0$ | $X'''(l) = 0$ | or $(N \tan N + M \tanh M = 0)$ |
| Sliding–pinned | $X'(0) = 0$ | $X(l) = 0$ | $\cos N = 0$ |
| | $X'''(0) = 0$ | $X''(l) = 0$ | |
| Sliding–sliding | $X'(0) = 0$ | $X'(l) = 0$ | $\sin N = 0$ |
| | $X'''(0) = 0$ | $X'''(l) = 0$ | |

***Solution.***   The frequency equation for a pinned–pinned beam is $\sin N = 0$, so $N = i\pi$, or

$$N = \sqrt{-U + \sqrt{U^2 + \Omega^2}} = i\pi$$

Because

$$U = \frac{Tl^2}{2EI} \quad \text{and} \quad \Omega = \omega l^2 \sqrt{\frac{m}{EI}}$$

the expression for $N$ leads to the exact expression for the frequency of vibration of the tensile simply supported beam

$$\omega_i = \frac{i^2 \pi^2}{l^2} \sqrt{\frac{EI}{m}} \sqrt{1 + \frac{T}{\dfrac{EIi^2\pi^2}{l^2}}}$$

***Example.*** Find the frequencies of vibration for a clamped–pinned tensile beam, if $T/T_E = 0.3$.

***Solution.*** The frequency equation is

$$M \cosh M \sin N - N \sinh M \cos N = 0$$

The first Euler critical force $T_E = \dfrac{\pi^2 EI}{(0.7l)^2}$, so the parameter

$$U = \frac{Tl^2}{2EI} = 0.3 T_E \frac{l^2}{2EI} = 3.0201$$

and the frequency equation becomes

$$\frac{\sqrt{3.0201 + \sqrt{3.0201^2 + \Omega^2}}}{\sqrt{-3.0201 + \sqrt{3.0201^2 + \Omega^2}}} \tan\left(\sqrt{-3.0201 + \sqrt{3.0201^2 + \Omega^2}}\right)$$
$$- \tanh\left(\sqrt{3.0201 + \sqrt{3.0201^2 + \Omega^2}}\right) = 0$$

The root of this equation is $\Omega = \omega l^2 \sqrt{\dfrac{m}{EI}} = 17.519$, so the fundamental frequency of vibration of a tensile clamped–pinned beam equals

$$\omega = \frac{17.519}{l^2} \sqrt{\frac{EI}{m}} = \frac{4.1845^2}{l^2} \sqrt{\frac{EI}{m}}$$

If $T = 0$, then $\omega = \dfrac{3.9266^2}{l^2} \sqrt{\dfrac{EI}{m}}$.

Parameters

$$M = \sqrt{3.0201 + \sqrt{3.0201^2 + \Omega^2}} = 4.5604$$
$$N = \sqrt{-3.0201 + \sqrt{3.0201^2 + \Omega^2}} = 3.84152$$

The mode shape coefficients are (Table 3.7)

$$C_1 = 1, \quad C_2 = -\tanh M = -0.99978,$$
$$C_3 = -\frac{M}{N} = -1.1871, \quad C_4 = \frac{M}{N} \tan N = 0.99973$$

The modal displacement and slope

$$X(x) = X(l\xi) = \sinh 4.5604\xi - 0.99978 \cosh 4.5604\xi - 1.1871 \sin 3.84152\xi$$
$$+ 0.99973 \cos 3.84152\xi$$
$$X'(l\xi) = 4.5604 \cosh 4.5604\xi - 4.5593 \sinh 4.5064\xi - 4.56026 \cos 3.84152\xi$$
$$- 3.84048 \sin 3.84152\xi$$

***Control.***   At the left clamped end

$$X(0) = -0.99978 + 0.99973 \approx 0$$
$$X'(0) = 4.5604 - 4.5602 \approx 0$$

At the right pinned end

$$X(l) = 47.80563 - 47.80557 + 0.76468 - 0.76468 \approx 0$$

### 3.4.3  Mode shape coefficients

Exact expressions for mode shape coefficients for a beam with different boundary conditions under axial tensile load are presented in Table 3.7.

### 3.4.4  Mode shape coefficients. The case of a large *U*

The frequency equation and expressions for mode shape coefficients for a beam with different boundary conditions may be simplified if the dimensionless tension parameter $U$ is greater than about 12. The approximate frequency equations are presented in Table 3.8 (Bokaian, 1990). Additional dimensionless parameters

$$\alpha = \frac{\Omega}{U}, \quad \delta = -1 + \sqrt{1 + \frac{\Omega^2}{U^2}}$$

***Example.***   Find a value of tensile load $T$ that acts on the clamped–pinned beam so that parameter $U$ would be so big it would be possible to use the approximate results presented in Table 3.8.

***Solution.***   Let $T/T_E = k$, where $k$ is unknown. Parameter

$$U = \frac{Tl^2}{2EI} = kT_E \frac{l^2}{2EI} = k \frac{\pi^2 EI}{(0.7l)^2} \frac{l^2}{2EI} = 3.07k$$

So parameter $U$ equals 12 (the case of large $U$) starting from $k = T/T_E = 1.2$.

***Example.***   Compare the frequency of vibration and mode shape coefficients for the clamped–pinned beam by using exact and approximate formulas; parameter $U = 12$.

***Solution***
   *Exact solution.*   Parameters $M$ and $N$ are

$$M = \sqrt{U + \sqrt{U^2 + \Omega^2}} = \sqrt{12 + \sqrt{144 + \Omega^2}}$$

$$N = \sqrt{-U + \sqrt{U^2 + \Omega^2}} = \sqrt{-12 + \sqrt{144 + \Omega^2}}$$

The frequency equation (Table 3.6) is

$$\frac{M}{N} \tan N - \tanh M = 0$$

The root of this equation is $\Omega = 22.572$, which leads to parameters

$$M = \sqrt{12 + \sqrt{144 + \Omega^2}} = 6.1289$$

$$N = \sqrt{-12 + \sqrt{144 + \Omega^2}} = 3.6828$$

**TABLE 3.7.** Uniform one-span beams with different boundary conditions under tensile axial load: critical buckling load, frequency parameters, and mode shape coefficients

| Beam type | $U_{mi} = \dfrac{T_{cr}l^2}{2EI}$ | $P_{cr}$ | $\Omega_{0i} = \dfrac{\omega_{0i}l^2}{\alpha}$ | $\Omega^* = f(U^*)$ | $C_1$ | $C_2$ | $C_3$ | $C_4$ |
|---|---|---|---|---|---|---|---|---|
| | | | | | | Mode shape coefficients | | |
| Free–free | $\dfrac{i^2\pi^2}{2}$ | $\dfrac{\pi^2 EI}{l^2}$ | $\dfrac{(2i+1)^2\pi^2}{4}$ | – | 1 | $\dfrac{N^3(-\cosh M + \cos N)}{N^3\sinh M + M^3\sin N}$ | $\dfrac{N}{M}$ | $\dfrac{M^2 N(\cos N - \cosh M)}{N^3\sinh M + M^3\sin N}$ |
| Sliding–free | $\dfrac{(2i-1)^2\pi^2}{8}$ | $\dfrac{\pi^2 EI}{4l^2}$ | $\dfrac{(4i-1)^2\pi^2}{16}$ | – | 0 | 1 | 0 | $\dfrac{N\sinh M}{M\sin N}$ |
| Clamped–free | $\dfrac{(2i-1)^2\pi^2}{8}$ | $\dfrac{\pi^2 EI}{4l^2}$ | $\dfrac{(2i-1)^2\pi^2}{4}$ | – | 1 | $-\dfrac{M^2\sinh M + MN\sin N}{M^2\cosh M + N^2\cos N}$ | $-\dfrac{M}{N}$ | $\dfrac{M^2\sinh M + MN\sin N}{M^2\cos M + N^2\cos N}$ |
| Pinned–free | $\dfrac{i^2\pi^2}{2}$ | $\dfrac{\pi^2 EI}{l^2}$ | $\dfrac{(4i+1)^2\pi^2}{16}$ | – | 1 | 0 | $\dfrac{M^2}{N^2}\dfrac{\sinh M}{\sin N}$ | 0 |
| Pinned–pinned | $\dfrac{i^2\pi^2}{2}$ | $\dfrac{\pi^2 EI}{l^2}$ | $i^2\pi^2$ | $\Omega^* = \sqrt{1+U^*}$ | 0 | 0 | 1 | 0 |
| Clamped–pinned | $\dfrac{(2i+1)^2\pi^2}{8}$ | $\dfrac{\pi^2 EI}{(0.7l)^2}$ | $\dfrac{(4i+1)^2\pi^2}{16}$ | – | 1 | $-\tanh M$ | $-\dfrac{M}{N}$ | $\dfrac{M}{N}\tan N$ |
| Clamped–clamped | $\dfrac{(i+1)^2\pi^2}{2}$ | $\dfrac{4\pi^2 EI}{l^2}$ | $\dfrac{(2i+1)^2\pi^2}{4}$ | – | 1 | $\dfrac{M\sin N - N\sinh M}{N(\cosh M - \cos N)}$ | $-\dfrac{M}{N}$ | $-\dfrac{M\sin N - N\sinh M}{N(\cosh M - \cos N)}$ |
| Clamped–sliding | $\dfrac{i^2\pi^2}{2}$ | $\dfrac{\pi^2 EI}{l^2}$ | $\dfrac{(4i-1)^2\pi^2}{16}$ | – | 1 | $-\dfrac{M(\cosh M - \cos N)}{M\sinh M + N\sin N}$ | $-\dfrac{M}{N}$ | $\dfrac{M(\cosh M - \cos N)}{M\sinh M + N\sin N}$ |
| Sliding–pinned | $\dfrac{(2i-1)^2\pi^2}{8}$ | $\dfrac{\pi^2 EI}{4l^2}$ | $\dfrac{(2i-1)^2\pi^2}{4}$ | $\Omega^* = \sqrt{1+U^*}$ | 0 | 0 | 0 | 1 |
| Sliding–sliding | $\dfrac{i^2\pi^2}{2}$ | $\dfrac{\pi^2 EI}{l^2}$ | $i^2\pi^2$ | $\Omega^* = \sqrt{1+U^*}$ | 0 | 0 | 0 | 1 |

$U_{mi}$ is the exact critical buckling load parameter in the $i$th mode.
$P_{cr}$ is the exact critical load in the first mode.

**TABLE 3.8** Uniform one-span beams with different boundary conditions under tensile axial load: approximate frequency equations and mode shape coefficients for tension parameter $U > 12$

| Beam type | Frequency equation | Mode shape coefficients | | | |
| --- | --- | --- | --- | --- | --- |
| | | $C_1$ | $C_2$ | $C_3$ | $C_4$ |
| Free–free | $\left[\tan^{-1}\left(-\dfrac{\alpha^3}{4+3\alpha^2}\right) + (i+1)\pi\right]^2 = U\delta$ | $1$ | $-1$ | $\dfrac{N}{M}$ | $-\dfrac{M^2}{N^2}$ |
| Sliding–free | $\tan\sqrt{U\delta} = -\dfrac{\delta^3}{\alpha^3}$ | – | – | – | – |
| Clamped–free | $\tan\sqrt{U\delta} = -\dfrac{2+\alpha^2}{\alpha}$ | $1$ | $-1$ | $-\dfrac{M}{N}$ | $1$ |
| Pinned–free | $\tan\sqrt{U\delta} = \dfrac{\alpha^3}{\delta^3}$ | – | – | – | – |
| Pinned–pinned | $\sin N = 0^\dagger$ | $0$ | $0$ | $1$ | $0$ |
| Clamped–pinned | $\tan\sqrt{U\delta} = \dfrac{\delta}{\alpha}$ | $1$ | $-1$ | $-\dfrac{M}{N}$ | $\dfrac{M}{N}\tan N$ |
| Clamped–clamped | $\tan\sqrt{U\delta} = \alpha$ | $1$ | $-1$ | $-\dfrac{M}{N}$ | $-1$ |
| Clamped–sliding | $\tan\sqrt{U\delta} = \dfrac{\delta}{\alpha}$ | $1$ | $-1$ | $-\dfrac{M}{N}$ | $1$ |
| Sliding–pinned | $\cos N = 0^\dagger$ | $0$ | $0$ | $0$ | $1$ |
| Sliding–sliding | $\sin N = 0^\dagger$ | $0$ | $0$ | $0$ | $1$ |

$^\dagger$Approximate and exact frequency vibrations coincide (Table 3.6).

The mode shape coefficients are (Table 3.7)

$$C_1 = 1, \quad C_2 = -\tanh M = -0.99999$$

$$C_3 = -\frac{M}{N} = -1.6642, \quad C_4 = \frac{M}{N}\tan N = 1.00030$$

*Approximate solution.*   The frequency equation (Table 3.7)

$$\tan\sqrt{U\delta} = \frac{U}{\Omega}\left(-1 + \sqrt{1 + \frac{\Omega^2}{U^2}}\right)$$

or

$$\tan\sqrt{12\left(-1+\sqrt{1+\frac{\Omega^2}{144}}\right)+\frac{12}{\Omega}\left(1-\sqrt{1+\frac{\Omega^2}{144}}\right)}=0$$

The root of this equation is $\Omega = 22.569$. Parameters $M$, $N$ and mode shape coefficients practically coincide with results that were obtained by using exact formulas.

### 3.4.5  Upper and lower bound approximation to the frequency of vibration

Table 3.9 gives the upper and lower bound approximation to the frequency of vibration of tensile beams with different boundary conditions. The parameters $\Omega^*$, $\Omega$ are given in terms of tension parameter $U$ (Bokaian, 1990).

$$U = Tl^2/2EI, \quad \Omega = \omega l^2/\alpha, \quad \alpha^2 = EI/\rho A, \quad U^* = T/T_E,$$
$$\Omega^* = \Omega/\Omega_0, \quad \Omega^* = (1 + \gamma U^*)^{1/2}$$

***Example.***   Find the fundamental frequency of vibration for a pinned–pinned tensile beam.

***Solution.***   The value of a dimensionless natural frequency parameter is

$$\Omega = \sqrt{2}\pi i\sqrt{U}, \quad \text{so} \quad \omega l^2\sqrt{\frac{m}{EI}} = \sqrt{2}\pi i\sqrt{\frac{Tl^2}{2EI}}, (i = 1, 2, 3, \ldots)$$

The frequency of vibration is $\omega = \dfrac{\pi i}{l}\sqrt{\dfrac{T}{m}}$. This formula coincides with the exact expression for the frequency of transversal vibration of the string for which $\omega = \sqrt{\dfrac{T}{m}}k$, where the wavenumber $kl = \pi i$ (Crawford, 1965).

The upper bound value for the normalized natural frequency parameter $\Omega^*$ in terms of normalized tension parameter $U^*$ is $\Omega^* = \sqrt{1 + U^*}$, so

$$\Omega^* = \frac{\omega}{\omega_0} = \sqrt{1+\frac{Tl^2}{\pi^2 EI}}$$

which leads to

$$\omega = \frac{\pi^2}{l^2}\sqrt{\frac{EI}{m}}\sqrt{1+\frac{Tl^2}{\pi^2 EI}}$$

***Example.***   Find the upper bound value for the fundamental frequency of vibration of a clamped–pinned beam. Parameter $T/T_E = 1.1915$; in this case parameter $U = 12$.

***Solution.***   The upper bound value for

$$\Omega^* = \sqrt{1 + 0.978U^*} = \sqrt{1 + 0.978 \times 1.1915} = 1.47149$$

So the frequency of vibration of the beam under tensile load is $\omega = 1.47149\omega_0$, where $\omega_0 = \dfrac{3.9266^2}{l^2}\sqrt{\dfrac{EI}{m}}$ is the frequency of vibration of the beam without tensile load.

So the the parameter $\lambda^2$ for the upper bound value for the fundamental frequency of vibration equals $3.9266^2 \cdot 1.47149 = 22.68771$.

The exact value is 22.572.

**TABLE 3.9.**   Uniform one-span beams with different boundary conditions under tensile axial load: upper and lower bounds of frequency parameters

| Beam type | Value of $\Omega$ for lower modes ($i = 1, 2, 3\ldots$) | $\gamma$ | Upper bound value for $\Omega^*$ (The Rayleigh quotient) | Lower bound value for $\Omega$ ($i = 1$) |
|---|---|---|---|---|
| Free–free | $(i+1)\pi\sqrt{2U}$ | 0.975 | $\sqrt{1+0.975U^*}$ | $2\sqrt{2}\pi\sqrt{U}$ |
| Sliding–free | $\pi i\sqrt{2U}$ | 0.925 | $\sqrt{1+0.925U^*}$ | $\sqrt{2}\pi\sqrt{U}$ |
| Clamped–free | $\dfrac{\pi}{\sqrt{2}}(2i-1)\sqrt{U}$ | 0.926 | $\sqrt{1+0.926U^*}$ | $\dfrac{\pi}{\sqrt{2}}\sqrt{U}$ |
| Pinned–free | $\dfrac{\pi}{\sqrt{2}}(2i+1)\sqrt{U}$ | 1.13 | $\sqrt{1+U^*}^{\dagger}$ | $\dfrac{3\pi}{\sqrt{2}}\sqrt{U}$ |
| Pinned–pinned | $\pi i\sqrt{2U}$ | 1.0 | $\sqrt{1+U^*}^{\dagger}$ | – |
| Clamped–pinned | $\dfrac{2\pi i U}{\sqrt{2U}-1} \approx \pi i\sqrt{2U}$ | 0.978 | $\sqrt{1+0.978U^*}$ | $\sqrt{2}\pi\sqrt{U}$ |
| Clamped–clamped | $\dfrac{2\pi i U}{\sqrt{2U}-2} \approx \pi i\sqrt{2U}$ | 0.97 | $\sqrt{1+0.97U^*}$ | $\sqrt{2}\pi\sqrt{U}$ |
| Clamped–sliding | $\dfrac{\pi}{\sqrt{2}}(2i-1)\sqrt{U}$ | 0.97 | $\sqrt{1+0.97U^*}$ | $\dfrac{\pi}{\sqrt{2}}\sqrt{U}$ |
| Sliding–pinned | $\pi i\sqrt{2U}$ | 1.0 | $\sqrt{1+U^*}^{\dagger}$ | – |
| Sliding–sliding | $\pi i\sqrt{2U}$ | 1.0 | $\sqrt{1+U^*}^{\dagger}$ | – |

$^*$A tensile force has the effect of increasing the frequency of vibration.
$^{\dagger}$These expressions are exact.
The extensive numerical results and their detailed analysis and discussion for beams with different boundary conditions are presented by Bokaian (1990).

## 3.5  VERTICAL CANTILEVER BEAMS. THE EFFECT OF SELF-WEIGHT

This section contains the frequency parameter $\lambda$ for the transversal vibration of a Bernoulli–Euler vertical uniform cantilever beam, with account taken of the effect of self-weight. The axial tension at coordinate $x$ is $T(x) = mg(l - x)$; the gravity parameter and frequency of vibration are

$$\gamma = \frac{mgl^3}{EI}, \qquad \omega = \frac{\lambda^2}{l^2}\sqrt{\frac{EI}{m}}$$

Squared frequency parameters $\lambda^2$ for clamped–free (CL-FR) and pin-guided–clamped (PG-CL) beams are presented in Table 3.10 (Naguleswaran, 1991).

The critical gravity parameter $\gamma$ (natural frequency is equal to zero) for standing (CL-FR) and (PG-CL) beams is shown in Table 3.11.

## 3.6  GAUGE FACTOR

The gauge factor, $G_{Nn}$, describes the sensitivy of the beam as a gauge, vibrating at $\omega = \omega_n$ to changes in the axial force $N$ in the vicinity of the operating force $T_0$ (Tilmans, 1993).

***Static linearity.***  The axial force is a result of an external effect on the beam

$$G_{Nn} = \left[\frac{1}{\omega_n}\frac{\partial\omega}{\partial T}\right]_{T=T_0} \tag{3.17}$$

For a beam with rectangular cross-section

$$G_{Nn} = \frac{1}{2}\frac{\gamma_n\dfrac{1-v^2}{Ebh}\left(\dfrac{l}{h}\right)^2}{1+\gamma_n\varepsilon_0(1-v^2)\left(\dfrac{l}{h}\right)^2} \tag{3.18}$$

***Static nonlinearity.***  The axial force is a result of an axial elongation of the beam caused by vibration (Chapter 5.1.1). In this case, it is more useful to define a gauge factor $G_{\varepsilon n}$, describing the sensitivity of mode $n$ to changes in the strain $\varepsilon$ in the vicinity of the operating point $\varepsilon_0 = (1 - v)\sigma_0/E$

$$G_{\varepsilon n} = \left[\frac{1}{\omega_n}\frac{\partial\omega_n}{\partial\varepsilon}\right]_{\varepsilon=\varepsilon_0} \tag{3.19}$$

For a beam with rectangular cross-section

$$G_{\varepsilon n} = \frac{1}{2}\frac{\gamma_n(1-v^2)\left(\dfrac{l}{h}\right)^2}{1+\gamma_n\varepsilon_0(1-v^2)\left(\dfrac{l}{h}\right)^2} \tag{3.20}$$

**TABLE 3.10.** Uniform beams under tensile or compress self-weight: frequency parameters $\lambda^2$

| $\gamma$ | Clamped–free beam | Mode | | | Pin-guided–clamped beam | Mode | | |
|---|---|---|---|---|---|---|---|---|
| | | 1 | 2 | 3 | | 1 | 2 | 3 |
| 1000 | | 38.69551 | 92.28009 | 160.70198 | | 27.39606 | 106.09241 | 188.31185 |
| 500 | | 27.58902 | 67.61250 | 123.47107 | | 20.93861 | 80.59374 | 146.26870 |
| 100 | | 12.86874 | 36.50917 | 78.98033 | | 11.77221 | 46.47428 | 91.28511 |
| 50 | | 9.46853 | 30.21719 | 70.97195 | | 9.44016 | 37.70100 | 79.00464 |
| 10 | | 5.29178 | 23.91200 | 63.68155 | | 5.98253 | 26.65655 | 65.76094 |
| 5 | | 4.49431 | 22.99353 | 62.69867 | | 5.05197 | 24.52076 | 63.77334 |
| 4 | | 4.31675 | 22.80513 | 62.49990 | | 4.81491 | 24.05477 | 63.36528 |
| 3 | | 4.13139 | 22.61507 | 62.30037 | | 4.55173 | 23.57404 | 62.95365 |
| 2 | | 3.93715 | 22.42331 | 62.10009 | | 4.25542 | 23.47769 | 62.53844 |
| 1 | | 3.73263 | 22.22980 | 61.89904 | | 3.91564 | 22.56483 | 62.11963 |
| $0^\dagger$ | | **3.51602** | **22.03449** | **61.69721** | | **3.51602** | **22.03449** | **61.69721** |
| 1 | | 3.28492 | 21.83734 | 61.49461 | | 3.02762 | 21.48573 | 61.27118 |
| 2 | | 3.03604 | 21.63830 | 61.29121 | | 2.38845 | 20.91759 | 60.84154 |
| 3 | | 2.76454 | 21.43732 | 61.08701 | | 1.38853 | 20.32915 | 60.40828 |
| 4 | | 2.46297 | 21.23433 | 60.88201 | | – | 19.71963 | 59.97142 |
| 5 | | 2.11849 | 21.02928 | 60.76619 | | – | 19.08839 | 59.53097 |
| 6 | | 1.70527 | 20.82211 | 60.46945 | | – | 18.43517 | 59.08695 |
| 7 | | 1.15155 | 20.61276 | 60.26206 | | – | 17.76018 | 58.63939 |
| 10 | | – | 19.97091 | 59.63452 | | – | 15.62022 | 57.27586 |
| 20 | | – | 17.65619 | 57.48446 | | – | 9.00946 | 52.52602 |
| 30 | | – | 14.97621 | 55.23622 | | – | 5.94514 | 47.55911 |
| 40 | | – | 11.69198 | 52.87665 | | – | – | 42.56732 |
| 50 | | – | 7.06487 | 50.38951 | | – | – | 37.73623 |
| 100 | | – | – | 35.04791 | | – | – | 15.15995 |
| 120 | | – | – | 26.37614 | | – | – | 6.70530 |
| 140 | | – | – | 13.48791 | | – | – | |

$^\dagger$Special cases and related formulas presented by Karnovsky and Lebed (2003).

**TABLE 3.11.** The critical gravity parameters

| Standing clamped–free beam mode | | | Standing pin-guided–clamped beam mode | | |
|---|---|---|---|---|---|
| 1 | 2 | 3 | 1 | 2 | 3 |
| 7.837347 | 55.97743 | 148.5083 | 3.476597 | 44.13849 | 129.25843 |

In both cases, the coefficient

$$
\gamma_n = \frac{12}{l^2}\frac{\int\limits_0^l [X_n'(x)]^2\,dx}{\int\limits_0^l [X_n''(x)]^2\,dx}
$$

where $X_n(x)$ is the shape function for a particular mode $n$. Integrals in the formula for $\gamma$ are presented by Karnovsky and Lebed (2003).

The analytical expression for the gauge factor considering the effects of lateral deformation

$$
G_{\varepsilon n} = -(2+v) + \frac{1}{2}[1 + 2\varepsilon_0(1+v)]\frac{\gamma_n(1-v^2)\left(\dfrac{l}{h}\right)^2}{1 + \gamma_n\varepsilon_0(1-v^2)\left(\dfrac{l}{h}\right)^2} \tag{3.21}
$$

### Numerical results

*Clamped–clamped beam, Static nonlinearity.* The gauge factor $G$ of the fundamental mode as a function of the residual tensile strain $\varepsilon_0$ is presented in Fig. 3.8 (Tilmans, 1993)

The gauge factors for a cantilever beam with a lumped mass and for a simply supported beam with a symmetrical distributed mass are studied by Lebed *et al.* (1996a and b).

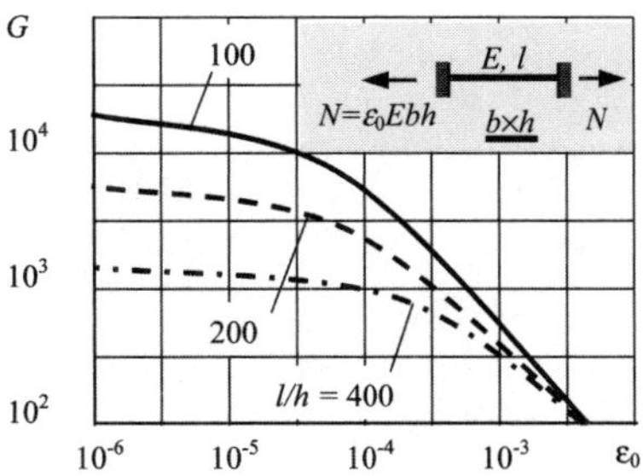


**FIGURE 3.8.** Gauge factor $G$ of the fundamental mode of vibration for a clamped–clamped beam for three values of the slenderness ratio $l/h$; a cross-section of the beam is rectangular, the width is $b$, height is $h$, $b > h$; and the Poisson ratio of material $v = 0.3$.

## *REFERENCES*

Amba-Rao, C.L. (1967) Effect of end conditions on the lateral frequencies of uniform straight columns. *Journal of the Acoustical Society of America,* **42,** 900–901.

Blevins, R.D. (1979) *Formulas for Natural Frequency and Mode Shape* (New York: Van Nostrand Reinhold).

Bokaian, A. (1988) Natural frequencies of beams under compressive axial loads. *Journal of Sound and Vibration,* **126**(1), 49–65.

Bokaian, A. (1990) Natural frequencies of beams under tensile axial loads. *Journal of Sound and Vibration,* **142**(3), 481–498.

Galef, A.E. (1968) Bending frequencies of compressed beams. *Journal of the Acoustical Society of America,* **44**(8), 643.

Karnovsky, I.A. and Lebed, O. (2003) Free vibrations of beams and frames. Eigenvalues and eigenfunctions. (New York: McGraw-Hill).

Liu, W.H. and Huang, C.C. (1988) Vibration of a constrained beam carrying a heavy tip body. *Journal of Sound and Vibration.* **123**(1), 15–19.

Liu, W.H. and Chen, K.S. (1989) Effects of lateral support on the fundamental natural frequencies and buckling coefficients. *Journal of Sound and Vibration,* **129**(1), 155–160.

Maurizi, M.J. and Belles, P.M. (1991) General equation of frequencies for vibrating uniform one-span beams under compressive axial loads. *Journal of Sound and Vibration* **145**(2), 345–347.

Naguleswaran, S. (1991) Vibration of a vertical cantilever with and without axial freedom at clamped end. *Journal of Sound and Vibration,* **146**(2), 191–198.

Tilmans, H.A.C. (1993) *Micro-Mechanical Sensors using Encapsulated Built-in Resonant Strain Gauges* (Enschede, The Netherlands: Febodruk) 310 p.

Timoshenko, S.P. and Gere, J.M. (1961) *Theory of Elastic Stability,* 2nd ed, (New York: McGraw-Hill).

Weaver, W., Timoshenko, S.P. and Young, D.H. (1990) *Vibration Problems in Engineering,* 5th edn (New York: Wiley).

## *FURTHER READING*

Crawford, F.S. (1965) *Waves.* Berkeley Physics Course, Vol. 3 (New York: McGraw-Hill).

Felgar, R.P. (1950) *Formulas for Integrals Containing Characteristic Functions of Vibrating Beams,* The University of Texas, Circular No.14.

Gorman, D.J. (1975) *Free Vibration Analysis of Beams and Shafts* (New York: Wiley).

Kim, Y.C. (1986) Natural frequencies and critical buckling loads of marine risers. *American Society of Mechanical Engineers, Fifth Symposium on Offshore Mechanics and Arctic Engineering,* pp. 442–449.

Kunukkasseril, V.X. and Arumugan, M. (1975) Transverse vibration of constrained rods with axial force fields. *Journal of the Acoustical Society of America,* **57**(1), 89–94.

Lebed, O.I., Karnovsky, I.A. and Chaikovsky, I. (1996) Limited displacement microfabricated beams and frames used as elastic elements in micromechanical devices. *Mechanics in Design,* University of Toronto, Ontario, Canada, Vol. 2, 1055–1061.

Lebed, O.I., Karnovsky, I.A. and Chaikovsky, I. (1996) Application of the mechanical impedance method to the definition of the mechanical properties of the thin film. *Mechanics in Design,* University of Toronto, vol. 2, 861–867.

Novacki, W. (1963) *Dynamics of Elastic Systems* (New York: Wiley.)

Paidoussis, M.P. and Des Trois Maisons, P.E. (1971) Free vibration of a heavy, damped, vertical cantilever. *Journal of Applied Mechanics,* **38,** 524–526.

Pilkington, D.F. and Carr, J.B. (1970) Vibration of beams subjected to end and axially distributed loading. *Journal of Mechanical Engineering Science*, **12**(1), 70–72.

Shaker, F.J. (1975) Effects of axial load on mode shapes and frequencies of beams. NASA Lewis Research Centre Report NASA-TN-8109.

Schafer, B. (1985) Free vibration of a gravity loaded clamped-free beam, *Ingenieur-Archiv*, **55**, 66–80.

Wittrick, W.H. (1985) Some observations on the dynamic equations of prismatic members in compression. *International Journal of Mechanical Science*, **27**(6), 375–382

Young, D. and Felgar, R.P., Jr. (1949) Tables of characteristic functions representing the normal modes of vibration of a beam. *The University of Texas Publication*, No. 4913.

# BRESS–TIMOSHENKO UNIFORM PRISMATIC BEAMS

Chapter 4 focuses on uniform Bress–Timoshenko beams. Eigenvalues and eigenfunctions for beams with classical boundary conditions are presented.

## 4.1 FUNDAMENTAL RELATIONSHIPS

### 4.1.1 Differential equations

The Bress–Timoshenko theory is used for describing the vibration of thick beams and for calculation of higher frequencies of vibrations. Timoshenko's equation is called the wave equation of the transverse vibration of a beam.

The slope of the deflection curve (DC) depends not only on the rotation of the cross-sections of the beam but also on the shearing deformations. Love and Bress–Timoshenko theories take into account the effects of rotatory inertia and shearing force (Chapter 1) (Timoshenko, 1922; Weaver *et al.* 1990). The free-body diagram of an element of a Timoshenko beam theory is presented in Fig. 4.1.

The angle $\psi$ denotes the slope of the deflection curve due to bending deformation alone, e.g., when the shearing force is neglected; the angle $\beta$ between the tangent to the deflection curve and the normal to the face denotes the shear deformation of the element (shear angle). Due to shear deformation, the tangent to the deflection curve will not be perpendicular to the face $a - b$.

The total slope, bending moment and shear force are

$$\frac{dy}{dx} = \psi + \beta$$

$$M = -EI\frac{d\psi}{dx} \tag{4.1}$$

$$Q = k\beta AG = kAG\left(\frac{dy}{dx} - \psi\right)$$

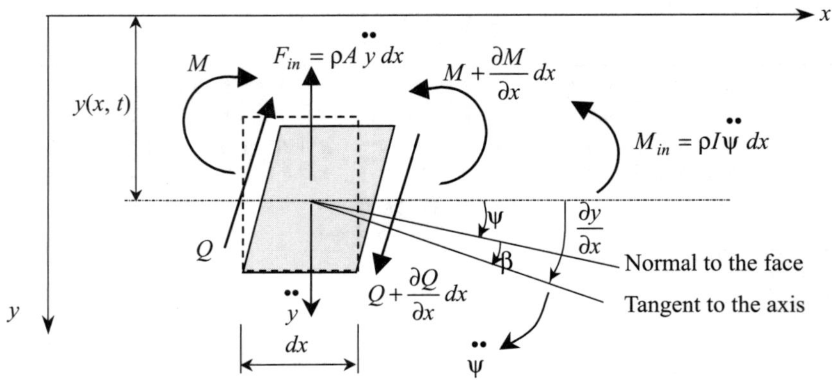


**FIGURE 4.1**   Notation and geometry of an element of a Timoshenko beam.

where $k$ is a shear coefficient depending on the shape of the cross-section, and $G$ is the modulus of elasticity in the shear. Shear coefficients for different cross-sections are presented in Table 4.1.

The higher order theories (Heyliger and Reddy, 1988; Stephen, 1978, 1983; Levinson 1981, 1982; Stephen and Levinson, 1979; Bickford, 1982; Murty, 1985, Ewing, 1990) correctly account for the stress-free conditions on the upper and lower surfaces of the beam. In this case the need for a shear correction coefficient is eliminated.

***Timoshenko equations. First form – coupled equations***

$$EI\frac{\partial^2 \psi}{\partial x^2} + kAG\left(\frac{\partial y}{\partial x} - \psi\right) - \rho I\frac{\partial^2 \psi}{\partial t^2} = 0$$

$$kAG\left(\frac{\partial^2 y}{\partial x^2} - \frac{\partial \psi}{\partial x}\right) - \rho A\frac{\partial^2 y}{\partial t^2} = 0$$

(4.2)

These equations describe coupled rotational and vertical displacements of the uniform beam. The shear coefficient $k$ for various cross sections is presented in Table 4.1.

Differential equations for non-uniform beams are presented by Thomson (1981).

## Timoshenko complete equations. Second form – separated equations

The system (4.2) may be presented in form of separated equations with respect of $y$ and $\psi$. The corresponding procedure is as follows. The second equation (4.2) may be rewritten as

$$\frac{\partial \psi}{\partial x} = \frac{\partial^2 y}{\partial x^2} - \frac{\rho A}{kAG}\frac{\partial^2 y}{\partial t^2}$$

(a)

Differentiation of the first equation of (4.2) with respect to $x$ leads to

$$EI\frac{\partial^3 \psi}{\partial x^3} + kAG\left(\frac{\partial^2 y}{\partial x^2} - \frac{\partial \psi}{\partial x}\right) - \rho I\frac{\partial^3 \psi}{\partial x\partial t^2} = 0$$

(b)

**TABLE 4.1.**   Shear coefficients for various cross-sections (Love, 1927; Cowper, 1966)

| Cross-section | Coefficient $k$ |
|---|---|
| Circle | $\dfrac{6(1+v)}{7+6v}$ |
| Hollow circle | $\dfrac{6(1+v)(1+m^2)^2}{(7+6v)(1+m^2)^2+(20+12v)m^2}, \quad m=\dfrac{b}{a}$ |
| Rectangular | $\dfrac{10(1+v)}{12+11v}$ |
| Ellipse | $\dfrac{12(1+v)a^2(3a^2+b^2)}{(40+37v)a^4+(16+10v)a^2b^2+vb^4}$ |
| Semicircle | $\dfrac{1+v}{1.305+1.273v}$ |
| Thin walled round tube | $\dfrac{2(1+v)}{4+3v}$ |
| Thin walled square tube | $\dfrac{20(1+v)}{48+39v}$ |
| Thin-walled I-section | $\dfrac{10(1+v)(1+3m)^2}{M_1+vM_2+5mn^2[6(1+m)+v(8+9m)]}, \quad m=\dfrac{2bt_f}{ht_w}, \quad n=\dfrac{b}{h}$ <br> $M_1=12+72m+150m^2+90m^3, \quad M_2=11+66m+135m^2+90m^3$ |
| Thin-walled box section | $\dfrac{10(1+v)(1+3m)^2}{M_1+vM_2+10mn^2(3+v+3m)}, \quad m=\dfrac{bt_1}{ht}; \quad n=\dfrac{b}{h}$ |
| Spar-and-web section | $\dfrac{10(1+v)(1+3m)^2}{M_1+vM_2}, \quad m=\dfrac{2F_s}{ht}, \quad F_s \text{ is area of one spar}$ |
| Thin-walled T-section | $\dfrac{10(1+v)(1+4m)^2}{M_3+vM_4+10mn^2(1+m)[3+(1+4m)]}, \quad m=\dfrac{bt_1}{ht}, \quad n=\dfrac{b}{h},$ <br> $M_3=12+96m+276m^2+192m^3, \quad M_4=11+88m+248m^2+216$ |

Substitution of (a) into (b) leads to

$$EI\frac{\partial^3\psi}{\partial x^3} + \rho A\frac{\partial^2 y}{\partial t^2} - \rho I\frac{\partial^3\psi}{\partial x\partial t^2} = 0 \qquad\text{(c)}$$

Differentiating twice equation (a) with respect to $x$ and $t$, leads to

$$\frac{\partial^3\psi}{\partial x^3} = \frac{\partial^4 y}{\partial x^4} - \frac{\rho A}{kAG}\frac{\partial^4 y}{\partial x^2\partial t^2} \qquad\text{(d)}$$

$$\frac{\partial^3\psi}{\partial x\partial t^2} = \frac{\partial^4 y}{\partial x^2\partial t^2} - \frac{\rho A}{kAG}\frac{\partial^4 y}{\partial t^4} \qquad\text{(e)}$$

Substitution of (d, e) into (c) leads to equation with respect to $y$

$$EI\frac{\partial^4 y}{\partial x^4} + \rho A\frac{\partial^2 y}{\partial t^2} - \rho I\left(1 + \frac{E}{kG}\right)\frac{\partial^4 y}{\partial x^2\partial t^2} + \frac{\rho^2 I}{kG}\frac{\partial^4 y}{\partial t^4} = 0 \qquad(4.3)$$

The equation with respect to bending slope $\psi$ (Cheng, 1970)

$$EI\frac{\partial^4\psi}{\partial x^4} + \rho A\frac{\partial^2\psi}{\partial t^2} - \rho I\left(1 + \frac{E}{kG}\right)\frac{\partial^4\psi}{\partial x^2\partial t^2} + \frac{\rho^2 I}{kG}\frac{\partial^4\psi}{\partial t^4} = 0 \qquad(4.3a)$$

This equation may be derived in a similar way.

The Timoshenko equation describes the disturbance of propagation in the axial direction of two waves with velocities $v_1 = \sqrt{\dfrac{E}{\rho}}$ and $v_2 = \sqrt{\dfrac{kG}{\rho}}$.

### 4.1.2  Kinetic and potential energy

The expressions for kinetic energy $T$ and potential energy $U$ may be written as follows (Bolotin, 1978; Yokoyama, 1991)

$$T = \frac{1}{2}\int_0^l m(x)\left(\frac{\partial y}{\partial t}\right)^2 dx + \frac{1}{2}\int_0^l \rho I\left(\frac{\partial^2 y}{\partial x\partial t} - \frac{\partial\beta}{\partial t}\right)^2 dx \qquad(4.4)$$

$$U = \frac{1}{2}\int_0^l EI\left(\frac{\partial^2 y}{\partial x^2} - \frac{\partial\beta}{\partial x}\right)^2 dx + \frac{1}{2}\int_0^l kGA\left(\frac{\partial y}{\partial x} - \psi\right)^2 dx + \frac{1}{2}\int_0^l k_{tr}y^2\, dx + \frac{1}{2}\int_0^l k_G\left(\frac{\partial y}{\partial x}\right)^2 dx \qquad(4.5)$$

where   $y$ = total transversal deflection
       $\psi$ = slope of the deflecting curve due to bending deformation alone
       $k_{tr}$ = Winkler foundation modulus
       $k_G$ = shear foundation modulus
       $k$ = shear coefficient (Table 4.1)
       $\beta$ = shear angle

The work $W$ done by a compressive axial force $N$ (positive for tension) is

$$W = -N\int_0^l \frac{1}{2}\left(\frac{\partial y}{\partial x}\right)^2 dx \qquad(4.6)$$

## *4.2  ANALYTICAL SOLUTION*

### 4.2.1  Frequency and normal mode equations

The differential equations of the transverse vibration of uniform beams are

$$
EI\frac{\partial^4 y}{\partial x^4} + \frac{\gamma A}{g}\frac{\partial^2 y}{\partial t^2} - \left(\frac{\gamma I}{g} + \frac{EI\gamma}{kgG}\right)\frac{\partial^4 y}{\partial x^2 \partial t^2} + \frac{\gamma I}{g}\frac{\gamma}{kgG}\frac{\partial^4 y}{\partial t^4} = 0
$$

$$
EI\frac{\partial^4 \psi}{\partial x^4} + \frac{\gamma A}{g}\frac{\partial^2 \psi}{\partial t^2} - \left(\frac{\gamma I}{g} + \frac{EI\gamma}{kgG}\right)\frac{\partial^4 \psi}{\partial x^2 \partial t^2} + \frac{\gamma I}{g}\frac{\gamma}{kgG}\frac{\partial^4 \psi}{\partial t^4} = 0
$$

$$(4.7)$$

where $y$, $\psi$ and $\dfrac{\partial y}{\partial x}$ are the transversal displacement, bending slope and slope of the center line of the beam, respectively.

***Solution.***   We assume the solutions in the form

$$
y = X(x)e^{j\omega t} \quad \text{and} \quad \psi = \Psi e^{j\omega t}
$$

Equations for normal functions of $X$ and $\Psi$ are (Huang, 1961)

$$
X^{IV} + b^2(r^2 + s^2)X'' - b^2(1 - b^2 r^2 s^2)X = 0
$$

$$
\Psi^{IV} + b^2(r^2 + s^2)\Psi'' - b^2(1 - b^2 r^2 s^2)\Psi = 0
$$

$$(4.8)$$

$$
b^2 = \frac{mL^4}{EI}\omega^2 \quad r^2 = \frac{I}{Al^2} \quad s^2 = \frac{EI}{kAGl^2} \quad m = \rho A, \quad \rho = \frac{\gamma}{g}
$$

where $r$ is the dimensionless radius of gyration.

***Eigenvalues.***   The frequencies of vibration may be calculated by the formula

$$
\omega = \frac{b}{l^2}\sqrt{\frac{EI}{m}}
$$

It is necessary to differentiate two cases.

   *Case 1.* This case would correspond to lower frequency vibrations

$$
\sqrt{(r^2 - s^2)^2 + \frac{4}{b^2}} > (r^2 + s^2) \quad \text{or} \quad b^2 r^2 s^2 = \frac{m\omega^2 I}{kA^2 G} < 1
$$

$$(4.9)$$

*Case 2.* This case would correspond to higher frequency vibrations

$$
\sqrt{(r^2 - s^2)^2 + \frac{4}{b^2}} < (r^2 + s^2) \quad \text{or} \quad b^2 r^2 s^2 = \frac{m\omega^2 I}{kA^2 G} > 1
$$

$$(4.10)$$

***Eigenfunctions.***   Let

$$\binom{\alpha}{\beta} = \frac{1}{\sqrt{2}}\sqrt{\mp(r^2 + s^2) + \sqrt{(r^2 - s^2)^2 + \frac{4}{b^2}}} \quad \text{for case 1}$$

$$\alpha = j\alpha' = \frac{j}{\sqrt{2}}\sqrt{(r^2 + s^2) - \sqrt{(r^2 - s^2)^2 + \frac{4}{b^2}}}, \quad j^2 = -1 \quad \text{for case 2}$$

Parameter $\beta$ for cases 1 and 2 are the same.

   *Case 1* ($\alpha$ is a real number, i.e. $b^2 r^2 s^2 < 1$)

$$X(\xi) = C_1 \cosh b\alpha\xi + C_2 \sinh b\alpha\xi + C_3 \cos b\alpha\xi + C_4 \sin b\alpha\xi, \quad \xi = x/l$$
$$\Psi = C_1' \sinh b\alpha\xi + C_2' \cosh b\alpha\xi + C_3' \sin b\alpha\xi + C_4' \cos b\alpha\xi$$

(4.11)

   *Case 2* ($\alpha$ is an imaginary number, i.e. $b^2 r^2 s^2 > 1$)

$$X(\xi) = C_1 \cosh b\alpha'\xi + jC_2 \sinh b\alpha'\xi + C_3 \cos b\beta\xi + C_4 \sin b\beta\xi, \quad \xi = x/l$$
$$\Psi = jC_1' \sin b\alpha\xi + C_2' \cos b\alpha\xi + C_3' \sin b\beta\xi + C_4' \cos b\beta\xi$$

(4.12)

Only one half of the constants are independent. They are related as follows:

$$C_1 = \frac{l}{b\alpha}[1 - b^2 s^2(\alpha^2 + r^2)]C_1' \qquad C_1' = \frac{b}{l}\frac{\alpha^2 + s^2}{\alpha}C_1$$

$$C_2 = \frac{l}{b\alpha}[1 - b^2 s^2(\alpha^2 + r^2)]C_2' \qquad C_2' = \frac{b}{l}\frac{\alpha^2 + s^2}{\alpha}C_2$$

$$\text{or}$$

$$C_3 = -\frac{l}{b\beta}[1 + b^2 s^2(\beta^2 - r^2)]C_2' \qquad C_3' = -\frac{b}{l}\frac{\beta^2 - s^2}{\beta}C_3$$

$$C_3 = \frac{l}{b\beta}[1 + b^2 s^2(\beta^2 - r^2)]C_4' \qquad C_4' = \frac{b}{l}\frac{\beta^2 - s^2}{\beta}C_4$$

The orthogonality condition of normal functions for the $m$th and $n$th modes

$$\int\limits_0^1 (X_m X_n + r_0^2 \Psi_m \Psi_n)\mathrm{d}\xi = 0, \quad m \neq n,$$

where $r_0$ is the radius of gyration of the cross-section around the principal axis, $r_0 = \sqrt{\dfrac{I}{A}}$.

   The frequency equation and normal modes for a beam with classical boundary conditions are presented in Tables 4.2 and 4.3. Additional notation:

$$\lambda = \frac{\alpha}{\beta} \quad \text{for case 1;} \quad \text{and} \quad \lambda' = \frac{\alpha'}{\beta} \quad \text{for case 2;}$$

$$\zeta = \frac{\alpha^2 + r^2}{\alpha^2 + s^2} = \frac{\beta^2 - s^2}{\beta^2 - r^2} = \frac{\alpha^2 + r^2}{\beta^2 - r^2} = \frac{\beta^2 - s^2}{\alpha^2 + s^2} \quad \text{for both cases}$$

***Special case (Bernoulli–Euler theory).***   In this case $r = 0, s = 0$ and

$$\alpha = \beta = \frac{1}{\sqrt{b}}, \quad \lambda = \frac{\alpha}{\beta} = 1, \quad \lambda' = \frac{\alpha'}{\beta} = j, \quad \zeta = \frac{\alpha^2 + r^2}{\beta^2 - r^2} = \frac{\beta^2 - s^2}{\alpha^2 + s^2} = 1.$$

**TABLE 4.2.** Uniform Bress–Timoshenko one-span beams with different boundary conditions: Frequency equations and mode shape expressions for Case 1 (Huang, 1961)



| Beam type | Frequency equations | Normal modes | Parameters |
|---|---|---|---|
| Pinned–pinned | $\sin b\beta = 0$ | $X = D \sin b\beta\xi$ <br> $\psi = H \sin b\beta\xi$ | |
| Free–free | $2 - 2\cosh b\alpha \cos b\beta + \dfrac{b}{\sqrt{1 - b^2 r^2 s^2}}$ <br> $\times\,[b^2 r^2 (r^2 - s^2)^2$ <br> $+ (3r^2 - s^2)]\sinh b\alpha \sin b\beta = 0$ | $X = \left[ D\cosh b\alpha\xi + \lambda\delta \sinh b\alpha\xi + \dfrac{1}{\zeta}\cos b\beta\xi + \delta \sin b\beta\xi \right]$ <br> $\Psi = H\left[ \cosh b\alpha\xi - \dfrac{\delta}{\lambda}\sinh b\alpha\xi + \zeta \cos b\beta\xi + \dfrac{1}{\delta}\sin b\beta\xi \right]$ | $\delta = \dfrac{\cosh b\alpha - \cos b\beta}{\lambda \sinh b\alpha - \zeta \sin b\beta}$ |
| Clamped–clamped | $2 - 2\cosh b\alpha \cos b\beta + \dfrac{b}{\sqrt{1 - b^2 r^2 s^2}}$ <br> $\times\,[b^2 s^2 (r^2 - s^2)^2$ <br> $+ (3r^2 - r^2)]\sinh b\alpha \sin b\beta = 0$ | $X = D\left[ \cosh b\alpha\xi + \lambda\zeta \sinh b\alpha\xi - \cos b\beta\xi + \delta \sin b\beta\xi \right]$ <br> $\Psi = H\left[ \cosh b\alpha\xi + \dfrac{\delta}{\lambda\zeta}\sinh b\alpha\xi - \cos b\beta\xi + \theta \sin b\beta\xi \right]$ | $\delta = \dfrac{-\cosh b\alpha + \cos b\beta}{\lambda\xi \sinh b\alpha + \sin b\beta}$ <br> $\theta = \dfrac{\lambda\zeta(-\cosh b\alpha + \cos b\beta)}{\sinh b\alpha + \lambda\zeta \sin b\beta}$ |
| Clamped–free | $2 + [b^2 (r^2 - s^2)^2 + 2]\cosh b\alpha \cos b\beta$ <br> $- \dfrac{b(r^2 + s^2)}{\sqrt{1 - b^2 r^2 s^2}}\sinh b\alpha \sin b\beta = 0$ | $X = D\left[ \cos b\alpha\xi - \lambda\xi\delta \sinh b\alpha\xi - \cos b\beta\xi + \delta \sin b\beta\xi \right]$ <br> $\Psi = H\left[ \cosh b\alpha\xi + \dfrac{\theta}{\lambda\zeta}\sinh b\alpha\xi - \cos b\beta\xi + \theta \sin b\beta\xi \right]$ | $\delta = \dfrac{\sinh b\alpha - \lambda \sin b\beta}{\lambda(\zeta \cosh b\alpha + \cos b\beta)}$ <br> $\theta = -\dfrac{\zeta(\lambda \sinh b\alpha + \sin b\beta)}{\cosh b\alpha + \cos b\beta}$ |
| Clamped–pinned | $\lambda\zeta \tanh b\alpha - \tan b\beta = 0$ | $X = D[\cosh b\alpha\xi - \coth b\alpha \sinh b\alpha\xi$ <br> $- \cos b\beta\xi + \cot b\beta \sin b\beta\xi]$ <br> $\Psi = H\left[ \cosh b\alpha\xi + \dfrac{\theta}{\lambda\zeta}\sinh b\alpha\xi - \cos b\beta\xi + \theta \sin b\beta\xi \right]$ | $\theta = -\dfrac{\zeta(\lambda \sinh b\alpha + \sin b\beta)}{\cosh b\alpha + \zeta \cos b\beta}$ |
| Pinned–free | $\lambda \tanh b\alpha - \zeta \tan b\beta = 0$ | $X = \lambda\dfrac{\cos b\beta}{\cosh b\alpha}\sinh b\alpha\xi + \sin b\beta\xi$ <br> $\Psi = \dfrac{1}{\lambda}\dfrac{\sin b\beta}{\sinh b\alpha}\cosh b\alpha\xi + \cos b\beta\xi$ | |

**TABLE 4.3.** Uniform Bress–Timoshenko one-span beams with different boundary conditions: Frequency equations and mode shape expressions for Case 2

| Beam type | Frequency equations | Normal modes | Parameters |
|---|---|---|---|
| Pinned–pinned | $\sin b\beta = 0$ | $X = D \sin b\beta\xi$<br>$\psi = H \sin b\beta\xi$ | |
| Free–free | $2 - 2\cos b\alpha' \cos b\beta + \dfrac{b}{\sqrt{b^2 r^2 s^2 - 1}}$<br>$\times [b^2 r^2 (r^2 - s^2)^2$<br>$+ (3r^2 - s^2)]\sin b\alpha' \sin b\beta = 0$ | $X = D\left[\cos b\alpha'\xi - \lambda'\eta \sin b\alpha'\xi + \dfrac{1}{\zeta}\cos b\beta\xi + \eta \sin b\beta\xi\right]$<br>$\Psi = H\left[\cos b\alpha'\xi - \dfrac{\eta}{\lambda'}\sin b\alpha'\xi + \zeta \cos b\beta\xi + \dfrac{1}{\eta}\sin b\beta\xi\right]$ | $\eta = -\dfrac{\cos b\alpha' - \cos b\beta}{\lambda' \sin b\alpha' + \zeta \sin b\beta}$ |
| Clamped–clamped | $2 - 2\cos b\alpha' \cos b\beta + \dfrac{b}{\sqrt{b^2 r^2 s^2 - 1}}$<br>$\times [b^2 s^2 (r^2 - s^2)^2$<br>$+ (3s^2 - r^2)]\sin b\alpha' \sin b\beta = 0$ | $X = D\cos b\alpha'\xi - \lambda'\zeta\eta \sin b\alpha'\xi - \cos b\beta\xi + \eta \sin b\beta\xi]$<br>$\Psi = H\left[\cos b\alpha'\xi + \dfrac{\mu}{\lambda'\zeta}\sinh b\alpha'\xi - \cos b\beta\xi + \mu \sin b\beta\xi\right]$ | $\eta = \dfrac{\cos b\alpha' - \cos b\beta}{\lambda'\xi \sin b\alpha' - \sin b\beta}$<br>$\mu = \dfrac{\lambda'\zeta(-\cos b\alpha' + \cos b\beta)}{\sinh b\alpha' + \lambda'\zeta \sin b\beta}$ |
| Clamped–free | $2 + [b^2 (r^2 - s^2)^2 + 2]\cos b\alpha' \cos b\beta$<br>$- \dfrac{b(r^2 + s^2)}{\sqrt{b^2 r^2 s^2 - 1}}\sinh b\alpha' \sin b\beta = 0$ | $X = D[\cos b\alpha'\xi + \lambda'\zeta\eta \sin b\alpha'\xi - \cos b\beta\xi + \eta \sin b\beta\xi]$<br>$\Psi = H\left[\cos b\alpha'\xi + \dfrac{\mu}{\lambda'\zeta}\sin b\alpha'\xi - \cos b\beta\xi + \mu \sin b\beta\xi\right]$ | $\eta = \dfrac{\sin b\alpha' - \lambda' \sin b\beta}{\lambda'(\zeta \cos b\alpha' + \cos b\beta)}$<br>$\mu = \dfrac{\zeta(\lambda' \sin b\alpha' - \sin b\beta)}{\cos b\alpha' + \cos b\beta}$ |
| Clamped–pinned | $\lambda'\zeta \tan b\alpha' + \tan b\beta = 0$ | $X = D[\cosh b\alpha'\xi - \cot b\alpha' \sin b\alpha'\xi$<br>$- \cos b\beta\xi + \cot b\beta \sin b \sin b\beta\xi]$<br>$\Psi = H\left[\cos b\alpha'\xi - \dfrac{\mu}{\lambda'\zeta}\sin b\alpha'\xi - \cos b\beta\xi + \mu \sin b\beta\xi\right]$ | $\mu = -\dfrac{\zeta(\lambda' \sinh b\alpha' - \sin b\beta)}{\cosh b\alpha' + \zeta \cos b\beta}$ |
| Pinned–free | $\lambda' \tan b\alpha' - \zeta \tan b\beta = 0$ | $X = -\lambda'\dfrac{\cos b\beta}{\cos b\alpha'}\sin b\alpha'\xi + \sin b\beta\xi$<br>$\Psi = -\dfrac{1}{\lambda'}\dfrac{\sin b\beta}{\sinh b\alpha'}\cos b\alpha'\xi + \cos b\beta\xi$ | |

***Example.*** For a pinned–free beam the frequency equations for cases 1 and 2 are

$$\lambda \tanh b\alpha - \zeta \tan b\beta = 0$$

$$\lambda' \tan b\alpha' + \zeta \tan b\beta = 0$$

In both cases, the frequency equations reduce to

$$\tanh \sqrt{b} - \tan \sqrt{b} = 0$$

***Solution of equation 4.8.*** The expression for the normal function may be presented as follows

$$X(\xi) = A_1 \sinh b\alpha\xi + A_2 \cosh b\alpha\xi + A_3 \sin b\beta\xi + A_4 \cos b\beta\xi$$

or, by using special functions, such as modified, or generalized Krylov's functions

$$X(\xi) = A_1 X_1(\xi) + A_2 X_2(\xi) + A_3 X_3(\xi) + A_4 X_4(\xi)$$

where

$$X_1(\xi) = \frac{1}{b^2(\alpha^2 + \beta^2)}(b^2\alpha^2 \cosh jb\beta\xi + b^2\beta^2 \cosh b\alpha\xi)$$

$$X_2(\xi) = \frac{1}{b^2(\alpha^2 + \beta^2)}\left(-\frac{jb\alpha^2}{\beta} \sinh jb\beta\xi + \frac{b\beta^2}{\alpha} \sinh b\alpha\xi\right)$$

$$X_3(\xi) = \frac{1}{b^2(\alpha^2 + \beta^2)}(\cosh b\alpha\xi - \cosh jb\beta\xi)$$

$$X_4(\xi) = \frac{1}{b^2(\alpha^2 + \beta^2)}\left(\frac{1}{b\alpha} \sinh b\alpha\xi + \frac{j}{b\beta} \sinh jb\beta\xi\right)$$

$$\binom{\alpha}{\beta} = \frac{1}{\sqrt{2}}\sqrt{\mp(r^2 + s^2) + \sqrt{(r^2 - s^2)^2 + \frac{4}{b^2}}}$$

The functions $X_i(\xi)$ and their derivatives result in the unit matrix at $\xi = 0$ (Karnovsky and Lebed O. 2003).

***Special case.*** For the Bernoulli–Euler theory, the parameters are

$$s = r = 0, \quad \alpha = \beta = \sqrt{\frac{1}{b}} = \frac{1}{\lambda}, \quad \lambda^4 = \frac{\omega^2 m l^4}{EI}$$

and functions $X_i$ transfer to Krylov functions. The normal functions and their derivatives for Bernoulli–Euler beams with different boundary conditions are presented in Appendix A.

## 4.2.2 State matrix. Dynamic stiffness matrix

The dynamic stiffness coefficients are presented as non-dimensional parameters corresponding to the effects of rotary inertia and of bending and shear deformation. The beam element and positive directions for the bending moment $M$, shear force $Q$, normal functions $X$ and $\psi$ are presented in Fig. 4.2.

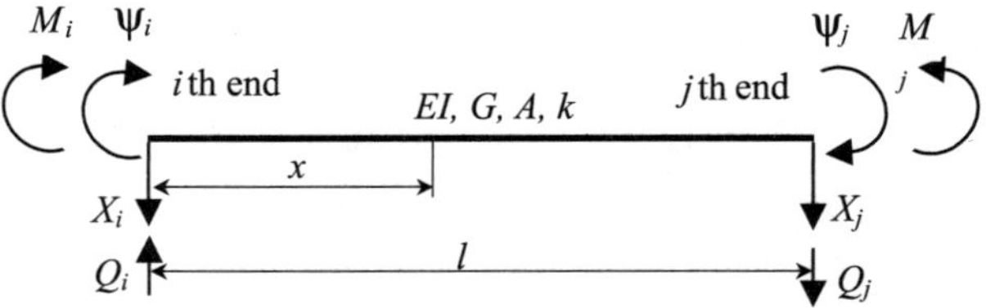



**FIGURE 4.2.**   Timoshenko beam, positive notation.

### *Conditions matrix*

The vector of condition $[X, \Psi, Q, M]^T$ for any section of a beam can be presented in the following form (Cheng, 1970)

$$
\begin{bmatrix} X(\xi) \\ \Psi(\xi) \\ Q(\xi) \\ M(\xi) \end{bmatrix} = A \times \begin{bmatrix} C_1 \\ C_2 \\ C_3 \\ C_4 \end{bmatrix}
\tag{4.13}
$$

where the matrix of the system is

$$
A = \begin{bmatrix}
\cosh b\alpha\xi & \sinh b\alpha\xi & \cos b\beta\xi & \sin b\beta\xi \\
T \sinh b\alpha\xi & T \cosh b\alpha\xi & -U' \sin b\beta\xi & U' \cos b\beta\xi \\
(\mu - kGAT)\sinh b\alpha\xi & (\mu - kGAT)\cosh b\alpha\xi & (-\eta + kAGU')\sin b\beta\xi & (\eta - kAGU')\cos b\beta\xi \\
-\Omega \cosh b\alpha\xi & -\Omega \sinh b\alpha\xi & \tau' \cos b\beta\xi & \tau' \sin b\beta\xi
\end{bmatrix}
$$

Parameters $b, r, s, \alpha, \beta$ are presented in Section 4.2.1. Additional parameters are

$$
T = \frac{b}{l}\frac{\alpha^2 + s^2}{\alpha} \qquad \mu = kaG\frac{\alpha b}{l} \qquad \eta = kAG\frac{\beta b}{l}
$$

$$
\Omega = EI\frac{\alpha b}{l}T \qquad \tau' = EI\frac{\beta b}{l}U' \qquad U' = \frac{b}{l}\frac{\beta^2 - s^2}{\beta}
$$

The vector of integration constants may be presented in terms of initial parameters as follows

$$
\begin{bmatrix} C_1 \\ C_2 \\ C_3 \\ C_4 \end{bmatrix} =
\begin{bmatrix}
1 & 0 & 1 & 0 \\
0 & T & 0 & U' \\
0 & (\mu - kGAT) & 0 & (\eta - kAGU') \\
-\Omega & 0 & \tau' & 0
\end{bmatrix}^{-1}
\times
\begin{bmatrix} Y(0) \\ \Psi(0) \\ Q(0) \\ M(0) \end{bmatrix}
$$

### Dynamic stiffness matrix

This matrix describes relationships between internal forces (bending moment and shear) and displacements (deflection and slope) for two different sections of a beam.

$$
\begin{bmatrix} M_i \\ M_j \\ Q_i \\ Q_j \end{bmatrix} =
\begin{bmatrix}
s(M\Psi)_1 & s(M\Psi)_2 & s(MX)_1 & s(MX)_2 \\
 & s(M\Psi)_1 & s(MX)_2 & s(MX)_1 \\
 & & s(QX)_1 & s(QX)_2 \\
\text{Symmetric} & & & s(QX)_1
\end{bmatrix}
\times
\begin{bmatrix} \Psi_i \\ \Psi_j \\ X_i \\ X_j \end{bmatrix}
\tag{4.14}
$$

The elements of the dynamic stiffness matrix for cases 1 and 2 are as follows

*Case 1.* $b^2 r^2 s^2 < 1$

$$
s(M\Psi)_1 = \frac{l[-\alpha(\beta^2 - s^2)\sinh b\alpha \cos b\beta + \beta(\alpha^2 + s^2)\cosh b\alpha \sin b\beta]}{kAGbs^2(\alpha^2 + \beta^2)D}
$$

$$
s(M\Psi)_2 = \frac{l[\alpha(\beta^2 - s^2)\sinh b\alpha - \beta(\alpha^2 + s^2)\sin b\beta]}{kAGbs^2(\alpha^2 + \beta^2)D}
$$

$$
s(MX)_1 = \frac{l^2[\alpha\beta(2s^2 + \alpha^2 - \beta^2)(1 - \cosh b\alpha \cos b\beta) - (2\alpha^2\beta^2 - \alpha^2 s^2 + \beta^2 s^2)\sinh b\alpha \sin b\beta]}{EIb^2(\alpha^2 + \beta^2)^2 \alpha\beta D}
$$

$$
s(MX)_2 = \frac{(\beta^2 - s^2)(\alpha^2 + s^2)[-\cosh b\alpha + \cos b\beta]}{kAGs^2(\alpha^2 + \beta^2)D}
$$

$$
s(QX)_1 = \frac{l[\alpha(\beta^2 - s^2)\cosh b\alpha \sin b\beta + \beta(\alpha^2 + s^2)\sinh b\alpha \cos b\beta]}{EIb\alpha\beta(\alpha^2 + \beta^2)D}
$$

$$
s(QX)_2 = \frac{l[\beta(\alpha^2 + s^2)\sinh b\alpha + \alpha(\beta^2 - s^2)\sin b\beta]}{EIb\alpha\beta(\alpha^2 + \beta^2)D}
$$

$$
D = \frac{l^2\{2\alpha\beta(\beta^2 - s^2)(\alpha^2 + s^2) - 2\alpha\beta(\beta^2 - s^2)(\alpha^2 + s^2)\cosh b\alpha \cos b\beta + [(\alpha^2 - \beta^2)(\alpha^2\beta^2 - s^4) + 4\alpha^2\beta^2 s^2]\sinh b\alpha \sin b\beta\}}{kAGEIb^2(\alpha^2 + \beta^2)^2 \alpha\beta s^2}
$$

$$
\tag{4.15}
$$

*Case 2.* $b^2 r^2 s^2 > 1,\ \alpha' = \mathrm{j}\alpha$

$$s(M\Psi)_1 = \frac{1}{bD}[\alpha'(\beta^2 - s^2)\sin b\alpha' \cos b\beta + \beta(\alpha^2 + s^2)\sin b\beta \cos b\alpha']\frac{EI}{l}$$

$$s(M\Psi)_2 = -\frac{1}{bD}[\alpha'(\beta^2 - s^2)\sin b\alpha' + \beta(\alpha^2 + s^2)\sin b\beta]\frac{EI}{l}$$

$$s(MX)_1 = \frac{1}{(\alpha^2 + \beta^2)D}(\alpha^2 + s^2)(\beta^2 - s^2)\left[(s^2 - r^2) - \frac{s^2(s^2 + r^2) + 2\beta^2\alpha^2}{\beta\alpha'}\sin b\alpha' \sin b\beta\right]\frac{EI}{l^2}$$

$$+ \frac{(\alpha^2 + s^2)(\beta^2 - s^2)}{(\alpha^2 + \beta^2)D}(r^2 - s^2)\cos b\alpha' \cos b\beta\frac{EI}{l^2}$$

$$s(MX)_2 = \frac{(\alpha^2 + s^2)(\beta^2 - s^2)}{D}[\cos b\beta - \cos b\alpha']\frac{EI}{l^2}$$

$$S(QX)_1 = \frac{1}{bD}\left[\frac{(\alpha^2 + s^2)}{\alpha'}\sin b\alpha' + \frac{(\beta^2 - s^2)}{\beta}\sin b\beta\right]\frac{EI}{l^3}$$

$$S(QX)_2 = \frac{1}{bD}\left[\frac{(\alpha^2 + s^2)}{\alpha'}\cos b\beta \sin b\alpha' + \frac{(\beta^2 - s^2)}{\beta}\sin b\beta \cos b\alpha'\right]\frac{EI}{l^3}$$

$$D = \frac{2(\alpha^2 + s^2)(\beta^2 - s^2)}{b^2(\alpha^2 + \beta^2)}[1 - \cos b\beta \cos b\alpha']$$

$$+ \frac{\beta^2(\alpha + s^2)^2 + \alpha'^2(\beta^2 - s^2)^2}{\alpha'\beta b^2(\alpha^2 + \beta^2)}\sin b\beta \sin b\alpha' \tag{4.16}$$

Another representation of the Timoshenko equation is presented by Genkin and Tarckhanov (1979).

The transitional matrix for the Timoshenko equation is presented by Ivovich (1981) and Pilkey (1994).

## 4.3 SOLUTIONS FOR THE SIMPLEST CASES

In this section, the results for the fundamental mode of vibration for simply supported, cantilever and clamped beams are presented.

### 4.3.1 Pinned–pinned beam. Frequency equation and frequency of vibration: exact solution

***Truncated differential equation.***     In this case, the term $\dfrac{\partial^4 y}{\partial t^4}$ in the Timoshenko equation is omitted.

*Effect of rotary motion and shearing force.*     Frequency equation (Filin, 1981)

$$\frac{n^2\pi^2}{l^2}\frac{m\omega^2}{EA}\left(1 + \frac{E}{kG}\right) + \frac{m\omega^2}{EI} = \frac{n^4\pi^4}{l^4} \tag{4.17}$$

Frequency of vibration

$$\omega = \frac{n^2\pi^2}{l^2}\sqrt{\frac{EI}{m}}R$$

where correction factor

$$R = \cfrac{1}{\sqrt{1 + \cfrac{n^2\pi^2}{l^2}\cfrac{I}{A}\left(1 + \cfrac{E}{kG}\right)}}$$

*Effect of shearing force.* The frequency equation and natural frequency of vibration are

$$\frac{n^2\pi^2}{l^2}\frac{m\omega^2}{kAG} + \frac{m\omega^2}{EI} = \frac{n^4\pi^4}{l^4}$$

$$\omega = \frac{n^2\pi^2}{l^2}\sqrt{\frac{EI}{m}}R$$

where the correction factor is

$$R = \cfrac{1}{\sqrt{1 + \cfrac{n^2\pi^2}{l^2}\cfrac{EI}{kAG}}}$$

## Complete differential equation. Frequency equation and frequency of vibration

*Effect of rotary motion and shearing force.* The frequency equation, and natural frequency of vibration are

$$\frac{n^2\pi^2}{l^2}\frac{m\omega^2}{EA}\left(1 + \frac{E}{Gk}\right) + \frac{m\omega^2}{EI}\left(1 - \frac{mr_0^2}{AGk}\omega^2\right) = \frac{n^4\pi^4}{l^4}$$

$$\omega_{1,2}^2 = \frac{b_1 \pm \sqrt{b_1^2 - 4ac}}{2a} \tag{4.18}$$

where

$$r_0^2 = \frac{I}{A}, \quad b_1 = \frac{m}{EI} + \frac{n^2\pi^2 m}{l^2 EA}\left(1 + \frac{E}{Gk}\right), \quad a = \frac{m^2 r_0^2}{EAGIk}, \quad c = \frac{n^4\pi^4}{l^4}$$

*Effect of shearing force.* The frequency equation, and natural frequency of vibration are

$$\frac{n^2\pi^2}{l^2}\frac{m\omega^2}{AGk} + \frac{m\omega^2}{EI}\left(1 - \frac{mr^2}{kAG}\omega^2\right) = \frac{n^4\pi^4}{l^4}$$

$$\omega_{1,2}^2 = \frac{b_2 \pm \sqrt{b_2^2 - 4ac}}{2a}$$

where

$$b_2 = \frac{m}{EI} + \frac{n^2\pi^2 m}{l^2 AGk}$$

*Effect of rotary motion.*   The frequency equation, and natural frequency of vibration are

$$\frac{n^2\pi^2}{l^2}\frac{m\omega^2}{EA} + \frac{m\omega^2}{EI} = \frac{n^4\pi^4}{l^4}$$

$$\omega = \frac{n^2\pi^2}{l^2}\sqrt{\frac{EI}{m}}\frac{1}{\sqrt{1 + r_0^2\,\dfrac{n^2\pi^2}{l^2}}}$$

*Technical (Bernoulli–Euler) theory.*   The effects of rotary motion and shearing force are neglected.

The natural frequency of vibration is

$$\omega = \frac{n^2\pi^2}{l^2}\sqrt{\frac{EI}{m}}$$

Different numerical approaches, vast numerical results and their detailed analysis are presented in Sekhniashvili (1960).

***Approximate solution.***   The approximate expressions for the fundamental frequency of vibration of pinned–pinned beam are obtained by using the Bubnov–Galerkin method from the following assumptions (Sekhniashvili, 1960).

**1.** Elastic curve $y = \sin \pi x/l$, $E/kG \ll 1$

$$\omega_1 = \frac{\pi^2}{l^2}\sqrt{\frac{EI}{m}}R \quad R = \left[1 - \frac{1}{2}\pi^2\frac{r_0^2}{l^2}\left(1 + \frac{E}{kG}\right)\right]$$

If correction factor $R < 0$ then one needs to use the following formula.
**2.** More precise formula; elastic curve $y = \sin \pi x/l$

$$\omega_1 = \frac{\pi^2}{l^2}\sqrt{\frac{EI}{m}}R$$

where the correction factor is

$$R = \sqrt{\frac{1}{1 + \pi^2\dfrac{r_0^2}{l^2}\left(1 + \dfrac{E}{kG}\right)}}$$

**3.** The governing equation is a four-term differential one, the elastic curve neglects the effects of rotary motion and shearing force. Elastic curve

$$y = \frac{ml^4}{24EI}(\xi - 2\xi^3 + \xi^4), \quad \xi = \frac{x}{l}$$

In this case the fundamental frequency of vibration

$$\omega_1 = \frac{9.876}{l^2}\sqrt{\frac{EI}{m}}R$$

where the correction factor is

$$R = \sqrt{\dfrac{1}{1 + 9.871\dfrac{r_0^2}{l^2}\left(1 + \dfrac{E}{kG}\right)}}$$

**4.** The elastic curve takes into account the effects of rotary motion and shearing force

$$y(x) = \frac{ml^4}{24EI}\left\{\left[\frac{24(1+v)r_0^2}{kl^2} + 1\right]\xi - 2\xi^3 + \xi^4\right\}$$

where $v$ is the Poisson coefficient.

Initial parameters are

$$y(0) = 0, \quad y'(0) = \frac{ml^3}{24EI} + \frac{ml}{2kGA}, \quad y''(0) = 0, \quad y'''(0) = -\frac{ml}{2}$$

The fundamental frequency of vibration

$$\omega_1 = \frac{9.876}{l^2}\sqrt{\frac{EI}{m}}R, \quad R = \sqrt{\frac{A_0}{B_0}} \quad A_0 = 1 + 60\frac{1}{k}(1+v)\frac{r_0^2}{l^2}$$

$$B_0 = 1 + \frac{3654}{31}\frac{1}{k}(1+v)\frac{r_0^2}{l^2} + \frac{120960}{31}\frac{1}{k^2}(1+v)^2\frac{r_0^2}{l^4}$$

$$+ \frac{7560}{31}\frac{i^2}{l^2}\left[1 + 2\frac{1}{k}(1+v)\right]\left[\frac{17}{420} + 2\frac{1}{k}(1+v)\frac{r_0^2}{l^2}\right]$$

**5.** Euler two-term differential equation; the elastic curve takes into account the effects of rotary motion and shearing force

$$y(x) = \frac{ml}{24EI}\left\{\left[\frac{24(1+v)r_0^2}{kl^2} + 1\right]\xi - 2\xi^3 + \xi^4\right\}$$

The fundamental frequency of vibration

$$\omega_1 = \frac{9.876}{l^2}\sqrt{\frac{EI}{m}}R, \quad R = \sqrt{\frac{A_0}{B_0}}; \quad A_0 = 1 + 60\frac{1+v}{k}\cdot\frac{r_0^2}{l^2}$$

$$B_0 = 1 + \frac{3654}{31}\frac{1+v}{k}\frac{r_0^2}{l^2} + \frac{120960}{31}\frac{(1+v)^2}{k^2}\frac{r_0^4}{l^4}$$

Correction factor $R$ for pinned–pinned–beam, which are calculated using different governing equations and assumptions concerning elastic curves, are presented in Table 4.4.

### 4.3.2 Cantilever beam. Approximate solution

The frequency of the fundamental mode of vibration may be calculated by the formula

$$\omega_1 = \frac{3.529}{l^2}\sqrt{\frac{EI}{m}}R$$

where the correction factor is

$$R = \sqrt{\frac{A_1}{B_1}}, \quad A_1 = 1 + \frac{40}{3k}(1+v)\frac{r_0^2}{l^2}$$

$$B_1 = 1 + \frac{21.96}{k}(1+v)\frac{r_0^2}{l^2} + \frac{132.82}{k^2}(1+v)^2\frac{r_0^4}{l^4}$$

$$- \frac{135}{26}\frac{r_0^2}{l^2}\left[1 + \frac{2}{k}(1+v)\right]\left[\frac{1}{7} - \frac{1.6}{k}(1+v)\frac{r_0^2}{l^2} - \frac{64}{k^2}(1+v)^2\frac{r_0^4}{l^4}\right]$$

Galerkin's method has been applied; the expression for the elastic curve takes into account the shear effect (Sekhniashvili, 1960).

**TABLE 4.4.**  Pinned–pinned beam. Correction factor $R$ for the fundamental mode of vibration including effects of rotary motion and shearing force. Beam with rectangular cross-section ($1/k = 3/2$, $1 + v = 1.25$)

| | | | | | $h/l$ | | | | | | |
|---|---|---|---|---|---|---|---|---|---|---|---|
| Expr | 0.05 | 0.1 | 0.2 | 0.3 | 0.4 | 0.5 | 0.6 | 0.7 | 0.8 | 0.9 | 1.0 |
| (1) | 0.9951 | 0.9805 | 0.9219 | 0.8244 | 0.6878 | 0.5122 | 0.2975 | 0.0438 | – | – | – |
| (2) | 0.9954 | 0.9811 | 0.9301 | 0.8603 | 0.7849 | 0.7114 | 0.6448 | 0.5860 | 0.5347 | 0.4902 | 0.4516 |
| (3) | 0.9953 | 0.9810 | 0.9300 | 0.8601 | 0.7844 | 0.7112 | 0.6446 | 0.5858 | 0.5345 | 0.4900 | 0.4514 |
| (4) | 0.9835 | 0.9411 | 0.8111 | 0.6776 | 0.5675 | 0.4808 | 0.4159 | 0.3647 | 0.3241 | 0.2912 | 0.2641 |
| (5) | 0.9870 | 0.9562 | 0.8409 | 0.7333 | 0.6257 | 0.5376 | 0.4679 | 0.4124 | 0.3676 | 0.3310 | 0.3008 |

1. Sign (–) means loss of physical meaning. In this case use expression (2).
2. Rows 1 and 2 have multiples $9.869/l^2(EI/m)^{1/2}$.
3. Rows 3,4,5 have multiples $9.876/l^2(EI/m)^{1/2}$.

### 4.3.3  Clamped beam. Approximate solution

The frequency of the fundamental mode of vibration may be calculated by the formula

$$\omega_1 = \frac{22.449}{l^2}\sqrt{\frac{EI}{m}}R$$

where the correction factor

$$R = \sqrt{\frac{A_2}{B_2}}, \quad A_2 = 1 + \frac{120}{k}(1+v)\frac{r_0^2}{l^2}$$

$$B_2 = 1 + \frac{216}{k}(1+v)\frac{r_0^2}{l^2} + \frac{12096}{k^2}(1+v)^2\frac{r_0^4}{l^4}$$

$$+ 1260\frac{r_0^2}{l^2}\left[1 + \frac{2}{k}(1+v)\right]\left[\frac{71}{105} + \frac{1.6}{k}(1+v)\frac{r_0^2}{l^2} + \frac{96}{k^2}(1+v)^2\frac{r_0^4}{l^4}\right]$$

Galerkin's method has been applied; the expression for the elastic curve takes into account the shear effect.

Correction factor $R$ for one-span beams with different values $h : l$ for prismatic beams or $r_0^2 : l^2$ for beams of any cross section are presented in Table 4.5. These data correspond to static elastic curves, which takes into account the effects of rotary motion and shear force.

**TABLE 4.5.** Correction factor $R$ for prismatic beams with different boundary conditions and values $h : l$ or $r_0^2 : l^2$

$$\left( r_0^2 = \frac{I}{A}; k = 2/3, v = 0.25 \right)$$

| $h : l$ or $r_0^2 : l^2$ | | | | |
|---|---|---|---|---|
| 0.05 or | 1/4800 | 0.9835 | 0.9987 | 0.9760 |
| 0.10 | 1/1200 | 0.9411 | 0.9976 | 0.9118 |
| 0.20 | 4/1200 | 0.8111 | 0.9784 | 0.7297 |
| 0.30 | 9/1200 | 0.6776 | 0.9502 | 0.5552 |
| 0.40 | 16/1200 | 0.5675 | 0.9096 | 0.4191 |
| 0.50 | 25/1200 | 0.4808 | 0.8572 | 0.3212 |
| 0.60 | 36/1200 | 0.4159 | 0.7954 | 0.2492 |
| 0.70 | 49/1200 | 0.3647 | 0.7275 | 0.1986 |
| 0.80 | 64/1200 | 0.3241 | 0.6580 | 0.1606 |
| 0.90 | 81/1200 | 0.2912 | 0.5902 | 0.1320 |
| 1.00 | 100/1200 | 0.2641 | 0.5268 | 0.1101 |

## 4.4  BEAMS WITH A LUMPED MASS AT THE MIDSPAN

The frequency equations for one-span Timoshenke beams with different boundary conditions and one additional lumped mass are presented.

### 4.4.1  Simply supported beam

The design diagram is presented in Fig. 4.3(a).
 Symmetric and antisymmetric vibration are considered separately.

***Symmetric modes of vibration.***   The frequency equation may be written as follows

$$b^4 \beta \alpha (\beta^2 + \alpha^2) \cosh \frac{b\beta}{2} \cosh \frac{b\alpha}{2}$$

$$+ nb^2 \left[ b\beta \left( d - \frac{s^2}{2} b^2 \beta^2 \right) \sinh \frac{b\alpha}{2} \cos \frac{b\beta}{2} - b\alpha \left( d - \frac{s^2}{2} b^2 \alpha^2 \right) \right] \sin \frac{b\beta}{2} \cosh \frac{b\alpha}{2} = 0 \quad (4.19)$$

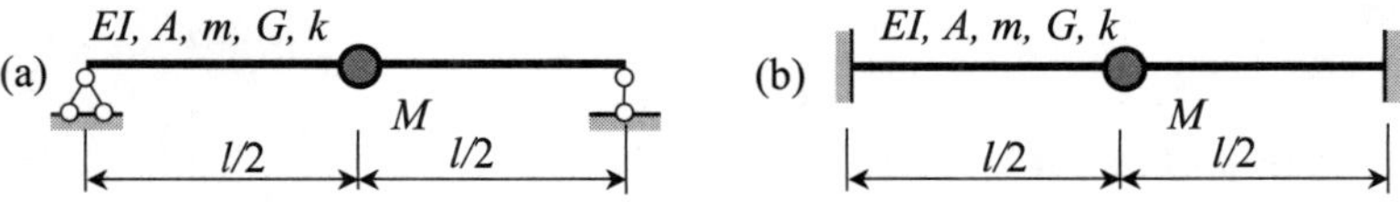


**FIGURE 4.3**   Uniform one-span beams with lumped mass in the middle of the span.

where

$$\begin{pmatrix} b\alpha \\ b\beta \end{pmatrix} = \frac{b}{\sqrt{2}}\sqrt{\mp(r^2 + s^2) + \sqrt{(r^2 - s^2)^2 + \frac{4}{b^2}}}$$

$$b^2 = \frac{ml^4}{EI}\omega^2, \quad r^2 = \frac{I}{Al^2} = \frac{r_0^2}{l^2}, \quad s^2 = \frac{EI}{kAl^2 G} = r^2\frac{E}{kG}$$

$$d = \frac{1}{2}\left[1 + b^2 r^4\left(\frac{E}{kG}\right)^2\right] = \frac{1}{2}(1 + b^2 s^2), \quad n = \frac{M}{ml}, \quad r_0 = \sqrt{\frac{I}{A}}$$

Here, $r_0$ is the radius of gyration of a cross-section, and $r$ is the non-dimensional radius of gyration, which is the reciprocal of slenderness.

The frequency parameters $b$ for the fundamental mode and different values $n = M/ml$ are presented in Table 4.6. Assume that $l/r_0 = r^{-1} = 20$, i.e. for rectangular cross-section $h : l \approx 1 : 5.8; k = 5/6; E/G = 2.33$ (Filippov, 1970).

**TABLE 4.6.**  Simply supported uniform beam with lumped mass at the middle of the span: Fundamental frequency parameter according to two theories

|  | | $n = M/ml$ | | |
|---|---|---|---|---|
| Frequency parameter $b$ | 0.045 | 0.638 | 2.270 | 4.370 |
| Timoshenko's theory | 9.00 | 6.25 | 4.00 | 3.00 |
| Technical theory ($b_0$) | 9.45 | 6.50 | 4.16 | 3.12 |
| ($b_0 - b$)100%/$b_0$ | 4.8 | 4.13 | 4.0 | 3.8 |

Vast numerical results for simply-supported beams with lumped mass along the span are presented in Maurizi and Belles (1991).

***Antisymmetric modes of vibration.***   The frequency equation may be written as

$$\sin\frac{b\beta}{2}\sinh\frac{b\alpha}{2} = 0 \tag{4.20}$$

***Special case:***   For Bernoulli–Euler theory, parameters $s = r = 0$, $\alpha = \beta = \dfrac{1}{\sqrt{b}}$. The frequency equation is $\sin\dfrac{\sqrt{b}}{2} = 0$ and frequency of vibration $\omega = \dfrac{4\pi^2}{l^2}\sqrt{\dfrac{EI}{m}}$.

### 4.4.2  Clamped beam

The design diagram is presented in Fig. 4.3(b).

***Symmetric modes of vibration.***   The frequency equation may be written as follows (Filippov, 1970)

$$
b^2(\beta^2 + \alpha^2)\left( b\beta \cosh\frac{b\alpha}{2}\sin\frac{b\beta}{2} + b\alpha \sinh\frac{b\alpha}{2}\cos\frac{b\beta}{2}\right)
$$

$$
+ nb^2\left\{ \left(\cos\frac{b\beta}{2}\cosh\frac{b\alpha}{2} - 1\right)\left[2d - b^2\frac{s^2}{2}(\beta^2 - \alpha^2)\right]\right.
$$

$$
\left. - \frac{1}{\beta\alpha}\sinh\frac{b\alpha}{2}\sin\frac{b\beta}{2}\left[\frac{b^2 s^2}{2}(\beta^4 + \alpha^4) - d(\beta^2 - \alpha^2)\right]\right\} = 0 \qquad (4.21)
$$

The frequency parameters $b$ for the fundamental mode and different values $n = M/ml$ are presented in Table 4.7. Assume that $l/r_0 = 20$, i.e. for rectangular cross-section $h : l \approx 1 : 5.8$; $k = 5/6$; $E/G = 23.33$ (Filippov, 1970).

**TABLE 4.7.**   Clamped uniform beam with lumped mass at the middle of the span: Fundamental frequency parameter according to two theories

|  | $n = M/ml$ | | | |
|---|---|---|---|---|
| Frequency parameter $b$ | 0.053 | 0.319 | 0.851 | 1.78 |
| $b$ (Timoshenko's theory) | 20 | 16 | 12 | 9 |
| $b_0$ (Technical theory) | 21.1 | 16.78 | 12.58 | 9.43 |
| $(b_0 - b)100\%/b_0$ | 5.22 | 4.65 | 4.60 | 4.55 |

***Antisymmetric modes of vibration.***   The frequency equation may be written as

$$
\frac{1}{\beta}\sin\frac{b\beta}{2}\cosh\frac{b\alpha}{2} - \frac{1}{\alpha}\sinh\frac{b\alpha}{2}\cos\frac{b\beta}{2} = 0 \qquad (4.22)
$$

## 4.5  CANTILEVER TIMOSHENKO BEAM OF UNIFORM CROSS-SECTION WITH TIP MASS AT THE FREE END

The frequency coefficients for the first five modes of vibration of the Timoshenko beam of uniform cross-section are presented in Table 4.8 (Rossi *et al.*, 1990).

***Notation***

$$
\Omega^2 = \rho A L^4 \omega^2 / EI; \quad n = M/mL; \quad r^2 = I/AL^2; \quad v = 0.3; \quad k = 5/6.
$$

## 4.6  UNIFORM SPINNING BRESS–TIMOSHENKO BEAMS

In this section, a free vibration analysis of a spinning, finite Timoshenko beam with general boundary conditions is presented. Analytical solutions of the frequency equations and mode shapes are given for six types of boundary conditions.

**TABLE 4.8**  Timoshenko cantilever uniform beam with tip mass at the free end: Frequency parameters for different mode shapes

| $r^2$ | $n$ | $\Omega_1$ | $\Omega_2$ | $\Omega_3$ | $\Omega_4$ | $\Omega_5$ |
|---|---|---|---|---|---|---|
| $10^{-7}$ | 0.0 | 3.51 | 22.03 | 61.69 | 120.89 | 199.85 |
| | 0.2 | 2.61 | 18.20 | 53.55 | 108.18 | 182.42 |
| | 0.4 | 2.16 | 17.17 | 52.06 | 106.45 | 180.53 |
| | 0.6 | 1.89 | 16.70 | 51.44 | 105.77 | 179.82 |
| | 0.8 | 1.70 | 16.42 | 51.10 | 105.41 | 179.45 |
| | 1.0 | 1.55 | 16.25 | 50.89 | 105.19 | 179.22 |
| 0.0004 | 0.0 | 3.50 | 21.47 | 58.14 | 109.02 | 171.29 |
| | 0.2 | 2.60 | 17.82 | 50.86 | 98.55 | 158.19 |
| | 0.4 | 2.16 | 16.83 | 49.48 | 97.04 | 156.66 |
| | 0.6 | 1.88 | 16.37 | 48.91 | 96.45 | 156.08 |
| | 0.8 | 1.69 | 16.10 | 48.60 | 96.14 | 155.77 |
| | 1.0 | 1.55 | 15.93 | 48.40 | 95.94 | 155.58 |
| 0.0016 | 0.0 | 3.46 | 20.01 | 50.56 | 88.19 | 129.98 |
| | 0.2 | 2.58 | 16.82 | 44.89 | 80.96 | 121.79 |
| | 0.4 | 2.14 | 15.92 | 43.74 | 79.81 | 120.69 |
| | 0.6 | 1.87 | 15.50 | 43.25 | 79.35 | 120.26 |
| | 0.8 | 1.68 | 15.26 | 42.99 | 79.10 | 120.03 |
| | 1.0 | 1.54 | 15.10 | 42.82 | 78.94 | 119.89 |
| 0.0036 | 0.0 | 3.40 | 18.14 | 42.89 | 70.84 | 100.38 |
| | 0.2 | 2.54 | 15.48 | 38.60 | 65.86 | 95.22 |
| | 0.4 | 2.12 | 14.70 | 37.65 | 64.96 | 94.39 |
| | 0.6 | 1.85 | 14.33 | 37.25 | 64.60 | 94.06 |
| | 0.8 | 1.66 | 14.11 | 37.02 | 64.40 | 93.89 |
| | 1.0 | 1.52 | 13.97 | 36.88 | 64.28 | 93.78 |
| 0.0064 | 0.0 | 3.32 | 16.23 | 36.53 | 57.94 | 79.68 |
| | 0.2 | 2.50 | 14.05 | 33.21 | 54.57 | 77.08 |
| | 0.4 | 2.08 | 13.39 | 32.42 | 53.88 | 76.47 |
| | 0.6 | 1.82 | 13.07 | 32.07 | 53.59 | 76.21 |
| | 0.8 | 1.64 | 12.88 | 31.88 | 53.43 | 76.07 |
| | 1.0 | 1.50 | 12.76 | 31.76 | 53.32 | 75.97 |
| 0.01 | 0.0 | 3.22 | 14.46 | 31.50 | 47.90 | 62.34 |
| | 0.2 | 2.44 | 12.70 | 28.88 | 46.09 | 61.24 |
| | 0.4 | 2.04 | 12.13 | 28.20 | 45.61 | 60.53 |
| | 0.6 | 1.78 | 11.86 | 27.90 | 45.39 | 60.16 |
| | 0.8 | 1.61 | 11.70 | 27.72 | 45.27 | 59.95 |
| | 1.0 | 1.47 | 11.59 | 27.62 | 45.19 | 59.82 |

The case of $n = 10^7$ corresponds, from a practical engineering viewpoint, to the Bernouilli–Euler theory.

### 4.6.1 Fundamental equations

A spinning beam and its frame of reference are presented in Fig. 4.4.

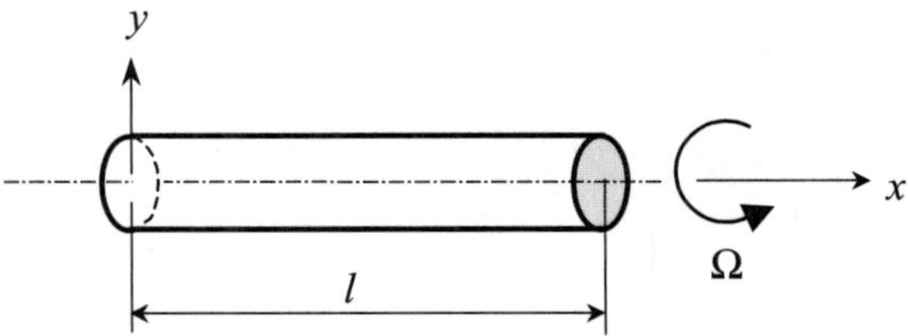

**FIGURE 4.4** Spinning Bress–Timoshenko beam (boundary conditions are not shown).

***Differential equations*** The transverse deflections along $0z$, $0y$ axis are represented by $u_y, u_z, y = u_z + ju_y$; and their corresponding bending angles by $\psi_y, \varphi_z$, so $\phi = \phi_z + j\phi_y$. Differential equation may be presented as follows (Zu and Han, 1992).

$$\frac{EI}{l^4}\frac{\partial^4 y}{\partial\xi^4} + \rho A \frac{\partial^2 y}{\partial t^2} - \frac{\rho I}{l^2}\left(1 + \frac{E}{kG}\right)\frac{\partial^4 y}{\partial\xi^2\partial t^2} + \frac{\rho^2 I}{kG}\frac{\partial^4 y}{\partial t^4} - j\Omega\frac{\rho J}{kG}\frac{\partial^3 y}{\partial t^3} + j\frac{\Omega J}{l^2}\frac{\partial^3 y}{\partial\xi^2\partial t} = 0$$

$$\frac{EI}{l^4}\frac{\partial^4 \psi}{\partial\xi^4} + \rho A \frac{\partial^2 \psi}{\partial t^2} - \frac{\rho I}{l^2}\left(1 + \frac{E}{kG}\right)\frac{\partial^4 \psi}{\partial\xi^2\partial t^2} + \frac{\rho^2 I}{kG}\frac{\partial^4 \psi}{\partial t^4} - j\Omega\frac{\rho J}{kG}\frac{\partial^3 \psi}{\partial t^3} + j\frac{\Omega J}{l^2}\frac{\partial^3 \psi}{\partial\xi^2\partial t} = 0$$

$$(4.23)$$

where $x$ is the spatial coordinate along the beam axis, $\xi = x/l$; $y$ and $\psi$ are the total deflection and bending slope; $E$, $G$, $\rho$ and $k$ are Young's modulus, shear modulus, mass density and shear coefficient; $A$, $l$ are the cross-sectional area and length of beam; $I$ and $J$ are the transverse moment of inertia of an axisymmetric cross-section and the polar mass moment of inertia, respectively; $\Omega$ is rotational speed; $j^2 = -1$.

***Solution***
The solution of these equations can be obtained using the separation of variables technique.

$$y = X(\xi)T(t) = X_0 \exp(jp\xi)\exp(j\omega t)$$
$$\psi = \Psi(\xi)T(t) = \Psi_0 \exp(j\psi)\exp(j\omega t)$$

where $X_0$ and $\Psi_0$ are complex amplitudes, $\omega$ is the natural frequency and $p$ is the coefficient characterizing the normal modes.

***Characteristic equation***

$$p^4 - \frac{B}{A}p^2 + \frac{C}{A} = 0 \qquad (4.24)$$

where

$$A = \frac{EI}{\rho A l^2}$$

$$B = \frac{I}{Al^2}\left(1 + \frac{E}{kG}\right)\omega^2 - \frac{\Omega J}{\rho A l^2}\omega$$

$$C = \frac{\rho I}{kAG}\omega^4 - \frac{\Omega J}{kAG}\omega^3 - \omega^2$$

**TABLE 4.9.**  Spinning Bress–Timoshenko uniform beam with different boundary conditions: frequency equation and mode shape expression for Case 1 (Zu and Han, 1992)

| Beam type | Frequency equations | Normal modes | Parameters |
|---|---|---|---|
| Pinned–pinned | $\sin s_2 = 0$ | $X = D \sin s_2 \xi$ <br> $\psi = H \cos s_2 \xi$ | |
| Free–free | $c_1 s_1 \left(\dfrac{s_1}{l} - c_1\right)(\cosh s_1 - \cos s_2)^2$ <br> $- \left(c_1 s_1 \sinh s_1 - \dfrac{s/l - c_1}{s_2/l + c_2} c_2 s_2 \sin s_2\right)$ <br> $\times \left[\left(\dfrac{s_1}{l} - c_1\right) \sinh s_1\right.$ <br> $\left. + \dfrac{c_1 s_1}{c_2 s_2}\left(\dfrac{s_2}{l} + c_2\right) \sin s_2\right] = 0$ | $X = D\left(\cosh s_1 \xi - \dfrac{d}{s_1/l - c_1} \sinh s_1 \xi\right.$ <br> $\left. - \dfrac{c_1 s_1}{c_2 s_2} \cos s_2 \xi + \dfrac{d}{s_2/l + c_2} \sinh s_2 \xi\right)$ <br> $\psi = H(c_1 \sinh s_1 \xi - \dfrac{dc_1}{s_1/l - c_1} \cosh s_1 \xi$ <br> $- \dfrac{c_1 s_1}{s_2} \sin s_2 \xi - \dfrac{dc_2}{s_2/l + c_2} \cos s_2 \xi$ | $d = \dfrac{\left(\dfrac{s_1}{l} - c_1\right) \sinh s_1 + \dfrac{c_1 s_1}{c_2 s_2}\left(\dfrac{s_2}{l} + c_2\right) \sin s_2}{\cosh s_1 - \cos s_2}$ |
| Clamped–clamped | $c_1(\cosh s_1 - \cos s_2)^2$ <br> $- (c_1 \sinh s_1 - c_2 \sin s_2)$ <br> $\times \left(\sinh s_1 + \dfrac{c_1}{c_2} \sin s_2\right) = 0$ | $X = D\left(\cosh s_1 \xi - \dfrac{d}{c_1} \sinh s_1 \xi - \cos s_2 \xi - \dfrac{d}{c_2} \sin s_2 \xi\right)$ <br> $\psi = H(c_1 \sinh s_1 \xi - d \cosh s_1 \xi - c_2 \sin s_2 \xi + d \sin s_2 \xi)$ | $d = \dfrac{c_1 \sinh s_1 - c_2 \sin s_2}{\cosh s_1 - \cos s_2}$ |
| Clamped–free | $(c_1 s_1 \cosh s_1 - c_2 s_2 \cos s_2)$ <br> $\times \left[\left(\dfrac{s_1}{l} - c_1\right) \cosh s_1 + \dfrac{c_1}{c_2}\left(\dfrac{s_2}{l} + c_2\right) \cos s_2\right]$ <br> $- c_1(s_1 \sinh s_1 + s_2 \sin s_2)$ <br> $\times \left[\left(\dfrac{s_1}{l} - c_1\right) \sinh s_1 + \left(\dfrac{s_2}{l} + c_2\right) \sin s_2\right] = 0$ | $X = D\left(\cosh s_1 \xi - \dfrac{d}{c_1} \sinh s_1 \xi - \cos s_2 \xi - \dfrac{d}{c_2} \sin s_2 \xi\right)$ <br> $\psi = H(c_1 \sinh s_1 \xi - d \cosh s_1 \xi - c_2 \sin s_2 \xi + d \cos s_2 \xi)$ | $d = \dfrac{c_1 s_1 \cosh s_1 - c_2 s_2 \cos s_2}{s_1 \sinh s_1 + s_2 \sin s_2}$ |
| Clamped–pinned | $c_1(\cosh s_1 - \cos s_2)(s_1 \sinh s_1 + s_2 \sin s_2)$ <br> $- \left(\sinh s_1 + \dfrac{c_1}{c_2} \sin s_2\right)$ <br> $\times (c_1 s_1 \cosh s_1 - c_2 s_2 \cos s_2) = 0$ | $X = D\left(\cosh s_1 \xi - \dfrac{d}{c_1} \sinh s_1 \xi - \cos s_2 \xi - \dfrac{d}{c_2} \sin s_2 \xi\right)$ <br> $\psi = H(c_1 \sinh s_1 \xi - d \cosh s_1 \xi - c_2 \sin s_2 \xi + d \cos s_2 \xi)$ | $d = \dfrac{c_1 s_1 \cosh s_1 - c_2 s_2 \cos s_2}{s_1 \sinh s_1 + s_2 \sin s_2}$ |
| Pinned–free | $c_1 s_1 \left(\dfrac{s_2}{l} + c_2\right) \sinh s_1 \cos s_2$ <br> $- c_2 s_2 \left(\dfrac{s_1}{l} - c_1\right) \cosh s_1 \sin s_2 = 0$ | $X = D\left(\sinh s_1 \xi - \dfrac{d}{c_2} \sin s_2 \xi\right)$ <br> $\psi = H(c_1 \cosh s_1 \xi + d \cos s_2 \xi)$ | $d = \dfrac{c_1 s_1 \sinh s_1}{s_2 \sin s_2}$ |

**TABLE 4.10.** Spinning Bress–Timoshenko uniform beam with different boundary conditions: Frequency equations and mode shape expression for Case 2 (Zu and Han, 1992)



| Beam type | Frequency equations | Normal modes | Parameters |
|---|---|---|---|
| Pinned–pinned | $\sin s_2 = 0$ | $X = D\sin s_2\xi$ <br> $\psi = H\cos s_2\xi$ | |
| Free–free | $c_1's_1'\left(\dfrac{s_1'}{l}+c_1'\right)(\cos s_1' - \cos s_2)^2$ <br> $+\left(c_1's_1'\sin s_1' - \dfrac{s_1'/l+c_1'}{s_2/l+c_2}c_2s_2\sin s_2\right)$ <br> $\times\left[\left(\dfrac{s_1'}{L}+c_1'\right)\sinh s_1 - \dfrac{c_1's_1'}{c_2s_2}\left(\dfrac{s_2}{L}+c_2\right)\sin s_2\right]=0$ | $X = D\Bigg(\cos s_1'\xi + \dfrac{d'}{s_1'/l+c_1'}\sin s_1'\xi$ <br> $-\dfrac{c_1's_1'}{c_2s_2}\cos s_2\xi - \dfrac{d'}{s_2/l+c_2}\sin s_2\xi\Bigg)$ <br> $\psi = H\Bigg(c_1'\sin s_1'\xi - \dfrac{d'c_1'}{s_1'/l+c_1'}\cos s_1'\xi$ <br> $-\dfrac{c_1's_1'}{s_2}\sin s_2\xi + \dfrac{d'c_2}{s_2/l+c_2}\cos s_2\xi\Bigg)$ | $d' = \dfrac{\left(\dfrac{s_1'}{l}+c_1'\right)\sin s_1' - \dfrac{c_1's_1'}{c_2s_2}\left(\dfrac{s_2}{l}+c_2\right)\sin s_2}{\cos s_1' - \cos s_2}$ |
| Clamped–clamped | $c_1'(\cos s_1' - \cos s_2)^2$ <br> $+(c_1'\sin s_1' - c_2\sin s_2)\left(\sin s_1' - \dfrac{c_1'}{c_2}\sin s_2\right)=0$ | $X = D\left(\cos s_1'\xi + \dfrac{d'}{c_1'}\sin s_1'\xi - \cos s_2\xi - \dfrac{d'}{c_2}\sin s_2\xi\right)$ <br> $\psi = H(c_1'\sin s_1'\xi - d'\cos s_1'\xi - c_2\sin s_2\xi + d'\sin s_2\xi)$ | $d' = \dfrac{c_1'\sin s_1' - c_2'\sin s_2}{\cos s_1' - \cos s_2}$ |
| Clamped–free | $(c_1's_1'\cos s_1' - c_2s_2\cos s_2)$ <br> $\times\left[\left(\dfrac{s_1'}{l}+c_1'\right)\cos s_1' - \dfrac{c_1'}{c_2}\left(\dfrac{s_2}{l}+c_2\right)\cos s_2\right]$ <br> $+c_1'(s_1'\sin s_1' - s_2\sin s_2)$ <br> $\times\left[\left(\dfrac{s_1'}{l}+c_1'\right)\sin s_1' - \left(\dfrac{s_2}{l}+c_2\right)\sin s_2\right]=0$ | $X = D\left(\cos s_1'\xi - \dfrac{d'}{c_1'}\sin s_1'\xi - \cos s_2\xi + \dfrac{d'}{c_2}\sin s_2\xi\right)$ <br> $\psi = H(c_1'\sin s_1'\xi + d'\cos s_1'\xi - c_2\sin s_2\xi - d'\cos s_2\xi)$ | $d' = \dfrac{c_1's_1'\cos s_1' - c_2s_2\cos s_2}{s_1'\sin s_1' - s_2\sin s_2}$ |
| Clamped–pinned | $c_1'(\cosh s_1' - \cos s_2)(s_1'\sinh s_1' + s_2\sin s_2)$ <br> $-\left(\sinh s_1' + \dfrac{c_1'}{c_2}\sin s_2\right)$ <br> $\times(c_1's_1'\cosh s_1' - c_2s_2\cos s_2)=0$ | $X = D\left(\cosh s_1'\xi - \dfrac{d'}{c_1'}\sinh s_1'\xi - \cos s_2\xi - \dfrac{d'}{c_2}\sin s_2\xi\right)$ <br> $\psi = H(c_1'\sinh s_1'\xi + d\cosh s_1'\xi - c_2\sin s_2\xi + d'\cos s_2\xi)$ | $d' = \dfrac{c_1's_1'\cos s_1' - c_2s_2\cos s_2}{s_1'\sin s_1' - s_2\sin s_2}$ |
| Pinned–free | $c_1's_1'\left(\dfrac{s_2'}{l}+c_2'\right)\sinh s_1'\cos s_2$ <br> $-c_2s_2\left(\dfrac{s_1'}{l}+c_1'\right)\cosh s_1'\sin s_2 = 0$ | $X = D\left(\sinh s_1'\xi - \dfrac{d'}{c_2}\sin s_2\xi\right)$ <br> $\psi = H(c_1'\cosh s_1'\xi + d'\cos s_2\xi)$ | $d' = \dfrac{c_1's_1'\sin s_1'}{s_2\sin s_2}$ |

The roots of the characteristic equation are $p_1 = \pm js_1, p_2 = \pm s_2$, where

$$\binom{s_1}{s_2} = \sqrt{\frac{\mp B + \sqrt{B^2 - 4AC}}{2A}} = \frac{1}{\sqrt{2}} \sqrt{\mp \frac{B}{A} + \sqrt{\frac{B^2}{A^2} - \frac{4C}{A}}}$$

The roots of the equation in terms of $b^2 = \dfrac{ml^4}{EI}\omega^2, r^2 = \dfrac{I}{Al^2}, s^2 = r^2\dfrac{E}{kG}$ and $f = \dfrac{J\omega^2}{kAG}$ are

$$\binom{s_1}{s_2} =$$

$$\frac{b}{\sqrt{2}} \sqrt{\mp \left[ (r^2 + s^2) - 2\frac{\Omega}{\omega}r^2 \right] + \sqrt{(r^2 - s^2)^2 - 4(r^2 + s^2)\frac{\Omega}{\omega}r^2 + 4\frac{\Omega^2}{\omega^2}r^4 + \frac{4}{b^2}\left( \frac{\Omega}{\omega}f + 1 \right)}}$$

If $\Omega = 0$, then the equation for $s_{1,2}$ reduces to the following formula (Section 4.3.1)

$$\binom{b\alpha}{b\beta} = \frac{b}{\sqrt{2}} \sqrt{\mp (r^2 + s^2) + \sqrt{(r^2 - s^2)^2 + \frac{4}{b^2}}}$$

It is necessary to differentiate between two cases.

*Case 1*

$$\sqrt{B^2 - 4AC} > B \quad \text{or} \quad C < 0$$

In this case the roots are $p_{1,2} = \pm js_1, \pm s_2$. This case would correspond to lower frequencies of vibrations.

*Case 2*

$$\sqrt{B^2 - 4AC} < B \quad \text{or} \quad C > 0$$

In this case, the roots are $p_1 = \pm s_1'$ in which $s_1' = js_2$. This case would correspond to higher frequencies of vibrations.

The frequency equation and mode shape of vibration are presented in Tables 4.9 and 4.10 for cases 1 and 2, respectively. Additional parameters are

$$c_1 = \frac{1}{s_1}\left( \frac{\rho l}{kG}\omega^2 + \frac{s_1^2}{l} \right) \quad c_2 = \frac{1}{s_2}\left( \frac{\rho l}{kG}\omega^2 - \frac{s_2^2}{l} \right) \quad c_1' = \frac{1}{s_1'}\left( \frac{\rho l}{kG}\omega^2 - \frac{s_1'^2}{l} \right)$$

Numerical results have been obtained, analysed and discussed by Zu and Han, (1992).

## REFERENCES

Abramovich, H. and Elishakoff, I. (1990) Influence of shear deformation and rotary inertia on vibration frequencies via Love's equations. *Journal of Sound and Vibration*, **137**(3), 516–522.

Bolotin, V.V. (1978) *Vibration of Linear Systems*, vol. 1, 1978. In *Handbook: Vibration in Tecnnik*, vols 1–6 (Moscow: Mashinostroenie) (in Russian).

Cheng, F.Y. (1970) Vibrations of Timoshenko beams and frameworks. *Journal of the Structural Division, Proceedings of the American Society of Civil Engineers*, March 551–571.

Cowper, G.R. (1966) The shear coefficients in Timoshenko's beam theory. *Journal of Applied Mechanics, ASME*, **33**, 335–340.

Filippov, A.P. (1970) *Vibration of Deformable Systems*. (Moscow: Mashinostroenie) (in Russian).

Genkin, M.D. and Tarkhanov, G.V. (1979) *Vibration of Machine-building Structures* (Moscow: Nauka) (in Russian).

Huang, T.C. (1958) Effect of rotatory inertia and shear on the vibration of beams treated by the approximate methods of Ritz and Galerkin. *ASME Proceedings of The Third US National Congress of Applied Mechanics*, pp. 189–194.

Huang, T.C. (1961) The effect of rotary inertia and of shear deformation on the frequency and normal mode equations of uniform beams with simple end conditions. *ASME Journal of Applied Mechanics*, December, pp. 579–584.

Karnovsky, I.A. and Lebed, O. (2003) Free vibrations of beams and frames. Eigenvalues and eigenfunctions. (New York: McGraw-Hill).

Love, E.A.H. (1927) *A Treatise on the Mathematical Theory of Elasticity*, 4th edn (Cambridge: Cambridge University Press), vol 1, 1892; vol 2, 1893.

Rayleigh, J.W.S. (1945) *The Theory of Sound* 2nd edn (New York: Dover Publication vol. 1–2).

Sekhniashvili, E.A. (1960) *Free Vibration of Elastic Systems*. (Tbilisi: Sakartvelo) (in Russian).

Timoshenko, S.P. (1921) On the correction for shear of the differential equation for transverse vibrations of prismatic bars. *Philosophical Magazine and Journal of Science,* Series 6, **41**, 744–746. See also: (1953) *The Collect Papers*. (New.York: McGraw Hill).

Timoshenko, S.P. (1922) On the transverse vibrations of bars of uniform cross sections. *Philosophical Magazine and Journal of Science,* Series 6, **43**, 125–131.

Thomson, W.T. (1981). Theory of vibration with application. (Prentice Hall).

Weaver, W., Timoshenko, S.P. and Young, D.H. (1990) *Vibration Problems in Engineering*, 5th edn (New York: Wiley).

Yokoyama, T. (1987) Vibrations and transient responses of Timoshenko beams resting on elastic foundations. *Ingenieur-Archiv*, **57**, 81–90.

Yokoyama, T. (1991) Vibrations of Timoshenko beam-columns on two-parameter elastic foundations. *Earthquake Engineering and Structural Dynamics*, **20**, 355–370.

Zu, J.W.-Z. and Han, R.P.S. (1992) Natural frequencies and normal modes of a spinning Timoshenko Beam with general boundary conditions. *ASME Journal of Applied Mechanics*, **59**, S197–S204.



## FURTHER READING


Aalami, B. and Atzori, B. (1974) Flexural vibrations and Timoshenko's beam theory. *American Institute of Aeronautics and Astronautics Journal*, **12**(5), 679–685.

Abbas, B.A.H. and Thomas, J. (1977) The secondary frequency spectrum of Timoshenko beams. *Journal of Sound and Vibration*, **51**(1), 309–326.

Berdichevskii, V.L. and Kvashnina S.S. (1976) On equations describing the transverse vibrations of elastic bars. *Applied Mathematics and Mechanics*, PMM vol. 40, N1, 120–135.

Bickford, W.B. (1982) A consistent higher order beam theory. *Developments in Theoretical and Applied Mechanics*, **11**, 137–150.

Bresse, M. (1859) *Cours de Mechanique Appliquee*. (Paris: Mallet-Bachelier).

Bruch, J.C. and Mitchell, T.P. (1987) Vibrations of a mass-loaded clamped-free Timoshenko beam. *Journal of Sound and Vibration,* **114**(2), 341–345.

Carr, J.B. (1970) The effect of shear flexibility and rotatory inertia on the natural frequencies of uniform beams. *The Aeronautical Quarterly*, **21**, 79–90.

Clough, R.W. and Penzien, J. (1975) *Dynamics of Structures*, (New York: McGraw-Hill).

Cowper, G.R. (1968) On the accuracy of Timoshenko's beam theory, *Proceedings of the American Society of Civil Engineering, Journal of the Engineering Mechanics Division*, EM6, December, pp. 1447–1453.

Dolph, C.L. (1954) On the Timoshenko beam vibrations, *Quarterly of Applied Mathematics*, **12**, 175–187.

Downs, B. (1976) Transverse vibrations of a uniform simply supported Timoshenko beam without transverse deflection. *ASME Journal of Applied Mechanics*, December, pp. 671–674.

Ewing, M.S. (1990) Another second order beam vibration theory: explicit bending warping flexibility and restraint. *Journal of Sound and Vibration,* **137**(1), 43–51.

Filin, A.P. (1981) *Applied Mechanics of a Solid Deformable Body*, Vol. 3, (Moscow: Nauka) (in Russian).

Flugge, W. (1975) *Viscoelasticity*, 2nd Edn. (New York: Springer-Verlag).

Heyliger, P.R. and Reddy, J.N. (1988) A higher order beam finite element for bending and vibration problems, *Journal of Sound and Vibration*, **126**(2), 309–326.

Huang, T.C. and Kung, C.S. (1963) New tables of eigenfunctions representing normal modes of vibration of Timoshenko beams. *Developments in Theoretical and Applied Mechanics* **1**, pp. 59–16, *Proceedings of the 1st Southeastern Conference on Theoretical and Applied Mechanics, Gatlinburg, Tennessee,* USA, 3–4 May (New York: Plenum Press).

Huang, T.C. (1964) Eigenvalues and modifying quotients of vibration of beams. Report **25,** Engineering Experiment Station, University of Wisconsin, Madison, Wisconsin.

Ivovich, V.A. (1981) *Transitional Matrices in Dyanmics of Elastic Systems*. Handbook (Moscow: Mashinostroenie) 181p. (in Russian).

Kaneko, T. (1975) On Timoshenko's correction for shear in vibrating beams. *Journal of Physics D: Applied Physics*, **8,** 1927–1936.

Leung, A.Y. (1990) An improved third beam theory. *Journal of Sound and Vibration*, **142**(3) 527–528.

Levinson, M. (1981) A new rectangular beam theory. *Journal of Sound and Vibration*, **74,** 81–87.

Levinson, M. and Cooke, D.W. (1982) On the two frequency spectra of Timoshenko beams. *Journal of Sound and Vibration,* **84**(3) 319–326.

Maurizi, M.J. and Belles P.M. (1991). Natural frequencies of the beam-mass system: comparison of the two fundamental theories of beam vibrations. *Journal of Sound and Vibration*, **150**(2) 330–334.

Murty, A.V.K. (1985) On the shear deformation theory for dynamic analysis of beams, *Journal of Sound and Vibration,* **101**(1), 1–12.

Pilkey, W.D. (1994) *Formulas for Stress, Strain, and Structural Matrices*. (Wiley).

Rossi, R.E., Laura, P.A.A. and Gutierrez, R.H. (1990) A note on transverse vibrations of a Timoshenko beam of non-uniform thickness clamped at one end and carrying a concentrated mass at the other. *Journal of Sound and Vibration*, **143**(3), 491–502.

Stephen N.G. (1978) On the variation of Timoshenko's shear coefficient with frequency. *Journal of Applied Mechanics,* **45**, 695-697.

Stephen, N.G. and Levinson, M. (1979) A second order beam theory. *Journal of Sound and Vibration*, **67**, 293–305.

Stephen N.G. (1982) The second frequency spectrum of Timoshenko beams. *Journal of Sound and Vibration*, **80**(4), 578–582.

Wang, J.T.S. and Dickson, J.N. (1979) Elastic beams of various orders. *American Institute of Aeronautics and Astronautics Journal*, **17,** 535–537.

Wang, C.M. (1995) Timoshenko beam–bendings solutions in terms of Euler–Bernoulli solutions. *Journal of Engineering Mechanics.* June, 763–765.

Wang, C.M., Yang, T.Q. and Lam, K.Y. (1997) Viscoelastic Timoshenko beam solutions from Euler–Bernoulli solutions. *Journal of Engineering Mechanics,* July, 746–748.

White, M.W.D. and Heppler, G.R. (1993) Vibration modes and frequencies of Timoshenko beams with attached rigid bodies. *Journal of Applied Mechanics,* **62**, 193–199.

Young, D. and Felgar, R.P., Jr. (1949) Tables of Characteristic functions representing the normal modes of vibration of a beam. *The University of Texas Publication*, No. 4913.

# CHAPTER 5

# NON-UNIFORM ONE-SPAN BEAMS

In this chapter, free vibration analyses of non-uniform one-span beams with different boundary conditions are presented. Continuous and stepped beams are investigated.

## 5.1 CANTILEVER BEAMS

### 5.1.1 Influence coefficients for clamped–free beam of non-uniform cross-sectional area

The distributed mass and the second moment of inertia

$$m(x) = m_0\left[1 - \left(1 - \frac{m_1}{m_0}\right)\frac{x}{l}\right], \quad EI(x) = EI_0\left(1 - \frac{x}{l}\right)^n$$

where $m_0$, $I_0$ are mass per unit length and moment of inertia at clamped support ($x = 0$),
   $m_1$ is mass per unit length at free end ($x = l$),
   $n$ is any integer or decimal number.

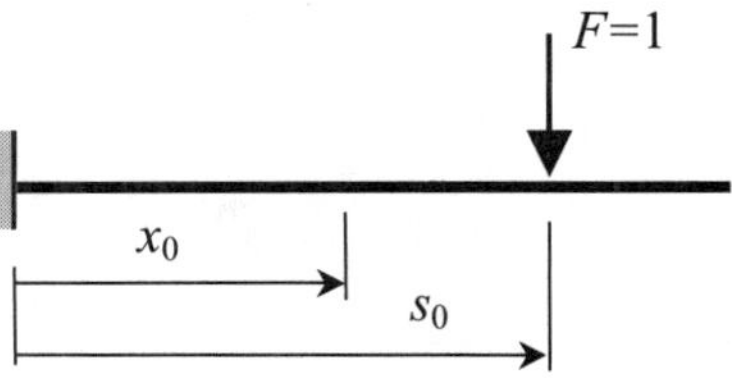


The unit force applied at $x = s_0$ and the position of any section is $x_0$, are as shown.
   The influence coefficient (Green's function) satisfies the Maxwell theorem, or the symmetry property

$$G(x, s) = G(s, x)$$

and may be presented in the form (Anan'ev, 1946)

$$G(x, s) = \frac{l^3}{(1 - n)(2 - n)EI_0}$$
$$\times \left[ (1 - s)^{2-n}(x - s) + (2 - n)xs - x - s - \frac{2}{3 - n}(1 - s)^{3-n} + \frac{2}{3 - n} \right] \text{ for } x \geq s$$

$$G(x, s) = \frac{l^3}{(1 - n)(2 - n)EI_0}$$
$$\times \left[ (1 - x)^{2-n}(s - x) + (2 - n)sx - s - x - \frac{2}{3 - n}(1 - x)^{3-n} + \frac{2}{3 - n} \right] \text{ for } x < s$$

where $x = \dfrac{x_0}{l}$, $s = \dfrac{s_0}{l}$ are non-dimensional parameters.

Expressions $G(x, s)$ have no singular points except $n = 1$, $n = 2$ and $n = 3$. For $n > 2$, $s \neq 1$, $x \neq 1$.

Special case: for a uniform cross-sectional area the parameter $n = 0$; in this case $G(s_{0_1}, s_0) = \dfrac{s_0^3}{3EJ}$.

### 5.1.2  Wedge and truncated wedge

A wedge and truncated wedge of length $l$ are presented in Figs. 5.1(a) and (b).

The differential equation of vibration accordingly to Bernoulli–Euler theory

$$\rho A(x) \frac{\partial^2 y(x, t)}{\partial t^2} = -\frac{\partial^2}{\partial x^2} \left[ EA(x) r^2(x) \frac{\partial^2 y(x, t)}{\partial x^2} \right] \tag{5.1}$$

where $r(x)$ is the radius of gyration of a cross-section about an axis through its centre parallel to the $z$ axis.

**Wedge.**  The natural frequency of vibration is

$$\omega = \frac{\lambda}{l^2} \sqrt{\frac{EI_0}{\rho A_0}} \tag{5.2}$$

where $A_0$ = cross-sectional area in the root section

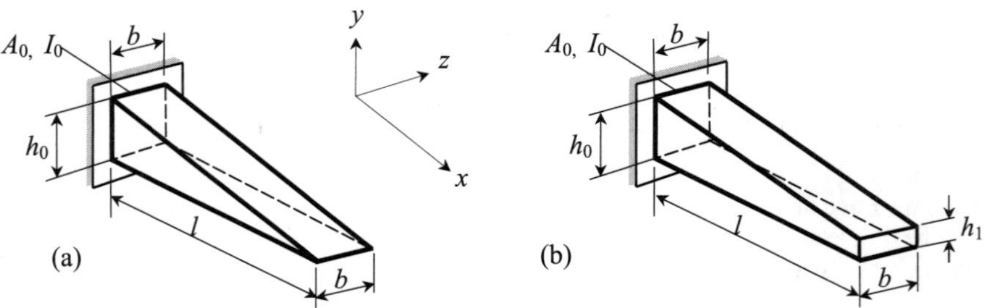


**FIGURE 5.1.**  Tapered cantilever beam: (a) wedge; (b) truncated wedge.

$I_0$ = second moment of inertia in the root section

$\lambda$ = frequency parameter.

*Bernoulli–Euler theory.* Exact values of $\lambda$ for the three lowest frequencies of vibrations are 5.315, 15.202 and 30.019, respectively.

*Timoshenko theory.* Approximate values of the frequency parameters, $\lambda$, in terms of the geometrical parameter, $\alpha = l/h_0$, are presented in Table 5.1, $E/G = 2.6$, shear coefficient $k = 0.833$ (bars of rectangular cross-section). The Rayleigh–Ritz method is applied (Gaines and Volterra, 1966).

**TABLE 5.1.** Frequency parameter $\lambda$ for wedge. Timoshenko theory

| | Mode 1 | | Mode 2 | | Mode 3 | |
|---|---|---|---|---|---|---|
| $\alpha$ | Upper bound | Lower bound | Upper bound | Lower bound | Upper bound | Lower bound |
| 3 | 4.9875 | 4.9362 | 13.2322 | 12.3542 | 24.0774 | 19.7426 |
| 4 | 5.1235 | 5.0846 | 13.9954 | 13.2577 | 26.1653 | 22.1151 |
| 5 | 5.1901 | 5.1575 | 14.3973 | 13.7426 | 27.3632 | 23.5295 |
| 10 | 5.2830 | 5.2586 | 14.9919 | 14.4585 | 29.2803 | 25.8134 |
| 15 | 5.3008 | 5.2776 | 15.1104 | 14.5977 | 29.6854 | 26.2810 |
| 20 | 5.3070 | 5.2843 | 15.1525 | 14.6466 | 29.8315 | 26.4463 |
| 50 | 5.3138 | 5.2915 | 15.1984 | 14.6993 | 29.9919 | 26.6251 |
| $\infty$ | 5.3151* | 5.2998 | 15.2072 | 14.8603 | 30.0199 | 27.5880 |

* See Table 5.7 for the following case: mode 1, $D = 0$, $H = 1$.

***Truncated wedge.*** The dimensionless frequency parameter $\lambda$, and the geometrical coefficient $\chi$ relating to vibration in the vertical plane are (Fig. 5.1(b))

$$\lambda^2 = \frac{\rho A_0 \omega^2 l^4}{E I_0}, \qquad \chi = \frac{h_1}{h_0}$$

**TABLE 5.2.** Frequency parameter $\lambda$ for truncated wedge. Bernoulli–Euler theory

| | Mode 1* | | Mode 2 | | Mode 3 | |
|---|---|---|---|---|---|---|
| $\chi = \dfrac{h_1}{h_0}$ | Upper bound | Lower bound | Upper bound | Lower bound | Upper bound | Lower bound |
| 0.0 | 5.3151 | 5.2998 | 15.2072 | 14.8603 | 30.0199 | 27.5880 |
| 0.1 | 4.6307 | 4.6246 | 14.9314 | 14.7291 | 32.8574 | 30.8563 |
| 0.2 | 4.2925 | 4.2891 | 15.7442 | 15.5782 | 36.9200 | 34.9201 |
| 0.3 | 4.0817 | 4.0794 | 16.6264 | 16.4745 | 40.6421 | 38.5641 |
| 0.4 | 3.9343 | 3.9326 | 17.4882 | 17.3449 | 44.0557 | 41.9052 |
| 0.5 | 3.8238 | 3.8225 | 18.3173 | 18.1797 | 47.2735 | 45.0479 |
| 0.6 | 3.7371 | 3.7361 | 19.1138 | 18.9799 | 50.3546 | 48.0489 |
| 0.7 | 3.6667 | 3.6659 | 19.8806 | 19.7493 | 53.3259 | 50.9370 |
| 0.8 | 3.6083 | 3.6076 | 20.6210 | 20.4915 | 56.2024 | 53.7292 |
| 0.9 | 3.5587 | 3.5581 | 21.3381 | 21.2099 | 58.9953 | 56.4382 |

* Fundamental parameter $\lambda$ for a nonlinear vibration of a cantilever tapered beam is presented in Table 7.3.

The frequency parameters according to the Bernoulli–Euler and Timoshenko theories are presented.

*Bernoulli–Euler theory.* Approximate values for the upper and lower bound of the frequency parameters $\lambda$ for different $\chi$ are presented in Table 5.2 (Gaines and Volterra, 1966).

*Timoshenko theory.* The approximate values for the upper and lower bounds of the frequency parameters $\lambda$ for $\chi = h_1/h_0 = 0.5$ and different parameters $\alpha = l/h_0$ are presented in Table 5.3; $E/G = 2.6$; $k = 0.833$ for bars of rectangular cross-section.

**TABLE 5.3.**  Frequency parameter for $\lambda$ for truncated wedge. Timoshenko theory

| | Mode 1 | | Mode 2 | | Mode 3 | |
|---|---|---|---|---|---|---|
| $\alpha$ | Upper bound | Lower bound | Upper bound | Lower bound | Upper bound | Lower bound |
| 3 | 3.5766 | 3.5597 | 13.9981 | 13.0774 | 29.5796 | 23.0213 |
| 4 | 3.6781 | 3.6670 | 15.4597 | 14.6962 | 34.3038 | 27.7793 |
| 5 | 3.7284 | 3.7204 | 16.3202 | 15.6882 | 37.5593 | 31.3497 |
| 10 | 3.7992 | 3.7957 | 17.7461 | 17.3985 | 44.1657 | 39.5036 |
| 15 | 3.8128 | 3.8101 | 18.0569 | 17.7793 | 45.9123 | 41.8514 |
| 20 | 3.8176 | 3.8152 | 18.1700 | 17.9173 | 46.5909 | 42.7687 |
| 50 | 3.8228 | 3.8207 | 18.2948 | 18.0685 | 47.3751 | 43.8163 |
| $\infty$ | 3.8238 | 3.8225 | 18.3173 | 18.1797 | 47.2753 | 45.0479 |

### 5.1.3  Cone and truncated cone

The clamped–free cone and truncated cone of length $l$ are presented in Figs. 5.2(a) and (b).

*Cone.*  The natural frequency of vibration is

$$\omega = \frac{\lambda}{l^2}\sqrt{\frac{EI_0}{\rho A_0}}$$

where $A_0 =$ cross-sectional area in the root section

$I_0 =$ second moment of inertia in the root section

$\lambda =$ frequency parameter.

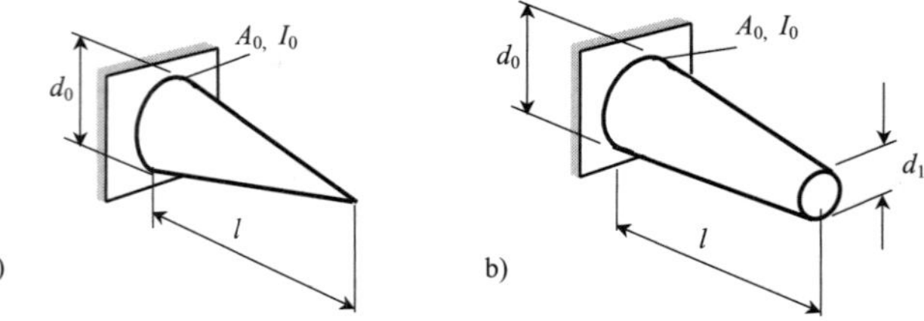


**FIGURE 5.2.**   Tapered cantilever beam: (a) cone; (b) truncated cone.

*Timoshenko theory.* Approximate values of frequency parameters $\lambda$ in terms of geometrical parameter $\alpha = l/d_0$ are presented in Table 5.4. The Rayleigh–Ritz method is applied; $E/G = 2.6$; $k = 0.9$ (Gaines and Volterra, 1966).

**TABLE 5.4.** Cantilevered cone: upper and lower bounds of frequency parameter $\lambda$ by Timoshenko theory

| | Mode 1 | | Mode 2 | | Mode 3 | |
|---|---|---|---|---|---|---|
| $\alpha$ | Upper bound | Lower bound | Upper bound | Lower bound | Upper bound | Lower bound |
| 3 | 8.1162 | 7.9585 | 18.5089 | 16.8389 | 31.4859 | 24.8727 |
| 4 | 8.3640 | 8.2390 | 19.5310 | 18.0814 | 33.9930 | 27.7074 |
| 5 | 8.4868 | 8.3783 | 20.0674 | 18.7478 | 35.4011 | 29.3633 |
| 10 | 8.6593 | 8.5728 | 20.8593 | 19.7301 | 37.6166 | 31.9877 |
| 15 | 8.6925 | 8.6096 | 21.0169 | 19.9209 | 38.0798 | 32.5202 |
| 20 | 8.7042 | 8.6225 | 21.0729 | 19.9880 | 38.2466 | 32.7085 |
| 50 | 8.7168 | 8.6365 | 21.1340 | 20.0605 | 38.4294 | 32.9124 |
| $\infty$ | **8.7193** | 8.6628 | 21.1457 | 20.3766 | 38.4540 | 34.3348 |

*Bernoulli–Euler theory.* Exact values of $\lambda$ of the three lowest frequency parameters of vibrations are 8.719 (see Table 5.5), 21.146 and 38.453, respectively (Volterra and Zachmanoglou, 1965).

### Truncated cone.

*Bernoulli–Euler theory. Analytical expression.* The fundamental frequency of vibration may be calculated by the formula (Dunkerley–Mikhlin estimates) (Brock, 1976)

$$\omega_1 \cong 8.72 \sqrt{\frac{1 - 0.016\delta}{1 + 5.053\delta} \frac{EI_0}{m_0 l^4}}, \quad m_0 = \pi \frac{d_0^2}{4}\rho, \quad \delta = \frac{d_1}{d_0} \tag{5.3}$$

where $\rho$ is mass density. There is a maximum error of 1.4% for $\delta = 0.1$. The exact results, for comparison, were calculated from formulas given by Conway *et al.* (1964).

*Numerical results.* For the beam presented in Fig. 5.2(b), the approximate upper and bound values of the frequency parameters $\lambda$ are presented in Table 5.5. The Rayleigh–Ritz method has been applied (Gaines and Volterra, 1966).

Natural frequency of vibration is

$$\omega = \frac{\lambda}{l^2} \sqrt{\frac{EI_0}{\rho A_0}}$$

where $A_0$ = cross-sectional area in the root section

      $I_0$ = second moment of inertia in the root section

      $\lambda$ = frequency parameter.

*Timoshenko theory.* The natural frequency of vibration is

$$\omega = \frac{\lambda}{l^2} \sqrt{\frac{EI_0}{\rho A_0}}$$

where $A_0$ = cross-sectional area in the root section

$I_0$ = second moment of inertia in the root section

$\lambda$ = frequency parameter.

**TABLE 5.5.**  Frequency parameter $\lambda$ for truncated cone. Bernoulli–Euler theory

| | Mode 1 | | Mode 2 | | Mode 3 | |
|---|---|---|---|---|---|---|
| $\delta = \dfrac{d_1}{d_0}$ | Upper bound | Lower bound | Upper bound | Lower bound | Upper bound | Lower bound |
| **0.0** | **8.7193** | 8.6628 | 21.1457 | 20.3766 | 38.4540 | 34.3348 |
| 0.1 | 7.2049 | 7.1827 | 18.6838 | 18.3071 | 37.2195 | 34.4834 |
| 0.2 | 6.1964* | 6.1863 | 18.3866 | 18.1268 | 39.8509 | 37.3952 |
| 0.3 | 5.5093 | 5.5037 | 18.6431 | 18.4313 | 42.8739 | 40.4715 |
| 0.4 | 5.0090* | 5.0056 | 19.0657 | 18.8807 | 45.7917 | 43.3934 |
| 0.5 | 4.6252 | 4.6229 | 19.5478 | 19.3807 | 48.6001 | 46.1909 |
| 0.6 | 4.3188* | 4.3172 | 20.0500 | 19.8956 | 51.3379 | 48.9003 |
| 0.7 | 4.0669 | 4.0658 | 20.5555 | 20.4104 | 54.0172 | 51.5382 |
| 0.8 | 3.8551* | 3.8543 | 21.0568 | 20.9189 | 56.6390 | 54.1102 |
| 0.9 | 3.6737 | 3.6730 | 21.5503 | 21.4182 | 59.2037 | 56.6205 |
| **1.0** | 3.5160[†] | 3.5155 | 22.0345 | 21.9072 | 61.7151 | 59.0746 |

    * Table 7.3 presents fundamental parameters for the case $\varepsilon = 1 - \delta$.

    [†] This case corresponds to a uniform cantilever beam.

    The above-mentioned reference present frequency parameters $\lambda$ for truncated-cone beams with different boundary conditions.

Approximate results for upper and lower values of the three lowest frequency parameters $\lambda$ for $\delta = \dfrac{d_1}{d_0} = 0.5$ and different parameters $\alpha$ are presented in Table 5.6; $E/G = 2.6$, $k = 0.9$ for bars of circular cross-section (Gaines and Volterra, 1966). The Rayleigh–Ritz method have been applied.

**TABLE 5.6.**  Frequency parameter $\lambda$ for truncated cone. Timoshenko theory

| | Mode 1 | | Mode 2 | | Mode 3 | |
|---|---|---|---|---|---|---|
| $\alpha = \dfrac{l}{d_0}$ | Upper bound | Lower bound | Upper bound | Lower bound | Upper bound | Lower bound |
| 3 | 4.3782 | 4.3555 | 15.9145 | 14.9203 | 33.3741 | 26.3384 |
| 4 | 4.4806 | 4.4657 | 17.2188 | 16.4251 | 37.8986 | 31.1577 |
| 5 | 4.5309 | 4.5199 | 17.9511 | 17.3019 | 40.8311 | 34.5640 |
| 10 | 4.6010 | 4.5956 | 19.1067 | 18.7301 | 46.3220 | 41.6202 |
| 15 | 4.6144 | 4.6100 | 19.3491 | 19.0332 | 47.6722 | 43.4606 |
| 20 | 4.6191 | 4.6150 | 19.4365 | 19.1418 | 48.1845 | 44.1573 |
| 50 | 4.6242 | 4.6205 | 19.5325 | 19.2601 | 48.7666 | 44.9396 |
| $\infty$ | 4.6252 | 4.6229 | 19.5478 | 19.3807 | 48.6001 | 46.1909 |

### 5.1.4  Doubly tapered beam

A doubly tapered clamped–free beam is presented in Fig. 5.3.

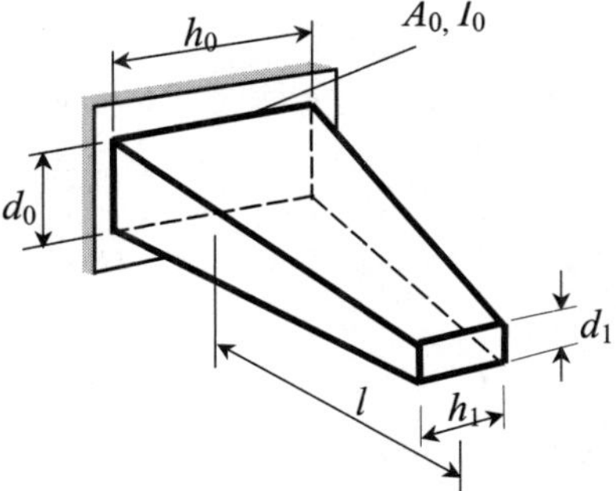


**FIGURE 5.3.**   Doubly-tapered cantilever beam.

The dimensionless tapered parameters are

$$D = \frac{d_1}{d_0}, \quad H = \frac{h_1}{h_0}$$

The natural frequency of vibration is

$$\omega = \frac{\lambda}{l^2} \sqrt{\frac{EI_0}{\rho A_0}}$$

where $A_0$ = cross-sectional area in the root section

$I_0$ = second moment of inertia in the root section

$\lambda$ = frequency parameter.

Frequency parameters, $\lambda$, for both Bernoulli–Euler and Timoshenko models of beams are presented in Table 5.7 (Downs, 1977). The results obtained by using the Bernoulli–Euler theory (identified by the letter E) can be applied to any material and slenderness ratio, but for the results based on Timoshenko theory (identified by the letter T), Poisson's ratio $v = 0.3$, the ratio of the radius of gyration at the cantilever root to beam length equals 0.08, and the shear coefficient of 0.85 applies.

### 5.1.5  Tapered beams with a tip mass

***Bernoulli–Euler theory.***   Tapered beams with a tip mass at the free end $(x = L_0)$ and clamped at $x = L_1 = L + L_0$ are presented in Fig. 5.4.

The flexural rigidity, $EI(x)$, and the area of cross-section, $A(x)$, are given by

$$EI(x) = \left(\frac{x}{L_1}\right)^4 EI_0 \tag{5.4}$$

$$A(x) = \left(\frac{x}{L_1}\right)^2 A_0 \tag{5.5}$$

**TABLE 5.7.** Frequency parameters $\lambda$ for a doubly-tapered cantilever beam

| Mode | $D$ | Theory | $H = h_1/h_0$ | | | | | |
|---|---|---|---|---|---|---|---|---|
| | | | 0.00 | 0.10 | 0.20 | 0.40 | 0.70 | 1.00 |
| | 0 | E | 8.71926 | 7.82581 | 7.21932 | 6.43567 | 5.74690 | **5.31511*** |
| | | T | 8.13372 | 7.36318 | 6.82660 | 6.11936 | 5.48694 | 5.08622 |
| | 0.1 | E | 8.24538 | 7.20487 | 6.53990 | 5.72513 | 5.04388 | 4.63072 |
| | | T | 7.68773 | 6.78850 | 6.19738 | 5.45828 | 4.82987 | 4.44487 |
| | 0.2 | E | 7.95834 | 6.87510 | 6.19639 | 5.37614 | 4.69913 | 4.29249 |
| 1 | | T | 7.40367 | 6.46638 | 5.86287 | 5.11942 | 4.49591 | 4.11769 |
| | 0.4 | E | 7.61278 | 6.51107 | 5.82882 | 5.00903 | 4.33622 | 3.93428 |
| | | T | 7.04337 | 6.09270 | 5.48871 | 4.74979 | 4.13389 | 3.76228 |
| | 0.7 | E | 7.32708 | 6.23078 | 5.55297 | 4.73721 | 4.06693 | 3.66675 |
| | | T | 6.72179 | 5.78358 | 5.18913 | 4.46107 | 3.85346 | 3.48689 |
| | 1.00 | E | 7.15648 | 6.07038 | 5.39759 | 4.58531 | 3.91603 | **3.51602**** |
| | | T | 6.51116 | 5.59053 | 5.00623 | 4.28829 | 3.68723 | 3.32405 |
| | 0.0 | E | 21.1457 | 18.6893 | 17.5871 | 16.4963 | 15.6876 | **15.2076*** |
| | | T | 18.5758 | 16.7222 | 15.8247 | 14.9034 | 14.2026 | 13.7798 |
| | 0.1 | E | 21.7596 | 18.6802 | 17.4387 | 16.2732 | 15.4298 | 14.9308 |
| | | T | 18.7124 | 16.4421 | 15.4497 | 14.4845 | 13.7683 | 13.3372 |
| 2 | 0.2 | E | 22.8572 | 19.6740 | 18.3855 | 17.1657 | 16.2744 | 15.7427 |
| | | T | 19.1968 | 16.9194 | 15.9230 | 14.9440 | 14.2069 | 13.7575 |
| | 0.4 | E | 25.1001 | 21.7763 | 20.3952 | 19.0649 | 18.0803 | 17.4879 |
| | | T | 20.1615 | 17.9264 | 16.9224 | 15.9107 | 15.1297 | 14.6449 |
| | 0.7 | E | 28.2077 | 24.6738 | 23.1578 | 21.6699 | 20.5555 | 19.8806 |
| | | T | 21.2941 | 19.1076 | 18.0864 | 17.0229 | 16.1766 | 15.6411 |
| | 1.0 | E | 31.0414 | 27.2989 | 25.6558 | 24.0211 | 22.7860 | **22.0345**** |
| | | T | 22.0961 | 19.9429 | 18.9026 | 17.7871 | 16.8752 | 16.2890 |
| | 0 | E | 38.4539 | 33.9593 | 32.5770 | 31.3715 | 30.5230 | 30.0241* |
| | | T | 31.6431 | 28.6418 | 27.5953 | 26.6492 | 25.9650 | 25.5528 |
| | 0.1 | E | 42.6956 | 37.1238 | 35.5883 | 34.2873 | 33.3747 | 32.8331 |
| | | T | 33.3657 | 29.8463 | 28.7514 | 27.7936 | 27.1025 | 26.6811 |
| 3 | 0.2 | E | 47.2173 | 41.4727 | 39.8336 | 38.4392 | 37.4635 | 36.8846 |
| | | T | 35.2211 | 31.8154 | 30.7258 | 29.7628 | 29.0643 | 28.6363 |
| | 0.4 | E | 55.2933 | 49.1472 | 47.3051 | 45.7384 | 44.6583 | 44.0248 |
| | | T | 38.1256 | 34.8498 | 33.7513 | 32.7692 | 32.0582 | 31.6243 |
| | 0.7 | E | 65.9035 | 59.1332 | 57.0157 | 55.2224 | 54.0152 | 53.3222 |
| | | T | 41.1711 | 37.9753 | 36.8520 | 35.8346 | 35.1026 | 34.6625 |
| | 1.00 | E | 75.4872 | 68.1145 | 65.7470 | 63.7515 | 62.4361 | **61.6972**** |
| | | T | 43.3089 | 40.1409 | 38.9854 | 37.9221 | 37.1588 | 36.7078 |

*Special cases*:
* $D = 0$, $H = 1$. This case corresponds to a clamped–free wedge (Table 5.1).
** $D = 1$, $H = 1$. This case corresponds to a cantilever uniform beam.

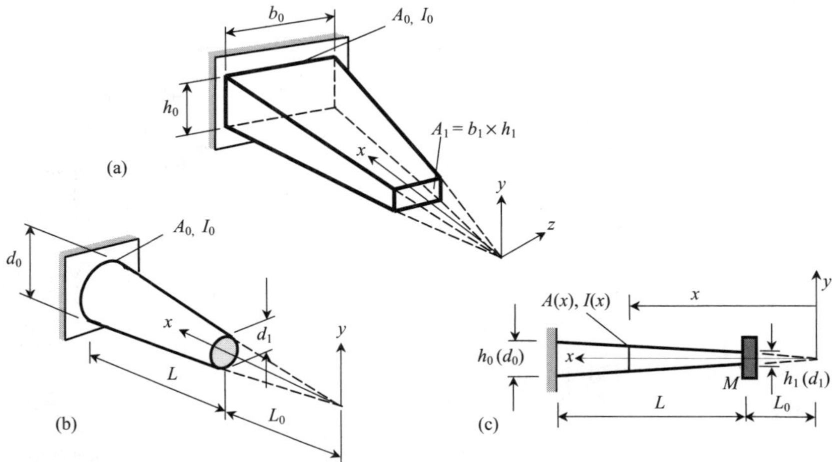

**FIGURE 5.4.**   (a) Truncated pyramid; (b) truncated cone; (c) general notation for a pyramid and cone with a tip mass.

where $A_0$ = the cross-section area at $x = L_1$ (root section)

$I_0$ = second moment of inertia at $x = L_1$.

Formulas (5.4) and (5.5) may be applied to doubly truncated pyramids (Fig. 5.4(a)) and truncated cones (Fig. 5.4(b),(c)).

The differential equation for mode shape $X$ is

$$\frac{1}{y_*^2}\frac{\mathrm{d}^2}{\mathrm{d}y_*^2}\left(y_*^4\frac{\mathrm{d}^2X}{\mathrm{d}y_*^2}\right) - \lambda^4 X = 0, \quad y_* = \frac{x}{L_1} \tag{5.6}$$

The general solution may be written by using Bessel functions

$$X = y_*^{-1}[C_1 J_2(z) + C_2 Y_2(z) + C_3 I_2(z) + C_4 K_2(z)], \quad z = 2\lambda\sqrt{y_*} \tag{5.7}$$

where $J_2$ and $Y_2$ are the second-order Bessel functions of the first and second kind, respectively; $J_2$ and $I_2$ and $K_2$ are the modified second-order Bessel functions of the first and second kind, respectively.

The frequency equation is presented by Lau (1984).

The natural frequency of vibration is

$$\omega = \frac{\lambda^2}{L^2}\sqrt{\frac{EI_0}{\rho A_0}} \tag{5.8}$$

Frequency coefficients, $\lambda$, for different modes of vibration in terms of dimensionless mass ratio $\eta = \dfrac{M}{\rho A_0 L}$ and geometry ratio $\dfrac{L_0}{L_1}$ are presented in Table 5.8 (Lau, 1984).

In this table Parameter $L_0/L_1 = 1$ corresponds to a uniform beam. Case * corresponds to a uniform beam without lumped mass. Case ** corresponds to a uniform beam with lumped mass.

**TABLE 5.8.**   Cantilevered tapered beams with lumped mass at the end: Frequency parameters $\lambda$ by Bernoulli–Euler theory

| $\eta$ | $L_0/L_1$ | Mode 1 | Mode 2 | Mode 3 | Mode 4 | Mode 5 |
|---|---|---|---|---|---|---|
| 0.0 | 0.1 | 2.68419 | 4.32206 | 6.09293 | 7.96900 | 9.90786 |
| | 0.2 | 2.48926 | 4.28783 | 6.31139 | 8.44048 | 10.62205 |
| | 0.3 | 2.34718 | 4.31754 | 6.54297 | 8.86120 | 11.22275 |
| | 0.4 | 2.23809 | 4.36633 | 6.76301 | 9.23817 | 11.74885 |
| | 0.5 | 2.15062 | 4.42127 | 6.96986 | 9.58190 | 12.22252 |
| | 0.6 | 2.07817 | 4.47772 | 7.16482 | 9.89975 | 12.65692 |
| | 0.7 | 2.01666 | 4.53382 | 7.34950 | 10.19682 | 13.06052 |
| | 0.8 | 1.96345 | 4.58876 | 7.52531 | 10.47678 | 13.43915 |
| | 0.9 | 1.91669 | 4.64222 | 7.69341 | 10.74233 | 13.79699 |
| | **1.0*** | 1.87510 | 4.69409 | 7.85476 | 10.99554 | 14.13717 |
| 0.2 | 0.1 | 1.10212 | 3.15637 | 5.11296 | 7.10001 | 9.10876 |
| | 0.2 | 1.29695 | 3.31638 | 5.45913 | 7.65757 | 9.88171 |
| | 0.3 | 1.41336 | 3.46089 | 5.75049 | 8.11207 | 10.50084 |
| | 0.4 | 1.48992 | 3.59643 | 6.01266 | 8.51076 | 11.03690 |
| | 0.5 | 1.54127 | 3.72491 | 6.25635 | 8.87359 | 11.51935 |
| | 0.6 | 1.57524 | 3.84671 | 6.48690 | 9.21107 | 11.96382 |
| | 0.7 | 1.59681 | 3.96180 | 6.70713 | 9.52936 | 12.37965 |
| | 0.8 | 1.60936 | 4.07015 | 6.91860 | 9.83234 | 12.77289 |
| | 0.9 | 1.61530 | 4.17182 | 7.12218 | 10.12251 | 13.14762 |
| | **1.0**** | 1.61640 | 4.26706 | 7.31837 | 10.40156 | 13.50670 |
| 0.4 | 0.1 | 0.92869 | 3.15207 | 5.11131 | 7.09909 | 9.10815 |
| | 0.2 | 1.09856 | 3.30136 | 5.45278 | 7.65382 | 9.87911 |
| | 0.3 | 1.20610 | 3.43098 | 5.73653 | 8.10341 | 10.49466 |
| | 0.4 | 1.28282 | 3.54948 | 5.98853 | 8.49516 | 11.02548 |
| | 0.5 | 1.34001 | 3.66057 | 6.22005 | 8.84917 | 11.50105 |
| | 0.6 | 1.38332 | 3.76604 | 6.43711 | 9.17624 | 11.93714 |
| | 0.7 | 1.41618 | 3.86680 | 6.64329 | 9.48294 | 12.34322 |
| | 0.8 | 1.44093 | 3.96331 | 6.84085 | 9.77360 | 12.72590 |
| | 0.9 | 1.45924 | 4.05581 | 7.03125 | 10.05118 | 13.08929 |
| | **1.0**** | 1.47241 | 4.14443 | 7.21549 | 10.31781 | 13.43668 |
| 0.6 | 0.1 | 0.83974 | 3.15064 | 5.11076 | 7.09879 | 9.10795 |
| | 0.2 | 0.99509 | 3.29631 | 5.45065 | 7.65256 | 9.87824 |
| | 0.3 | 1.09534 | 3.42067 | 5.73179 | 8.10049 | 10.49259 |
| | 0.4 | 1.16882 | 3.53275 | 5.98021 | 8.48985 | 11.02160 |
| | 0.5 | 1.22551 | 3.63669 | 6.20725 | 8.84072 | 11.49478 |
| | 0.6 | 1.27031 | 3.73472 | 6.41907 | 9.16396 | 11.92787 |
| | 0.7 | 1.30608 | 3.82810 | 6.61942 | 9.46622 | 12.33047 |
| | 0.8 | 1.33471 | 3.91761 | 6.81076 | 9.75189 | 12.70895 |
| | 0.9 | 1.35757 | 4.00369 | 6.99473 | 10.02406 | 13.06777 |
| | **1.0**** | 1.37567 | 4.08665 | 7.17252 | 10.28498 | 13.41021 |

**TABLE 5.8.** *(Continued)*

| $\eta$ | $L_0/L_1$ | Mode 1 | Mode 2 | Mode 3 | Mode 4 | Mode 5 |
|---|---|---|---|---|---|---|
| | 0.1 | 0.78174 | 3.14993 | 5.11048 | 7.09864 | 9.10785 |
| | 0.2 | 0.92717 | 3.29377 | 5.44958 | 7.65193 | 9.87781 |
| | 0.3 | 1.02193 | 3.41545 | 5.72941 | 8.09903 | 10.49154 |
| 0.8 | 0.4 | 1.09233 | 3.52417 | 5.97599 | 8.48716 | 11.01966 |
| | 0.5 | 1.14759 | 3.62425 | 6.20071 | 8.83643 | 11.49161 |
| | 0.6 | 1.19217 | 3.71808 | 6.40976 | 9.15769 | 11.92316 |
| | 0.7 | 1.22867 | 3.80712 | 6.60695 | 9.45760 | 12.32390 |
| | 0.8 | 1.25876 | 3.89227 | 6.79480 | 9.74059 | 12.70022 |
| | 0.9 | 1.28360 | 3.97412 | 6.97506 | 10.00979 | 13.05659 |
| | **1.0**** | 1.30409 | 4.05308 | 7.14898 | 10.26749 | 13.39631 |
| | 0.1 | 0.73947 | 3.14950 | 5.11032 | 7.09854 | 9.10779 |
| | 0.2 | 0.87751 | 3.29225 | 5.44894 | 7.65155 | 9.87755 |
| | 0.3 | 0.96797 | 3.41230 | 5.72798 | 8.09815 | 10.49092 |
| 1.0 | 0.4 | 1.03572 | 3.51895 | 5.97345 | 8.48555 | 11.01848 |
| | 0.5 | 1.08945 | 3.61661 | 5.19674 | 8.83384 | 11.48970 |
| | 0.6 | 1.13335 | 3.70777 | 6.40407 | 9.15389 | 11.92031 |
| | 0.7 | 1.16983 | 3.79396 | 6.59928 | 9.45234 | 12.31991 |
| | 0.8 | 1.20042 | 3.87617 | 6.78492 | 9.73366 | 12.69489 |
| | 0.9 | 1.22619 | 3.95507 | 6.96277 | 10.00099 | 13.04974 |
| | **1.0**** | 1.24792 | 4.03114 | 7.13413 | 10.25662 | 13.38776 |
| | 0.1 | 0.56219 | 3.14836 | 5.10988 | 7.09830 | 9.10763 |
| | 0.2 | 0.66808 | 3.28818 | 5.44723 | 7.65054 | 9.87685 |
| | 0.3 | 0.73854 | 3.40382 | 5.72413 | 8.09579 | 10.48924 |
| 3.0 | 0.4 | 0.79247 | 3.50478 | 5.96659 | 8.48121 | 11.01534 |
| | 0.5 | 0.83643 | 3.59561 | 6.18599 | 8.82686 | 11.48456 |
| | 0.6 | 0.87359 | 3.67898 | 6.38855 | 9.14358 | 11.91262 |
| | 0.7 | 0.90570 | 3.75657 | 6.57814 | 9.43802 | 12.30910 |
| | 0.8 | 0.93389 | 3.82956 | 6.75736 | 9.71464 | 12.68037 |
| | 0.9 | 0.95889 | 3.89876 | 6.92806 | 9.97659 | 13.03093 |
| | **1.0**** | 0.98123 | 3.96482 | 7.09160 | 10.22621 | 13.36409 |
| | 0.1 | 0.49484 | 3.14813 | 5.10979 | 7.09825 | 9.10760 |
| | 0.2 | 0.58822 | 3.28736 | 5.44688 | 7.65034 | 9.87671 |
| | 0.3 | 0.65054 | 3.40212 | 5.72336 | 8.09532 | 10.48891 |
| 5.0 | 0.4 | 0.69844 | 3.50190 | 5.96521 | 8.48034 | 11.01471 |
| | 0.5 | 0.73771 | 3.59130 | 6.18381 | 8.82546 | 11.48353 |
| | 0.6 | 0.77113 | 3.67299 | 6.38538 | 9.14149 | 11.91107 |
| | 0.7 | 0.80024 | 3.74867 | 6.57379 | 9.43510 | 12.30690 |
| | 0.8 | 0.82603 | 3.81953 | 6.75163 | 9.71073 | 12.67741 |
| | 0.9 | 0.84913 | 3.88642 | 6.92076 | 9.97155 | 13.02707 |
| | **1.0**** | 0.87002 | 3.94998 | 7.08254 | 10.21986 | 13.35920 |

*(Continued)*

**TABLE 5.8.** *(Continued)*

| | | Mode | | | | |
|---|---|---|---|---|---|---|
| $\eta$ | $L_0/L_1$ | 1 | 2 | 3 | 4 | 5 |
| | 0.1 | 0.45494 | 3.14804 | 5.10975 | 7.09823 | 9.10758 |
| | 0.2 | 0.54085 | 3.28702 | 5.44673 | 7.65025 | 9.87665 |
| | 0.3 | 0.59827 | 3.40138 | 5.72303 | 8.09511 | 10.48876 |
| 7.0 | 0.4 | 0.64248 | 3.50066 | 5.96462 | 8.47997 | 11.01444 |
| | 0.5 | 0.67882 | 3.58944 | 6.18287 | 8.82485 | 11.48308 |
| | 0.6 | 0.70982 | 3.67040 | 6.38401 | 9.14059 | 11.91040 |
| | 0.7 | 0.73693 | 3.74524 | 6.57191 | 9.43384 | 12.30596 |
| | 0.8 | 0.76103 | 3.81516 | 6.74915 | 9.70905 | 12.67614 |
| | 0.9 | 0.78272 | 3.88101 | 6.91759 | 9.96937 | 13.02541 |
| | **1.0**** | 0.80243 | 3.94344 | 7.07860 | 10.21712 | 13.35709 |
| | 0.1 | 0.42725 | 3.14798 | 5.10973 | 7.09822 | 9.10757 |
| | 0.2 | 0.50797 | 3.28682 | 5.44665 | 7.65021 | 9.87662 |
| | 0.3 | 0.56195 | 3.40098 | 5.72285 | 8.09500 | 10.48868 |
| 9.0 | 0.4 | 0.60356 | 3.49997 | 5.96429 | 8.47976 | 11.01429 |
| | 0.5 | 0.63780 | 3.58840 | 6.18235 | 8.82451 | 11.48284 |
| | 0.6 | 0.66707 | 3.66895 | 6.38325 | 9.14009 | 11.91003 |
| | 0.7 | 0.69270 | 3.74332 | 6.57086 | 9.43314 | 12.30544 |
| | 0.8 | 0.71555 | 3.81271 | 6.74777 | 9.70811 | 12.67543 |
| | 0.9 | 0.73616 | 3.87797 | 6.91582 | 9.96815 | 13.02448 |
| | **1.0**** | 0.75494 | 3.93976 | 7.07639 | 10.21558 | 13.35591 |

***Timoshenko theory.*** A cantilevered tapered beam with a tip mass at the free end is presented in Fig. 5.5. The width of the cross-section is assumed to be constant.

The governing equations and frequency equation are presented by Rossi *et al.* (1990). The natural frequency of vibration is

$$\omega = \frac{\lambda^2}{L^2}\sqrt{\frac{EI_0}{\rho A_0}}$$

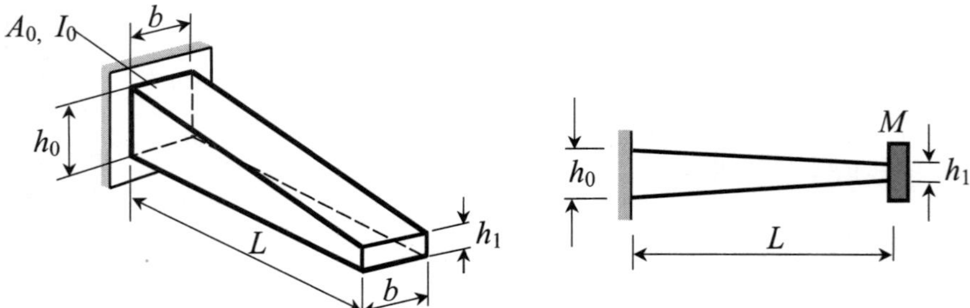


**FIGURE 5.5.** Cantilever tapered beam with end mass.

where $A_0$ = cross-sectional area in the root section

$I_0$ = second moment of inertia in the root section

$\lambda$ = frequency parameter.

Table 5.9 shows the first four frequency coefficients $\lambda$ for different combinations of the geometry ratio $h_1/h_0$, dimensionless parameters $\eta = I_0/(A_0 L)^2$ and mass ratio $\mu = M/M_b$ ($M_b$ is total beam mass). The finite element method has been applied for Poisson's coefficient $v = 0.3$, and shear coefficient $k = 5/6$ (Rossi *et al.*, 1990).

### 5.1.6  Haunched beam with a tip mass at the free end

A haunched beam with a tip mass at the free end is shown in Fig. 5.6. The width of the cross-section is assumed to be constant.

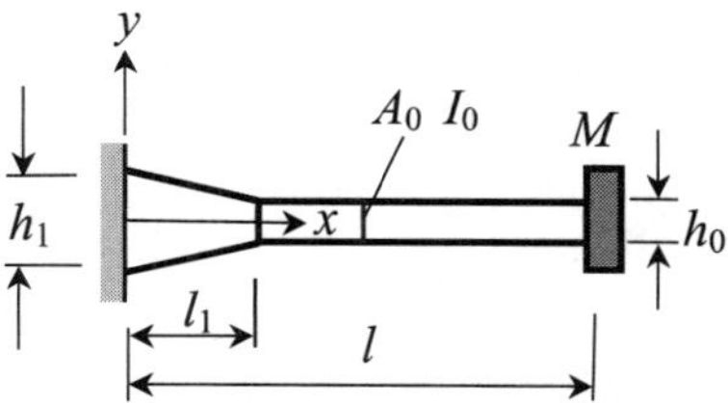


**FIGURE 5.6.**  Cantilevered haunched beam.

In the range from $x = 0$ (clamped end) to $x = l_1$ the area and moment of inertia are

$$A(x) = A_0\left[1 + k\left(1 - \frac{x}{l_1}\right)\right] \tag{5.9}$$

$$I(x) = I_0\left[1 + k\left(1 - \frac{x}{l_1}\right)\right]^3 \tag{5.10}$$

where the dimensionless parameters are

$$k = \frac{2a}{h_0}, \quad a = \frac{h_1 - h_0}{2}$$

The natural fundamental frequency of vibration, according to the Bernoulli–Euler theory, may be calculated by

$$\omega = \frac{3.57}{l_r^2}\sqrt{\frac{EI_0}{\left(1 + \frac{140}{33}\frac{M}{\rho A_0 l}\right)\rho A_0}} \tag{5.11}$$

where the reduced length of the beam is $l_r = \phi l$; $A_0$ and $I_0$ are the cross-sectional area and moment of inertia of any cross-section at $l_1 \leq x \leq l$.

Parameter $\phi$ is presented in Table 5.10 in terms of geometry ratios $n = h_1/h_0$ and $\eta = l_1/l$. The Rayleigh–Ritz method has been applied. If $\eta$ and $k$ are small, then the kinetic energy of the haunched part of the beam is neglected (Filippov, 1970).

**TABLE 5.9.** Cantilevered tapered beams with lumped mass at the end: Frequency parameters $\lambda^2$ by Timoshenko theory

| | | | Mode | | | |
|---|---|---|---|---|---|---|
| $\eta$ | $h_1/h_0$ | $\mu$ | 1 | 2 | 3 | 4 |
| $10^{-8}$ | 0.8 | 0.0 | 3.61 | 20.61 | 56.17 | 109.27 |
| | | 0.2 | 2.61 | 16.73 | 48.24 | 97.15 |
| | | 0.4 | 2.15 | 15.78 | 46.95 | 95.69 |
| | | 0.6 | 1.87 | 15.36 | 46.44 | 94.55 |
| | | 0.8 | 1.67 | 15.12 | 46.17 | 94.72 |
| | | 1.0 | 1.53 | 14.96 | 45.99 | 94.65 |
| | 0.6 | 0.0 | 3.73 | 19.09 | 50.31 | 96.92 |
| | | 0.2 | 2.61 | 15.09 | 42.55 | 85.34 |
| | | 0.4 | 2.11 | 14.25 | 41.51 | 84.18 |
| | | 0.6 | 1.82 | 13.88 | 41.10 | 83.82 |
| | | 0.8 | 1.63 | 13.68 | 40.90 | 83.51 |
| | | 1.0 | 1.48 | 13.55 | 40.76 | 83.43 |
| 0.0004 | 0.8 | 0.0 | 3.59 | 20.17 | 53.48 | 100.32 |
| | | 0.2 | 2.61 | 16.44 | 46.23 | 89.97 |
| | | 0.4 | 2.14 | 15.52 | 45.03 | 88.63 |
| | | 0.6 | 1.86 | 15.11 | 44.54 | 88.22 |
| | | 0.8 | 1.67 | 14.87 | 44.28 | 87.85 |
| | | 1.0 | 1.52 | 14.72 | 44.12 | 87.75 |
| | 0.6 | 0.0 | 3.72 | 18.77 | 48.36 | 90.46 |
| | | 0.2 | 2.60 | 14.89 | 41.13 | 80.20 |
| | | 0.4 | 2.11 | 14.06 | 40.13 | 79.23 |
| | | 0.6 | 1.82 | 13.70 | 39.74 | 78.89 |
| | | 0.8 | 1.62 | 13.51 | 39.55 | 78.50 |
| | | 1.0 | 1.48 | 13.38 | 39.42 | 78.49 |
| 0.0016 | 0.8 | 0.0 | 3.56 | 19.01 | 47.43 | 83.48 |
| | | 0.2 | 2.59 | 15.67 | 41.56 | 75.84 |
| | | 0.4 | 2.13 | 14.82 | 40.52 | 74.82 |
| | | 0.6 | 1.85 | 14.43 | 40.10 | 74.37 |
| | | 0.8 | 1.66 | 14.21 | 39.86 | 74.27 |
| | | 1.0 | 1.51 | 14.07 | 39.72 | 74.09 |
| | 0.6 | 0.0 | 3.69 | 17.88 | 43.74 | 77.32 |
| | | 0.2 | 2.58 | 14.32 | 37.65 | 69.47 |
| | | 0.4 | 2.09 | 13.54 | 36.76 | 68.67 |
| | | 0.6 | 1.81 | 13.20 | 36.43 | 68.26 |
| | | 0.8 | 1.61 | 13.02 | 36.24 | 68.13 |
| | | 1.0 | 1.47 | 12.89 | 36.12 | 68.09 |
| | 0.8 | 0.0 | 3.49 | 17.47 | 40.96 | 68.43 |
| | | 0.2 | 2.55 | 14.59 | 36.24 | 62.92 |
| | | 0.4 | 2.10 | 13.84 | 35.46 | 62.08 |
| | | 0.6 | 1.83 | 13.49 | 35.10 | 61.75 |
| | | 0.8 | 1.64 | 13.29 | 34.91 | 61.56 |
| | | 1.0 | 1.50 | 13.17 | 34.78 | 61.52 |

**TABLE 5.9.** (*Continued*)

| $\eta$ | $h_1/h_0$ | $\mu$ | Mode 1 | Mode 2 | Mode 3 | Mode 4 |
|---|---|---|---|---|---|---|
| 0.0036 | | | | | | |
| | 0.6 | 0.0 | 3.62 | 16.65 | 38.49 | 64.77 |
| | | 0.2 | 2.55 | 13.51 | 33.52 | 58.80 |
| | | 0.4 | 2.08 | 12.79 | 32.76 | 58.09 |
| | | 0.6 | 1.79 | 12.48 | 32.46 | 57.79 |
| | | 0.8 | 1.60 | 12.31 | 32.29 | 57.67 |
| | | 1.0 | 1.46 | 12.20 | 32.20 | 57.55 |
| | 0.8 | 0.0 | 3.42 | 15.84 | 35.35 | 56.91 |
| | | 0.2 | 2.51 | 13.42 | 31.66 | 52.81 |
| | | 0.4 | 2.07 | 12.76 | 30.92 | 52.13 |
| | | 0.6 | 1.80 | 12.45 | 30.60 | 51.84 |
| | | 0.8 | 1.62 | 12.28 | 30.42 | 51.74 |
| | | 1.0 | 1.48 | 12.16 | 30.32 | 51.64 |
| 0.0064 | | | | | | |
| | 0.6 | 0.0 | 3.55 | 15.30 | 33.69 | 54.75 |
| | | 0.2 | 2.52 | 12.58 | 29.63 | 50.08 |
| | | 0.4 | 2.05 | 11.94 | 28.96 | 49.48 |
| | | 0.6 | 1.77 | 11.66 | 28.69 | 49.24 |
| | | 0.8 | 1.58 | 11.50 | 28.55 | 49.13 |
| | | 1.0 | 1.44 | 11.40 | 28.46 | 49.06 |
| | 0.8 | 0.0 | 3.33 | 14.29 | 30.79 | 48.05 |
| | | 0.2 | 2.46 | 12.27 | 27.78 | 45.15 |
| | | 0.4 | 2.03 | 11.69 | 27.13 | 44.59 |
| | | 0.6 | 1.77 | 11.42 | 26.84 | 44.39 |
| | | 0.8 | 1.59 | 11.26 | 26.69 | 44.25 |
| | | 1.0 | 1.46 | 11.16 | 26.59 | 44.18 |
| 0.01 | | | | | | |
| | 0.6 | 0.0 | 3.47 | 13.98 | 29.68 | 46.97 |
| | | 0.2 | 2.47 | 11.64 | 26.27 | 43.27 |
| | | 0.4 | 2.01 | 11.07 | 25.67 | 42.76 |
| | | 0.6 | 1.74 | 10.81 | 25.43 | 42.55 |
| | | 0.8 | 1.56 | 10.67 | 25.30 | 42.45 |
| | | 1.0 | 1.42 | 10.58 | 25.22 | 42.38 |

**TABLE 5.10.** Parameter $\phi$ for a reduced length of a cantilever haunched beam

| $\eta = \dfrac{l_1}{l}$ | Parameter $n = h_1/h_0$ | | | |
|---|---|---|---|---|
| | 1.2 | 1.3 | 1.4 | 1.5 |
| 0.1 | 0.97 | 0.96 | 0.95 | 0.92 |
| 0.2 | 0.96 | 0.93 | 0.91 | 0.89 |

Expression for mode shape of vibration is

$$a_1 X(x) = a_1 \frac{x^2(3l - x)}{2l^3} \tag{5.12}$$

## 5.2  STEPPED BEAMS

### 5.2.1  Bernoulli–Euler beams with different boundary conditions

A beam with a discontinuous variation of thickness is presented in Fig. 5.7.

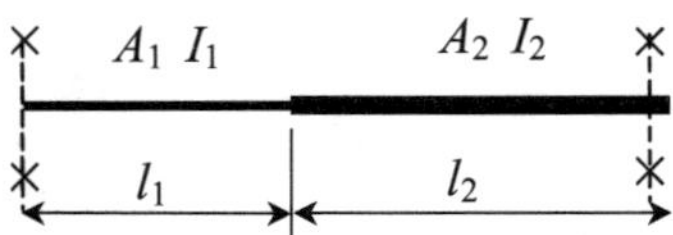

FIGURE 5.7. Stepped-beam geometry. The types of supports are not shown.

The natural frequency of vibrations is

$$\omega = \frac{\lambda^2}{l^2} \sqrt{\frac{EI_1}{\rho A_1}}, \quad l = l_1 + l_2$$

The exact frequency equations for beams with different boundary conditions are presented in Table 5.11. The dimensionless frequency and geometry parameters are

$$k = \frac{k_2}{k_1}, \quad k_i^4 = \frac{\rho A_i}{EI_i} \omega^2, \quad i = 1, 2 \quad \text{and} \quad I = \frac{I_2}{I_1}$$

Notation:

$$S1 = \sin k_1 l_1, \qquad S2 = \sin k_2 l_2, \qquad C1 = \cos k_1 l_1, \qquad C2 = \cos k_2 l_2,$$
$$SH1 = \sinh k_1 l_1, \quad SH2 = \sinh k_2 l_2, \quad CH1 = \cosh k_1 l_1, \quad CH2 = \cosh k_2 l_2.$$

Table 5.11 also shows the exact solution, $\lambda^2$, of the fundamental mode of vibration for beams with circular cross-section with $l_1 = l_2 = l/2$, $A_2 = \alpha A_1$, $I = I_2/I_1 = \alpha^2$ (Jano and Bert, 1989).

### 5.2.2  Stepped–cantilever Bernoulli–Euler beam

A stepped clamped–free beam is presented in Fig. 5.8.

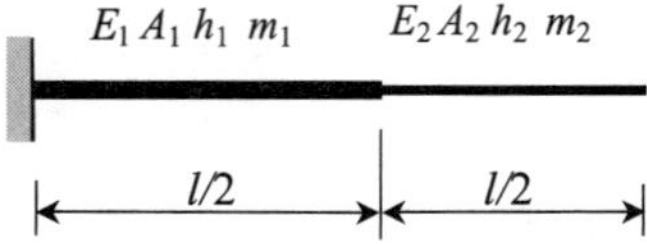

FIGURE 5.8.  Stepped–cantilever beam.

**TABLE 5.11** Frequency parameter $\lambda^2$ for stepped beams with different boundary conditions

| Type beam | Frequency equation (common case $l_1$ and $l_2$) | $I$ | $\lambda^2$ |
|---|---|---|---|
| Clamped–clamped | $\begin{bmatrix} S1-SH1 & C1-CH1 & -S2+SH2 & -C2+CH2 \\ C1-CH1 & -S1-SH1 & k(C2-CH2) & -k(S2+SH2) \\ -S1-SH1 & -C1-CH1 & k^2I(S2+SH2) & k^2I(C2+CH2) \\ -C1-CH1 & S1-SH1 & -k^3I(C2+CH2) & k^3I(S2-SH2) \end{bmatrix} = 0$ | **1***<br>5<br>10<br>20<br>40 | **22.3733**<br>25.9591<br>27.6807<br>30.3213<br>34.3252 |
| Pinned–pinned | $\begin{bmatrix} S1 & SH1 & -S2 & -SH2 \\ C1 & CH1 & kC2 & kCH2 \\ -S1 & SH1 & k^2IS2 & -k^2ISH2 \\ -C1 & CH1 & -k^3IC2 & k^3ICH2 \end{bmatrix} = 0$ | **1***<br>5<br>10<br>20<br>40 | **9.8696**<br>10.4129<br>9.8781<br>9.0747<br>8.1369 |
| Clamped–free | $\begin{bmatrix} S1-SH1 & C1-CH1 & -S2-SH2 & -C2-CH2 \\ C1-CH1 & -S1-SH1 & k(C2+CH2) & k(-S2+SH2) \\ -S1-SH1 & -C1-CH1 & k^2I(S2-SH2) & k^2I(C2-CH2) \\ -C1-CH1 & S1-SH1 & -k^3I(C2-CH2) & k^3I(S2+SH2) \end{bmatrix} = 0$ | **1***<br>5<br>10<br>20<br>40 | **3.5160**<br>2.4373<br>2.0629<br>1.7418<br>1.4685 |
| Clamped–pinned | $\begin{bmatrix} S1-SH1 & C1-CH1 & -S2 & -SH2 \\ C1-CH1 & -S1-SH1 & kC2 & -kCH2 \\ -S1-SH1 & -C1-CH1 & k^2IS2 & -k^2ISH2 \\ -C1-CH1 & S1-SH1 & -k^3IC2 & k^3ICH2 \end{bmatrix} = 0$ | **1***<br>5<br>10<br>20<br>40 | **15.4182**<br>16.2811<br>15.5129<br>14.2568<br>12.7501 |

(continued)

**TABLE 5.11**   (*continued*)

| Type beam | Frequency equation (common case $l_1$ and $l_2$) | $I$ | $\lambda^2$ |
|---|---|---|---|
| Free–free | $\begin{bmatrix} S1+SH1 & C1+CH1 & -S2-SH2 & -C2-CH2 \\ C1+CH1 & -S1+SH1 & k(C2+CH2) & k(-S2+SH2) \\ -S1+SH1 & -C1+CH1 & k^2I(S2-SH2) & k^2I(C2-CH2) \\ -C1+CH1 & S1+SH1 & -k^3I(C2-CH2) & k^3I(S2+SH2) \end{bmatrix} = 0$ | **1*** <br> 5 <br> 10 <br> 20 <br> 40 | **22.3733** <br> 24.1650 <br> 23.5459 <br> 22.4725 <br> 21.1907 |
| Guided–guided | $\begin{bmatrix} C1 & CH1 & -C2 & -CH2 \\ -S1 & SH1 & kS2 & kSH2 \\ -C1 & CH1 & k^2IC2 & -k^2ICH2 \\ S1 & SH1 & k^3IS2 & k^3ISH2 \end{bmatrix} = 0$ | **1*** <br> 5 <br> 10 <br> 20 <br> 40 | **9.8696** <br> 13.5124 <br> 15.9066 <br> 18.2949 <br> 20.1954 |
| Guided–pinned | $\begin{bmatrix} C1 & CH1 & -S2 & -SH2 \\ -S1 & SH1 & kC2 & kCH2 \\ -C1 & CH1 & -k^2IS2 & -k^2ISH2 \\ S1 & SH1 & -k^3IC2 & k^3ICH2 \end{bmatrix} = 0$ | **1*** <br> 5 <br> 10 <br> 20 <br> 40 | **2.4674** <br> 2.4372 <br> 2.3292 <br> 2.1841 <br> 2.0122 |

**TABLE 5.11** *(continued)*

| Type beam | Frequency equation (common case $l_1$ and $l_2$) | $I$ | $\lambda^2$ |
|---|---|---|---|
| Clamped–guided | $\begin{bmatrix} S1 - SH1 & C1 - CH1 & -C2 & -CH2 \\ C1 - CH1 & -S1 - SH1 & -kS2 & kSH2 \\ -S1 - SH1 & -C1 - CH1 & k^2 IC2 & -k^2 ICH2 \\ -C1 - CH1 & S1 - SH1 & k^3 IS2 & k^3 ISH2 \end{bmatrix} = 0$ | **1*** <br> 5 <br> 10 <br> 20 <br> 40 | **5.5933** <br> 5.6912 <br> 5.6321 <br> 5.3573 <br> 4.8913 |
| Free–guided | $\begin{bmatrix} S1 + SH1 & C1 + CH1 & -C2 & -CH2 \\ C1 + CH1 & -S1 + SH1 & -kS2 & kSH2 \\ -S1 + SH1 & -C1 + CH1 & k^2 IC2 & -k^2 ICH2 \\ -C1 + CH1 & S1 + SH1 & k^3 IS2 & k^3 ISH2 \end{bmatrix} = 0$ | **1*** <br> 5 <br> 10 <br> 20 <br> 40 | **5.5933** <br> 9.3624 <br> 11.0519 <br> 12.4070 <br> 13.2947 |
| Free–pinned | $\begin{bmatrix} S1 + SH1 & C1 + CH1 & -S2 & -SH2 \\ C1 + CH1 & -S1 + SH1 & k^2 IS2 & kCH2 \\ -S1 + SH1 & -C1 + CH1 & -k^2 IS2 & -k^2 ISH2 \\ -C1 + CH1 & S1 + SH1 & -k^3 IC2 & k^3 ICH2 \end{bmatrix} = 0$ | **1*** <br> 5 <br> 10 <br> 20 <br> 40 | **15.4182** <br> 18.6102 <br> 18.7641 <br> 18.4031 <br> 17.7778 |

*Special case*:
* Frequency parameters for uniform beams (Table 4.3, Karnovsky, Lebed, 2003).

Natural frequencies of vibration may be calculated by the formula

$$\omega_i = \frac{\lambda_i}{l^2}\sqrt{\frac{EI_1}{m_1}} \tag{5.12}$$

The frequency parameters $\lambda_i$ are presented in Table 5.12 for the dimensionless geometry ratio $H = \dfrac{h_2}{h_1}$, mass ratio $\dfrac{m_2}{m_1} = H^2$; $\dfrac{EI_2}{EI_1} = H^4$ for $H = 0.2$.

Finite element method higher degree polynomials have been used (Balasubramanian *et al.*, 1990).

**TABLE 5.12.**   Frequency parameter $\lambda$ for a stepped–cantilever Bernoulli–Euler beam

| | | Mode | | | |
|---|---|---|---|---|---|
| 1 | 2 | 3 | 4 | 5 | 6 |
| 2.78514 | 13.1938 | 18.5068 | 49.3539 | 85.8066 | 99.1000 |

### 5.2.3  Bernoulli–Euler beams with guided mass

Details of a guided–clamped stepped beam with a mass are shown in Fig. 5.9.

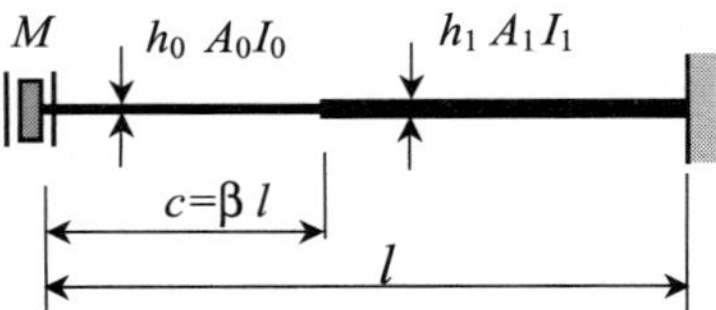


**FIGURE 5.9.**   Guided–clamped stepped beam with concentrated mass $M$.

The natural frequencies of vibration may be calculated by

$$\omega = \frac{\lambda^2}{l^2}\sqrt{\frac{EI_0}{\rho A_0}}$$

Frequency parameters $\lambda_i^2$ for different modes of vibration of a beam with rectangular cross-sections of constant width and $h_0/h_1 = 0.8$ are listed in Table 5.13. Dimensionless parameters are

$$\mu = \frac{M}{M_{\text{beam}}}, \quad \beta = \frac{c}{l}$$

The finite element method has been applied ([Laura *et al.*, 1989; Bambill and Laura, 1989).

**TABLE 5.13.** Guided–clamped stepped beam with guided mass: Frequency parameters $\lambda^2$ by Bernoulli–Euler theory

| $\beta = \dfrac{c}{l}$ | Mode | $\mu = M/M_{\text{beam}}$ | | | | | | | |
|---|---|---|---|---|---|---|---|---|---|
| | | 0.0 | 0.2 | 0.4 | 0.6 | 0.8 | 1.0 | 5.0 | 10.0 |
| | 1 | 6.70 | 5.45 | 4.71 | 4.20 | 3.832 | 3.54 | 1.792 | 1.28 |
| | 2 | 35.61 | 31.07 | 29.33 | 28.41 | 27.85 | 27.47 | 25.99 | 25.77 |
| 1/3 | 3 | 85.85 | 76.68 | 74.18 | 73.04 | 72.39 | 71.97 | 70.45 | 70.24 |
| | 4 | 159.5 | 147.3 | 144.6 | 143.4 | 142.8 | 142.4 | 140.9 | 140.8 |
| | 5 | 259.5 | 240.4 | 236.9 | 235.4 | 234.6 | 234.1 | 232.4 | 232.2 |
| | 1 | 6.81 | 5.51 | 4.74 | 4.22 | 3.84 | 3.55 | 1.79 | 1.28 |
| | 2 | 33.90 | 29.68 | 28.09 | 27.26 | 26.75 | 26.41 | 25.09 | 24.89 |
| 1/2 | 3 | 83.11 | 75.42 | 73.23 | 72.21 | 71.62 | 71.24 | 69.86 | 69.67 |
| | 4 | 154.0 | 140.7 | 137.7 | 136.4 | 135.7 | 135.3 | 133.7 | 133.5 |
| | 5 | 248.8 | 231.8 | 228.7 | 227.4 | 226.7 | 226.2 | 224.7 | 224.5 |
| | 1 | 6.791 | 5.46 | 4.69 | 4.17 | 3.79 | 3.504 | 1.760 | 1.26 |
| | 2 | 32.67 | 29.00 | 27.59 | 26.86 | 26.41 | 26.11 | 24.92 | 24.74 |
| 2/3 | 3 | 80.58 | 72.59 | 70.29 | 69.23 | 68.61 | 68.22 | 66.77 | 66.56 |
| | 4 | 150.1 | 138.0 | 135.3 | 134.1 | 133.4 | 133.0 | 131.5 | 131.3 |
| | 5 | 237.1 | 220.7 | 217.6 | 216.4 | 215.7 | 215.3 | 213.8 | 213.6 |

## 5.2.4 Cantilever Timoshenko beam with tip mass

Details of a stepped clamped–free beam having constant width $b$ and a mass $M$ at the tip are shown in Fig. 5.10.

Dimensionless parameters are

$$\gamma_A = \frac{A_2}{A_1}, \quad \gamma_L = \frac{L_1}{L}, \quad \eta = \frac{I_1}{A_1 L_1^2}, \quad \mu = \frac{M}{M_b}$$

where $M_b$ is the total beam mass.

Natural frequencies of vibration may be calculated by

$$\omega = \frac{\lambda}{L^2} \sqrt{\frac{EI_1}{\rho A_1}}$$

The frequency parameters, $\lambda$, for beams with shear factor $k = 5/6$, Poisson ratio $v = 0.3$ and different combinations of the parameters $\eta$, $\gamma_L$, $\gamma_A$ and $\mu$ are listed in Table 5.14 (Rossi

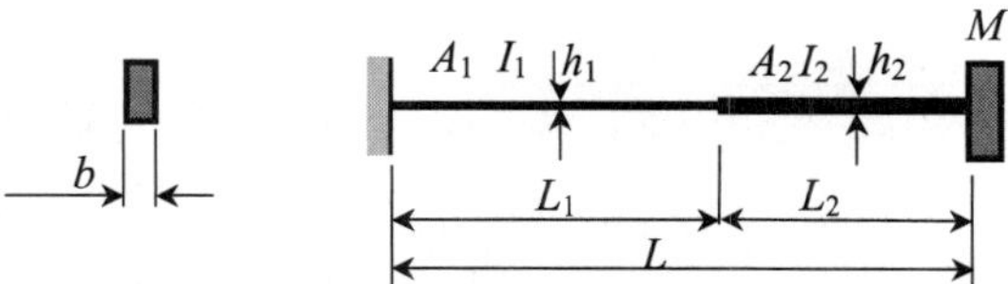


**FIGURE 5.10.** Clamped–free beam with concentrated mass $M$.

**TABLE 5.14.**   Clamped–free stepped beam with lumped mass at the end: Frequency parameters $\lambda$ by Timoshenko theory

| $\eta = \dfrac{I_1}{A_1 L_1^2}$ | $\gamma_L = \dfrac{L_1}{L}$ | $\gamma_A = \dfrac{A_2}{A_1}$ | $\mu = \dfrac{M}{M_b}$ | Mode 1 | 2 | 3 | 4 | 5 |
|---|---|---|---|---|---|---|---|---|
| $10^{-7}$ | 2/3 | 0.8 | 0.0 | 3.83 | 21.57 | 56.34 | 112.33 | 185.60 |
|  |  |  | 0.2 | 2.76 | 17.03 | 48.32 | 100.58 | 167.59 |
|  |  |  | 0.4 | 2.26 | 15.99 | 47.15 | 99.25 | 166.02 |
|  |  |  | 0.6 | 1.96 | 15.54 | 46.69 | 98.74 | 165.45 |
|  |  |  | 0.8 | 1.75 | 15.28 | 46.45 | 98.48 | 165.15 |
|  |  |  | 1.0 | 1.60 | 15.12 | 46.30 | 98.31 | 164.97 |
|  |  | 0.6 | 0.0 | 4.23 | 19.94 | 49.55 | 103.32 | 162.83 |
|  |  |  | 0.2 | 2.90 | 14.62 | 43.11 | 90.63 | 146.74 |
|  |  |  | 0.4 | 2.33 | 13.71 | 42.40 | 89.51 | 145.76 |
|  |  |  | 0.6 | 2.00 | 13.33 | 42.14 | 89.10 | 145.41 |
|  |  |  | 0.8 | 1.78 | 13.13 | 42.00 | 88.89 | 145.24 |
|  |  |  | 1.0 | 1.62 | 13.01 | 41.91 | 88.76 | 145.13 |
|  | 1/3 | 0.8 | 0.0 | 3.62 | 19.35 | 53.12 | 104.97 | 171.00 |
|  |  |  | 0.2 | 2.59 | 16.00 | 45.78 | 93.68 | 156.22 |
|  |  |  | 0.4 | 2.11 | 15.18 | 44.56 | 92.28 | 154.79 |
|  |  |  | 0.6 | 1.83 | 14.82 | 44.07 | 91.75 | 154.26 |
|  |  |  | 0.8 | 1.64 | 14.61 | 43.81 | 91.47 | 153.98 |
|  |  |  | 1.0 | 1.49 | 14.48 | 43.64 | 91.29 | 153.81 |
|  |  | 0.6 | 0.0 | 3.49 | 16.87 | 42.60 | 86.25 | 142.40 |
|  |  |  | 0.2 | 2.34 | 14.03 | 36.63 | 76.07 | 130.09 |
|  |  |  | 0.4 | 1.88 | 13.43 | 35.78 | 75.00 | 129.05 |
|  |  |  | 0.6 | 1.61 | 13.18 | 35.45 | 74.59 | 128.67 |
|  |  |  | 0.8 | 1.43 | 13.04 | 35.27 | 74.38 | 128.47 |
|  |  |  | 1.0 | 1.30 | 12.95 | 35.16 | 74.25 | 128.35 |
| 0.0004 | 2/3 | 0.8 | 0.0 | 3.82 | 21.35 | 55.04 | 107.50 | 173.62 |
|  |  |  | 0.2 | 2.75 | 16.90 | 47.34 | 96.69 | 157.79 |
|  |  |  | 0.4 | 2.26 | 15.87 | 46.21 | 95.44 | 156.36 |
|  |  |  | 0.6 | 1.96 | 15.42 | 45.76 | 94.97 | 155.83 |
|  |  |  | 0.8 | 1.75 | 15.17 | 45.33 | 94.72 | 155.56 |
|  |  |  | 1.0 | 1.60 | 15.01 | 45.37 | 94.56 | 155.39 |
|  |  | 0.6 | 0.0 | 4.22 | 19.78 | 48.57 | 99.47 | 154.33 |
|  |  |  | 0.2 | 2.89 | 14.54 | 42.33 | 87.74 | 139.58 |
|  |  |  | 0.4 | 2.32 | 13.63 | 41.64 | 86.69 | 138.66 |
|  |  |  | 0.6 | 2.00 | 13.26 | 41.38 | 86.30 | 138.33 |
|  |  |  | 0.8 | 1.78 | 13.06 | 41.25 | 86.10 | 138.16 |
|  |  |  | 1.0 | 1.62 | 12.93 | 41.16 | 85.97 | 138.05 |
|  | 1/3 | 0.8 | 0.0 | 3.62 | 19.30 | 52.84 | 103.92 | 168.34 |
|  |  |  | 0.2 | 2.59 | 15.97 | 45.57 | 92.84 | 154.00 |
|  |  |  | 0.4 | 2.11 | 15.16 | 44.37 | 91.47 | 152.59 |
|  |  |  | 0.6 | 1.83 | 14.79 | 43.88 | 90.95 | 152.07 |
|  |  |  | 0.8 | 1.64 | 14.59 | 43.62 | 90.67 | 151.80 |
|  |  |  | 1.0 | 1.49 | 14.45 | 43.45 | 90.50 | 151.63 |

**TABLE 5.14.** (*Continued*)

| $\eta = \dfrac{I_1}{A_1 L_1^2}$ | $\gamma_L = \dfrac{L_1}{L}$ | $\gamma_A = \dfrac{A_2}{A_1}$ | $\mu = \dfrac{M}{M_b}$ | Mode | | | | |
|---|---|---|---|---|---|---|---|---|
| | | | | 1 | 2 | 3 | 4 | 5 |
| | | | 0.0 | 3.49 | 16.84 | 42.45 | 85.67 | 140.87 |
| | | | 0.2 | 2.34 | 14.01 | 36.52 | 75.62 | 128.72 |
| | 1/3 | 0.6 | 0.4 | 1.88 | 13.42 | 35.68 | 74.56 | 127.70 |
| | | | 0.6 | 1.61 | 13.16 | 35.34 | 74.16 | 127.32 |
| | | | 0.8 | 1.43 | 13.02 | 35.17 | 73.95 | 127.13 |
| | | | 1.0 | 1.30 | 12.93 | 35.06 | 73.82 | 127.01 |
| | | | 0.0 | 3.80 | 20.72 | 51.68 | 96.39 | 148.97 |
| | | | 0.2 | 2.74 | 16.51 | 44.77 | 87.51 | 137.04 |
| | | 0.8 | 0.4 | 2.25 | 15.52 | 43.72 | 86.44 | 135.87 |
| | | | 0.6 | 1.95 | 15.08 | 43.31 | 86.02 | 135.44 |
| | | | 0.8 | 1.75 | 14.84 | 43.08 | 85.81 | 135.21 |
| | | | 1.0 | 1.59 | 14.68 | 42.94 | 85.67 | 135.08 |
| | 2/3 | | 0.0 | 4.20 | 19.32 | 45.98 | 90.28 | 135.74 |
| | | | 0.2 | 2.88 | 14.28 | 40.25 | 80.62 | 123.72 |
| | | 0.6 | 0.4 | 2.32 | 13.40 | 39.61 | 79.71 | 122.91 |
| | | | 0.6 | 1.99 | 13.04 | 39.36 | 79.37 | 122.62 |
| | | | 0.8 | 1.77 | 12.84 | 39.24 | 79.19 | 122.47 |
| | | | 1.0 | 1.61 | 12.72 | 39.16 | 70.09 | 122.38 |
| 0.0016 | | | 0.0 | 3.61 | 19.18 | 52.03 | 100.98 | 161.15 |
| | | | 0.2 | 2.58 | 15.88 | 44.98 | 90.50 | 147.92 |
| | | 0.8 | 0.4 | 2.11 | 15.08 | 43.80 | 89.18 | 146.60 |
| | | | 0.6 | 1.83 | 14.72 | 43.32 | 88.67 | 146.11 |
| | | | 0.8 | 1.64 | 14.51 | 43.06 | 88.40 | 145.85 |
| | | | 1.0 | 1.49 | 14.38 | 42.90 | 88.24 | 145.69 |
| | 1/3 | | 0.0 | 3.48 | 16.75 | 42.01 | 84.00 | 136.16 |
| | | | 0.2 | 2.34 | 13.96 | 36.19 | 74.33 | 124.88 |
| | | 0.6 | 0.4 | 1.87 | 13.36 | 35.36 | 73.30 | 123.91 |
| | | | 0.6 | 1.61 | 13.11 | 35.03 | 72.91 | 123.56 |
| | | | 0.8 | 1.43 | 12.97 | 34.86 | 72.70 | 123.37 |
| | | | 1.0 | 1.30 | 12.88 | 34.75 | 72.58 | 123.26 |
| | | | 0.0 | 3.77 | 19.80 | 47.35 | 84.14 | 125.06 |
| | | | 0.2 | 2.73 | 15.92 | 41.36 | 77.08 | 116.30 |
| | | 0.8 | 0.4 | 2.24 | 14.98 | 40.42 | 76.17 | 115.36 |
| | | | 0.6 | 1.94 | 14.57 | 40.04 | 75.82 | 115.01 |
| | | | 0.8 | 1.74 | 14.34 | 39.83 | 75.63 | 114.82 |
| | | | 1.0 | 1.59 | 14.19 | 39.71 | 75.52 | 114.71 |
| 0.0036 | 2/3 | | 0.0 | 4.16 | 18.62 | 42.54 | 79.68 | 116.41 |
| | | | 0.2 | 2.86 | 13.89 | 37.44 | 72.05 | 106.99 |
| | | 0.6 | 0.4 | 2.30 | 13.04 | 36.84 | 71.29 | 106.30 |
| | | | 0.6 | 1.98 | 12.69 | 36.62 | 71.00 | 106.05 |
| | | | 0.8 | 1.76 | 12.50 | 36.50 | 70.85 | 105.92 |
| | | | 1.0 | 1.60 | 12.38 | 36.43 | 70.76 | 105.84 |

(*continued*)

**TABLE 5.14.**   *(Continued)*

| $\eta = \dfrac{I_1}{A_1 L_1^2}$ | $\gamma_L = \dfrac{L_1}{L}$ | $\gamma_A = \dfrac{A_2}{A_1}$ | $\mu = \dfrac{M}{M_b}$ | Mode 1 | Mode 2 | Mode 3 | Mode 4 | Mode 5 |
|---|---|---|---|---|---|---|---|---|
| 0.0036 | 1/3 | 0.8 | 0.0 | 3.60 | 18.97 | 50.78 | 96.63 | 151.24 |
|  |  |  | 0.2 | 2.58 | 15.74 | 44.04 | 86.99 | 139.41 |
|  |  |  | 0.4 | 2.11 | 14.95 | 42.90 | 85.75 | 138.19 |
|  |  |  | 0.6 | 1.83 | 14.59 | 42.44 | 85.27 | 137.73 |
|  |  |  | 0.8 | 1.63 | 14.39 | 42.19 | 85.02 | 137.49 |
|  |  |  | 1.0 | 1.49 | 14.26 | 42.03 | 84.86 | 137.35 |
|  |  | 0.6 | 0.0 | 3.48 | 16.62 | 41.32 | 81.43 | 129.43 |
|  |  |  | 0.2 | 2.34 | 13.86 | 35.67 | 72.33 | 119.24 |
|  |  |  | 0.4 | 1.87 | 13.28 | 34.86 | 71.34 | 118.34 |
|  |  |  | 0.6 | 1.61 | 13.02 | 34.54 | 70.97 | 118.01 |
|  |  |  | 0.8 | 1.43 | 12.89 | 34.36 | 70.77 | 117.84 |
|  |  |  | 1.0 | 1.30 | 12.80 | 34.26 | 70.65 | 117.73 |
| 0.0064 | 2/3 | 0.8 | 0.0 | 3.73 | 18.70 | 42.91 | 73.22 | 105.69 |
|  |  |  | 0.2 | 2.70 | 15.19 | 37.77 | 67.55 | 99.16 |
|  |  |  | 0.4 | 2.22 | 14.33 | 36.92 | 66.78 | 98.40 |
|  |  |  | 0.6 | 1.93 | 13.94 | 36.58 | 66.48 | 98.11 |
|  |  |  | 0.8 | 1.73 | 13.72 | 36.39 | 66.32 | 97.96 |
|  |  |  | 1.0 | 1.58 | 13.58 | 36.28 | 66.22 | 97.86 |
|  |  | 0.6 | 0.0 | 4.11 | 17.77 | 38.92 | 69.94 | 99.89 |
|  |  |  | 0.2 | 2.84 | 13.40 | 34.39 | 63.86 | 92.56 |
|  |  |  | 0.4 | 2.29 | 12.59 | 33.85 | 63.22 | 91.97 |
|  |  |  | 0.6 | 1.97 | 12.26 | 33.65 | 62.97 | 91.75 |
|  |  |  | 0.8 | 1.75 | 12.07 | 33.54 | 62.85 | 91.64 |
|  |  |  | 1.0 | 1.59 | 11.96 | 33.47 | 62.77 | 91.57 |
|  | 1/3 | 0.8 | 0.0 | 3.59 | 18.69 | 49.18 | 91.48 | 140.31 |
|  |  |  | 0.2 | 2.57 | 15.55 | 42.84 | 82.76 | 129.89 |
|  |  |  | 0.4 | 2.11 | 14.78 | 41.74 | 81.61 | 128.77 |
|  |  |  | 0.6 | 1.83 | 14.43 | 41.30 | 81.17 | 128.35 |
|  |  |  | 0.8 | 1.63 | 14.23 | 41.06 | 80.93 | 128.13 |
|  |  |  | 1.0 | 1.49 | 14.10 | 40.91 | 80.78 | 127.99 |
|  |  | 0.6 | 0.0 | 3.47 | 16.43 | 40.40 | 78.22 | 121.64 |
|  |  |  | 0.2 | 2.33 | 13.73 | 34.99 | 69.80 | 112.58 |
|  |  |  | 0.4 | 1.87 | 13.16 | 34.19 | 68.87 | 111.75 |
|  |  |  | 0.6 | 1.60 | 12.91 | 33.88 | 68.51 | 111.45 |
|  |  |  | 0.8 | 1.43 | 12.77 | 33.71 | 68.33 | 111.29 |
|  |  |  | 1.0 | 1.30 | 12.68 | 33.61 | 68.21 | 111.20 |
|  | 2/3 | 0.8 | 0.0 | 3.67 | 17.52 | 38.81 | 64.15 | 90.53 |
|  |  |  | 0.2 | 2.67 | 14.40 | 34.38 | 59.52 | 85.64 |
|  |  |  | 0.4 | 2.20 | 13.60 | 33.60 | 58.85 | 85.02 |
|  |  |  | 0.6 | 1.91 | 13.24 | 33.29 | 58.59 | 84.77 |
|  |  |  | 0.8 | 1.71 | 13.04 | 33.12 | 58.45 | 84.65 |
|  |  |  | 1.0 | 1.56 | 12.91 | 33.01 | 58.36 | 84.57 |

**TABLE 12.14.**   *(Continued)*

| $\eta = \dfrac{I_1}{A_1 L_1^2}$ | $\gamma_L = \dfrac{L_1}{L}$ | $\gamma_A = \dfrac{A_2}{A_1}$ | $\mu = \dfrac{M}{M_b}$ | Mode | | | | |
|---|---|---|---|---|---|---|---|---|
| | | | | 1 | 2 | 3 | 4 | 5 |
| 0.01 | 2/3 | 0.6 | 0.0 | 4.05 | 16.82 | 35.49 | 61.68 | 86.54 |
| | | | 0.2 | 2.81 | 12.83 | 31.46 | 56.72 | 80.81 |
| | | | 0.4 | 2.27 | 12.08 | 30.96 | 56.17 | 80.30 |
| | | | 0.6 | 1.95 | 11.76 | 30.77 | 55.96 | 80.12 |
| | | | 0.8 | 1.74 | 11.59 | 30.67 | 55.85 | 80.02 |
| | | | 1.0 | 1.58 | 11.48 | 30.60 | 55.78 | 79.96 |
| | 1/3 | 0.8 | 0.0 | 3.58 | 18.36 | 47.35 | 86.02 | 129.48 |
| | | | 0.2 | 2.57 | 15.32 | 41.44 | 78.21 | 120.34 |
| | | | 0.4 | 2.10 | 14.56 | 40.40 | 77.15 | 119.33 |
| | | | 0.6 | 1.82 | 14.22 | 39.97 | 76.74 | 118.94 |
| | | | 0.8 | 1.63 | 14.03 | 39.74 | 76.52 | 118.74 |
| | | | 1.0 | 1.49 | 13.91 | 39.60 | 76.38 | 118.61 |
| | | 0.6 | 0.0 | 3.46 | 16.20 | 39.33 | 74.64 | 113.61 |
| | | | 0.2 | 2.33 | 13.57 | 34.17 | 66.94 | 105.57 |
| | | | 0.4 | 1.87 | 13.01 | 33.40 | 66.06 | 104.82 |
| | | | 0.6 | 1.60 | 12.76 | 33.10 | 65.72 | 104.54 |
| | | | 0.8 | 1.43 | 12.63 | 32.93 | 65.55 | 104.39 |
| | | | 1.0 | 1.29 | 12.54 | 32.83 | 65.44 | 104.30 |

*et al.*, 1990). The case of $\eta = 10^{-7}$ corresponds, from a practical engineering viewpoint, to the classical Bernoulli–Euler theory of vibrating beams.

## 5.3   ELASTICALLY RESTRAINED BEAMS

### 5.3.1   Tapered Bernoulli–Euler beam with one end spring-hinged and a tip body at the free end

Consider a rectangular beam of depth $h$, tapered linearly in the vertical plane and of constant width $b$ in the horizontal plane (Fig. 5.11). The rotational spring constant is $k$; a body attached at the free end has a mass $M$ and a rotatory inertia $J$.

The tapered parameter of the beam is $\alpha = (h_0 - h_1)/h_0$. The cross sectional area, flexural rigidity and the total beam mass $M_b$ are given by

$$A(x) = A_0\left(1 - \alpha\frac{x}{l}\right), \quad A_0 = bh_0 \tag{5.13}$$

$$EI(x) = EI_0\left(1 - \alpha\frac{x}{l}\right)^3, \quad I_0 = \frac{bh_0^3}{12} \tag{5.14}$$

$$M_b = \rho A_0 l\left(1 - \frac{\alpha}{2}\right) \tag{5.15}$$

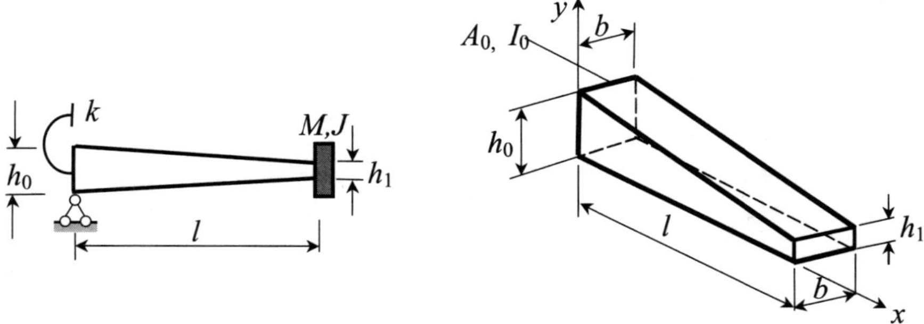

**FIGURE 5.11.**   Elastic–clamped tapered beam with a body attached at the free end.

The differential equation for eigenfunctions $X$ is given by

$$\frac{d^2}{d\xi^2}\left(\xi^3 \frac{d^2 X}{d\xi^2}\right) - \beta^4 \xi X = 0 \tag{5.16}$$

where $\xi = 1 - \alpha\frac{x}{l}$.   Frequency parameter

$$\beta^4 = \frac{\rho A_0}{E I_0}\omega^2 \frac{l^4}{\alpha^4} = \frac{\lambda^4}{\alpha^4} \tag{5.17}$$

The boundary conditions are:

At $\xi = 1$ (left end):     $X = 0$   and   $\dfrac{kl}{E I_0}\dfrac{dX}{d\xi} + \alpha\dfrac{d^2 X}{d\xi^2} = 0 \tag{5.18}$

At $\xi = 1 - \alpha$ (right end):

$$\frac{J}{M_b l^2}\lambda_n^4\left(1 - \frac{\alpha}{2}\right)\frac{dX}{d\xi} + \alpha(1-\alpha)^3 \frac{d^2 X}{d\xi^2} = 0$$

$$\frac{M}{M_b}\lambda_n^4\left(1 - \frac{\alpha}{2}\right)X - \alpha^3(1-\alpha)^2\left[3\frac{d^2 X}{d\xi^2} + (1-\alpha)\frac{d^3 X}{d\xi^3}\right] = 0 \tag{5.19}$$

The general expression for the vibration mode is expressed as

$$X(\xi) = \xi^{-1/2}[C_1 J_1(Z) + C_2 Y_1(Z) + C_3 I_1(Z) + C_4 K_1(Z)], \quad Z = 2\beta\sqrt{\xi} \tag{5.20}$$

where $J$, $Y$, $I$ and $K$ are Bessel functions of the first order with argument $Z$.

The frequency equation can be obtained by using the expression for $X(\xi)$ and the boundary conditions. The complete frequency equation is presented by Lee (1976).

The natural frequency of vibration of the $i$th mode is

$$\omega = \frac{\lambda^2}{l^2}\sqrt{\frac{E I_0}{\rho A_0}}$$

The frequency parameters $\lambda$ for various values of the spring stiffness, end mass, rotary inertia and taper parameter are shown in Table 5.15.

Elastic–clamped tapered Bernoulli–Euler beams with a tip body at the free end and with translational and rotational springs along the space are studied by Yang (1990).

**TABLE 5.15.** Fundamental frequency coefficients $\lambda$ for a tapered Bernoulli–Euler beam with one end spring-hinged and a tip body

$$\beta = \frac{kl}{EI_0}, \quad \mu = \frac{M}{M_b}, \quad \psi = \frac{J}{M_b l^2}, \quad \alpha = \frac{h_0 - h_1}{h_0}$$

| | | $\alpha = 0.2$ | | | | $\alpha = 0.4$ | | | | $\alpha = 0.6$ | | | |
|---|---|---|---|---|---|---|---|---|---|---|---|---|---|
| $\beta$ | $\mu$ | $\psi = 0.0$ | 1.0 | 10 | 100 | $\psi = 0.0$ | 1.0 | 10 | 100 | $\psi = 0.0$ | 1.0 | 10 | 100 |
| 0.1 | 0 | 0.76581 | 0.52648 | 0.31212 | 0.17655 | 0.80329 | 0.53786 | 0.31630 | 0.17873 | 0.85244 | 0.54214 | 0.31460 | 0.17749 |
| | 1 | 0.53447 | 0.46056 | 0.30581 | 0.17617 | 0.55190 | 0.47240 | 0.31044 | 0.17838 | 0.57229 | 0.48247 | 0.30994 | 0.17722 |
| | 10 | 0.31911 | 0.31147 | 0.26704 | 0.17291 | 0.32815 | 0.32008 | 0.27299 | 0.17537 | 0.33836 | 0.32954 | 0.27733 | 0.17484 |
| | 100 | 0.18068 | 0.18022 | 0.17624 | 0.15066 | 0.18570 | 0.18521 | 0.18102 | 0.15396 | 0.19134 | 0.19081 | 0.18624 | 0.15631 |
| 1.0 | 0 | 1.29285 | 0.80143 | 0.46224 | 0.26060 | 1.34860 | 0.77587 | 0.44305 | 0.24953 | 1.42170 | 0.71316 | 0.40350 | 0.22704 |
| | 1 | 0.88824 | 0.72881 | 0.45708 | 0.26031 | 0.90651 | 0.71950 | 0.43949 | 0.24932 | 0.92322 | 0.68145 | 0.40172 | 0.22694 |
| | 10 | 0.52778 | 0.51119 | 0.41814 | 0.25768 | 0.53565 | 0.51643 | 0.41034 | 0.24752 | 0.54128 | 0.51617 | 0.38579 | 0.22604 |
| | 100 | 0.29864 | 0.29763 | 0.28901 | 0.23562 | 0.30288 | 0.30173 | 0.29175 | 0.23108 | 0.30577 | 0.30431 | 0.29128 | 0.21709 |
| 10 | 0 | 1.75935 | 0.91673 | 0.52059 | 0.29303 | 1.80561 | 0.85139 | 0.48138 | 0.27085 | 1.86889 | 0.74992 | 0.42260 | 0.23769 |
| | 1 | 1.16386 | 0.86721 | 0.51759 | 0.29286 | 1.15829 | 0.81761 | 0.47944 | 0.27074 | 1.14177 | 0.73252 | 0.42163 | 0.23764 |
| | 10 | 0.68517 | 0.65103 | 0.49228 | 0.29135 | 0.67763 | 0.63679 | 0.46237 | 0.26976 | 0.66244 | 0.60760 | 0.41288 | 0.23715 |
| | 100 | 0.38725 | 0.38517 | 0.36753 | 0.27709 | 0.38270 | 0.38024 | 0.35919 | 0.26016 | 0.37376 | 0.37056 | 0.34235 | 0.23223 |
| 100 | 0 | 1.89955 | 0.93482 | 0.52973 | 0.29812 | 1.93315 | 0.86185 | 0.48677 | 0.27385 | 1.98350 | 0.75447 | 0.42501 | 0.23904 |
| | 1 | 1.23624 | 0.89117 | 0.52713 | 0.29797 | 1.21827 | 0.83202 | 0.48506 | 0.27375 | 1.18790 | 0.73884 | 0.42414 | 0.23899 |
| | 10 | 0.72545 | 0.68353 | 0.50490 | 0.29665 | 0.71062 | 0.66136 | 0.46999 | 0.27289 | 0.68747 | 0.62301 | 0.41627 | 0.23855 |
| | 100 | 0.40985 | 0.40728 | 0.38571 | 0.28414 | 0.40119 | 0.39821 | 0.37290 | 0.26441 | 0.38777 | 0.38396 | 0.35093 | 0.23415 |

*Special cases*:
Parameters $\mu = 0$ and $\psi = 0$ correspond to a tapered beam without a tip body.
Parameter $\psi = 0$ corresponds to a tapered beam with a tip mass but without a rotational effect.

### 5.3.2  Tapered simply supported Timoshenko beam with ends elastically restrained against rotation

A simply supported non-uniform beam with two rotational springs is presented in Fig. 5.12. The thickness $b$ of the beam is constant and the depth is tapered linearly in the plane of vibration. The parameters $h$, $A$ and $I$ are the depth, cross-sectional area and the moment of inertia; subscripts 0 and 1 denote the values at the left and right-hand supports, respectively; $\phi_1$ and $\phi_2$ are the flexibility constants of the rotational springs.

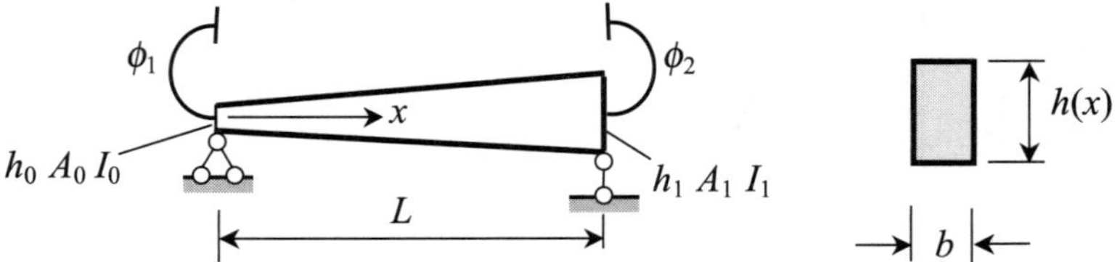


**FIGURE 5.12.**   Tapered Timoshenko beam with ends elastically restrained against rotation.

The cross-sectional area and flexural rigidity are given by

$$A(x) = A_0\left(1 + \frac{\alpha x}{L}\right) \tag{5.31}$$

$$EI(x) = EI_0\left(1 + \frac{\alpha x}{L}\right)^3 \tag{5.32}$$

The taper parameter and the dimensionless parameters of the beam are

$$\alpha = \frac{h_1 - h_0}{h_0} \geq 0, \quad \eta_0 = \frac{I_0}{A_0 L^2}, \quad \phi'_1 = \phi_1 \frac{EI_0}{L}; \quad \phi'_2 = \phi_2 \frac{EI_1}{L},$$

The natural frequency of vibration is

$$\omega = \frac{\lambda}{L^2}\sqrt{\frac{EI_0}{\rho A_0}}$$

**TABLE 5.16(a).**   Clamped–clamped beam. Fundamental frequency coefficients $\lambda_1$ ($\phi'_1 = \phi'_2 = 0$)

| | $\alpha = 0.0^*$ | | $\alpha = 0.05$ | | $\alpha = 0.10$ | | $\alpha = 0.15$ | | $\alpha = 0.20$ | |
|---|---|---|---|---|---|---|---|---|---|---|
| $\eta_0$ | (1) | (2) | (1) | (3) | (1) | (3) | (1) | (3) | (1) | (3) |
| 0.0009 | 21.263 | 20.872 | 21.659 | 21.325 | 22.073 | 21.765 | 22.503 | 22.197 | 22.949 | 22.621 |
| 0.0016 | 20.254 | 19.901 | 20.633 | 20.294 | 21.024 | 20.673 | 21.401 | 21.043 | 21.728 | 21.403 |
| 0.0025 | 19.309 | 18.837 | 19.643 | 19.173 | 19.931 | 19.492 | 20.227 | 19.801 | 20.530 | 20.101 |
| 0.0036 | 18.311 | 17.749 | 18.596 | 18.031 | 18.866 | 18.297 | 19.120 | 18.552 | 19.377 | 18.799 |
| 0.0049 | 17.313 | 16.682 | 17.539 | 16.919 | 17.772 | 17.138 | 18.010 | 17.348 | 18.247 | 17.549 |
| 0.0064 | 16.342 | 15.666 | 16.552 | 15.864 | 16.753 | 16.045 | 16.946 | 16.217 | 17.124 | 16.381 |

**TABLE 5.16(b).** Clamped–pinned beam. Fundamental frequency coefficients $\lambda_1$ ($\phi'_1 = 0$; $\phi'_2 = \infty$)

| | $\alpha = 0.0*$ | | $\alpha = 0.05$ | | $\alpha = 0.10$ | | $\alpha = 0.15$ | | $\alpha = 0.20$ | |
|---|---|---|---|---|---|---|---|---|---|---|
| $\eta_0$ | (1) | (2) | (1) | (3) | (1) | (3) | (1) | (3) | (1) | (3) |
| 0.0009 | 14.989 | 14.793 | 15.234 | 15.035 | 15.453 | 15.271 | 15.676 | 15.502 | 15.906 | 15.728 |
| 0.0016 | 14.517 | 14.358 | 14.739 | 14.578 | 14.922 | 14.789 | 15.102 | 14.996 | 15.289 | 15.197 |
| 0.0025 | 13.985 | 13.856 | 14.178 | 14.050 | 14.372 | 14.236 | 14.551 | 14.417 | 14.735 | 14.592 |
| 0.0036 | 13.478 | 13.311 | 13.671 | 13.480 | 13.849 | 13.641 | 13.988 | 13.796 | 14.131 | 13.945 |
| 0.0049 | 12.984 | 12.746 | 13.113 | 12.892 | 13.247 | 13.029 | 13.386 | 13.160 | 13.531 | 13.285 |
| 0.0064 | 12.439 | 12.178 | 12.571 | 12.303 | 12.708 | 12.418 | 12.838 | 12.527 | 12.946 | 12.663 |

**TABLE 5.16(c).** Elastically clamped–pinned beam. Fundamental frequency coefficients $\lambda_1$ ($\phi'_1 = 10$; $\phi'_2 = \infty$)

| | $\alpha = 0.0*$ | | $\alpha = 0.05$ | | $\alpha = 0.10$ | | $\alpha = 0.15$ | | $\alpha = 0.20$ | |
|---|---|---|---|---|---|---|---|---|---|---|
| $\eta_0$ | (1) | (2) | (1) | (3) | (1) | (3) | (1) | (3) | (1) | (3) |
| 0.0009 | 9.842 | 9.789 | 10.082 | 10.019 | 10.325 | 10.244 | 10.540 | 10.465 | 10.756 | 10.682 |
| 0.0016 | 9.703 | 9.657 | 9.922 | 9.878 | 10.129 | 10.094 | 10.338 | 10.305 | 10.548 | 10.512 |
| 0.0025 | 9.534 | 9.498 | 9.732 | 9.707 | 9.932 | 9.912 | 10.133 | 10.111 | 10.336 | 10.307 |
| 0.0036 | 9.342 | 9.315 | 9.532 | 9.513 | 9.723 | 9.705 | 9.916 | 9.892 | 10.106 | 10.074 |
| 0.0049 | 9.138 | 9.114 | 9.320 | 9.300 | 9.502 | 9.479 | 9.676 | 9.653 | 9.836 | 9.822 |
| 0.0064 | 8.926 | 8.901 | 9.097 | 9.074 | 9.256 | 9.241 | 9.413 | 9.402 | 9.565 | 9.556 |

**TABLE 5.16(d).** Pinned–pinned beam. Fundamental frequency coefficients $\lambda_1$ ($\phi'_1 = \phi'_2 = \infty$)

| | $\alpha = 0.0*$ | | $\alpha = 0.05$ | | $\alpha = 0.10$ | | $\alpha = 0.15$ | | $\alpha = 0.20$ | |
|---|---|---|---|---|---|---|---|---|---|---|
| $\eta_0$ | (1) | (2) | (1) | (3) | (1) | (3) | (1) | (3) | (1) | (3) |
| 0.0009 | 9.748 | 9.695 | 9.990 | 9.927 | 10.235 | 10.154 | 10.458 | 10.377 | 10.676 | 10.597 |
| 0.0016 | 9.611 | 9.567 | 9.835 | 9.790 | 10.045 | 10.007 | 10.067 | 10.221 | 10.469 | 10.430 |
| 0.0025 | 9.446 | 9.411 | 9.648 | 9.623 | 9.850 | 9.830 | 10.054 | 10.031 | 10.259 | 10.229 |
| 0.0036 | 9.257 | 9.232 | 9.450 | 9.433 | 9.645 | 9.627 | 9.840 | 9.816 | 10.033 | 10.001 |
| 0.0049 | 9.057 | 9.036 | 9.242 | 9.224 | 9.428 | 9.406 | 9.604 | 9.582 | 9.767 | 9.754 |
| 0.0064 | 8.850 | 8.827 | 9.026 | 9.003 | 9.189 | 9.172 | 9.346 | 9.336 | 9.500 | 9.494 |

*Special case*:
* The case $\alpha = 0$ corresponds to a uniform beam.

The governing functional is presented by Magrab (1979). The Ritz minimization procedure is presented by Guttierrez *et al.* (1991). The fundamental frequency parameters, $\lambda$, for tapered Timoshenko beams with different boundary conditions and parameters $\phi$ are presented in Tables 5.16(a)–(d). Calculations have been performed with Poisson coefficient $v = 0.3$ and shear coefficient $k = 0.833$.

In the tables are depicted numerical results obtained by (1) the optimized Ritz approach; (2) exact value and (3) by applying the finite element method, if the element was subdivided into 20 slices of constant thickness.

### 5.3.3 Stepped Timoshenko beam elastically restrained at one end and guided at the other

The stepped beam with translational and rotational springs at one end and a guided support at the other is presented in Fig. 5.13, where $K$ is the stiffness constant for a translational spring and $\phi$ is the flexibility constant for a rotational spring.

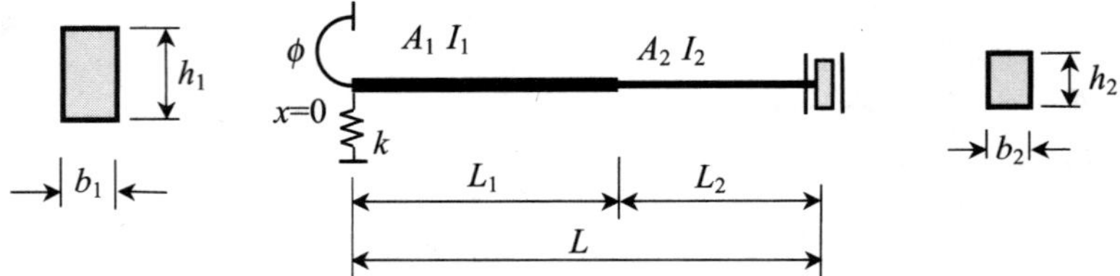


**FIGURE 5.13.**   Stepped beam with an elastic support at one end and a guided support at the other.

The natural frequency of vibration is

$$\omega = \frac{\lambda}{L^2}\sqrt{\frac{EI_1}{\rho A_1}}$$

The first five natural frequency coefficients, $\lambda$, for stepped beams with different boundary conditions in terms of the dimensionless parameters are presented in Tables 5.17(a)–(d). The dimensionless parameters are

$$k' = \frac{kL^3}{EI_1}; \quad \phi' = \phi\frac{EI_1}{L}; \quad \eta = \frac{I_1}{A_1 L_1^2}; \quad \gamma_L = \frac{L_1}{L}; \quad \gamma_b = \frac{b_2}{b_1}; \quad \gamma_h = \frac{h_2}{h_1}$$

Calculations have been performed with the shear factor $k = 0.866$, and the Poisson coefficient $v = 0.3$ (Guttierrez *et al.*, 1990). This article also contains the governing Timoshenko differential equation, boundary and compatibility conditions as well as expressions for eigenfunctions.

**TABLE 5.17(a).** Clamped–guided beam ($\phi' = 0$, $\kappa' = \infty$). Natural frequency coefficients $\lambda$

| $\dfrac{I_1}{A_1 L_1^2}$ | $\dfrac{L_1}{L}$ | $\dfrac{b_2}{b_1}$ | $\dfrac{h_2}{h_1}$ | Mode 1 | Mode 2 | Mode 3 | Mode 4 | Mode 5 |
|---|---|---|---|---|---|---|---|---|
| $10^{-7}$ | 0.25 | 1.0 | 0.8 | 5.319 | 26.304 | 62.996 | 117.694 | 189.343 |
| | | | 0.6 | 4.783 | 22.647 | 50.514 | 92.775 | 151.862 |
| | | 0.8 | 0.8 | 5.606 | 27.035 | 63.107 | 117.344 | 189.658 |
| | | | 0.6 | 4.957 | 23.452 | 50.831 | 92.123 | 151.544 |
| | 0.5 | 1.0 | 0.8 | 5.458 | 27.150 | 66.568 | 123.364 | 199.284 |
| | | | 0.6 | 5.484 | 22.616 | 58.752 | 102.704 | 173.226 |
| | | 0.8 | 0.8 | 5.822 | 26.966 | 66.977 | 122.722 | 200.092 |
| | | | 0.6 | 5.883 | 22.274 | 59.411 | 101.990 | 173.622 |
| 0.0036 | 0.25 | 1.0 | 0.8 | 5.299 | 25.938 | 61.172 | 111.798 | 174.952 |
| | | | 0.6 | 4.770 | 22.416 | 49.510 | 89.745 | 144.102 |
| | | 0.8 | 0.8 | 5.584 | 26.658 | 61.289 | 111.472 | 175.194 |
| | | | 0.6 | 4.943 | 23.209 | 49.805 | 89.126 | 143.872 |
| | 0.50 | 1.0 | 0.8 | 5.374 | 25.620 | 58.883 | 102.013 | 151.955 |
| | | | 0.6 | 5.411 | 21.687 | 52.979 | 88.517 | 138.784 |
| | | 0.8 | 0.8 | 5.731 | 25.453 | 59.171 | 101.453 | 152.436 |
| | | | 0.6 | 5.802 | 21.369 | 53.513 | 87.762 | 139.274 |
| 0.01 | 0.25 | 1.0 | 0.8 | 5.263 | 25.328 | 58.331 | 103.389 | 156.477 |
| | | | 0.6 | 4.747 | 22.024 | 47.887 | 85.099 | 133.002 |
| | | 0.8 | 0.8 | 5.546 | 26.029 | 58.456 | 103.086 | 156.643 |
| | | | 0.6 | 4.919 | 22.796 | 48.149 | 84.527 | 132.884 |
| | 0.50 | 1.0 | 0.8 | 5.236 | 23.475 | 50.325 | 82.550 | 117.386 |
| | | | 0.6 | 5.289 | 20.304 | 46.059 | 74.074 | 110.051 |
| | | 0.8 | 0.8 | 5.581 | 23.326 | 50.512 | 82.010 | 117.771 |
| | | | 0.6 | 5.666 | 20.017 | 46.452 | 73.323 | 110.493 |

**TABLE 5.17(b).**   Pinned–guided beam. Natural frequency coefficients $\lambda$ ($\phi' = \kappa' = \infty$)

| $\dfrac{I_1}{A_1 L_1^2}$ | $\dfrac{L_1}{L}$ | $\dfrac{b_2}{b_1}$ | $\dfrac{h_2}{h_1}$ | Mode 1 | Mode 2 | Mode 3 | Mode 4 | Mode 5 |
|---|---|---|---|---|---|---|---|---|
| $10^{-7}$ | 0.25 | 1.0 | 0.8 | 1.980 | 18.237 | 52.086 | 103.140 | 168.986 |
|  |  |  | 0.6 | 1.483 | 13.757 | 40.404 | 82.627 | 137.813 |
|  |  | 0.8 | 0.8 | 1.975 | 18.055 | 51.726 | 103.278 | 169.724 |
|  |  |  | 0.6 | 1.477 | 13.517 | 39.843 | 82.356 | 138.639 |
|  | 0.5 | 1.0 | 0.8 | 2.025 | 20.192 | 54.457 | 108.396 | 177.675 |
|  |  |  | 0.6 | 1.519 | 17.089 | 47.531 | 90.470 | 155.485 |
|  |  | 0.8 | 0.8 | 1.997 | 20.114 | 54.657 | 108.029 | 178.246 |
|  |  |  | 0.6 | 1.482 | 16.832 | 48.154 | 89.558 | 156.643 |
| 0.0036 | 0.25 | 1.0 | 0.8 | 1.979 | 18.110 | 51.055 | 99.210 | 158.951 |
|  |  |  | 0.6 | 1.483 | 13.696 | 39.899 | 80.547 | 132.112 |
|  |  | 0.8 | 0.8 | 1.974 | 17.931 | 50.697 | 99.292 | 159.585 |
|  |  |  | 0.6 | 1.476 | 13.457 | 39.343 | 80.271 | 132.853 |
|  | 0.50 | 1.0 | 0.8 | 2.018 | 19.540 | 50.254 | 93.489 | 142.828 |
|  |  |  | 0.6 | 1.515 | 16.677 | 44.553 | 81.045 | 129.834 |
|  |  | 0.8 | 0.8 | 1.991 | 19.449 | 50.443 | 93.074 | 143.240 |
|  |  |  | 0.6 | 1.479 | 16.421 | 45.079 | 80.230 | 130.606 |
| 0.01 | 0.25 | 1.0 | 0.8 | 1.976 | 17.892 | 49.494 | 93.379 | 145.386 |
|  |  |  | 0.6 | 1.481 | 13.591 | 39.053 | 77.267 | 123.773 |
|  |  | 0.8 | 0.8 | 1.971 | 17.717 | 49.035 | 93.387 | 145.904 |
|  |  |  | 0.6 | 1.475 | 13.352 | 38.507 | 76.984 | 124.399 |
|  | 0.50 | 1.0 | 0.8 | 2.007 | 18.546 | 44.978 | 78.401 | 113.996 |
|  |  |  | 0.6 | 1.509 | 16.022 | 40.596 | 70.271 | 106.318 |
|  |  | 0.8 | 0.8 | 1.980 | 18.437 | 45.157 | 77.946 | 114.327 |
|  |  |  | 0.6 | 1.473 | 15.770 | 41.012 | 69.538 | 106.823 |

**TABLE 5.17(c).**   Beam elastically restrained at one end and guided at the other. Natural frequency coefficients $\lambda$. The beam at the left-hand end is rigidly restrained against rotation ($\phi' = 0$) and elastically restrained in translation ($\kappa' = 10$)

| $\dfrac{I_1}{A_1 L_1^2}$ | $\dfrac{L_1}{L}$ | $\dfrac{b_2}{b_1}$ | $\dfrac{h_2}{h_1}$ | Mode 1 | Mode 2 | Mode 3 | Mode 4 | Mode 5 |
|---|---|---|---|---|---|---|---|---|
| $10^{-7}$ | 0.25 | 1.0 | 0.8 | 3.010 | 9.696 | 34.010 | 74.430 | 132.341 |
| | | | 0.6 | 3.168 | 8.190 | 28.309 | 60.305 | 103.823 |
| | | 0.8 | 0.8 | 3.249 | 9.664 | 34.315 | 74.645 | 131.803 |
| | | | 0.6 | 3.385 | 8.117 | 28.630 | 60.997 | 103.439 |
| | 0.5 | 1.0 | 0.8 | 2.958 | 10.046 | 34.993 | 80.145 | 139.666 |
| | | | 0.6 | 3.122 | 9.323 | 28.811 | 69.948 | 118.328 |
| | | 0.8 | 0.8 | 3.124 | 10.165 | 34.688 | 80.570 | 139.175 |
| | | | 0.6 | 3.284 | 9.499 | 28.501 | 70.123 | 118.482 |
| 0.0036 | 0.25 | 1.0 | 0.8 | 3.007 | 9.668 | 33.571 | 72.393 | 126.111 |
| | | | 0.6 | 3.164 | 8.176 | 28.054 | 59.179 | 100.629 |
| | | 0.8 | 0.8 | 3.245 | 9.635 | 33.862 | 72.607 | 125.617 |
| | | | 0.6 | 3.381 | 8.103 | 28.364 | 59.839 | 100.245 |
| | 0.50 | 1.0 | 0.8 | 2.946 | 9.919 | 33.205 | 71.469 | 116.725 |
| | | | 0.6 | 3.110 | 9.225 | 27.746 | 63.908 | 102.720 |
| | | 0.8 | 0.8 | 3.112 | 10.030 | 32.926 | 71.741 | 116.362 |
| | | | 0.6 | 3.272 | 9.391 | 27.449 | 64.020 | 102.697 |
| 0.01 | 0.25 | 1.0 | 0.8 | 3.001 | 9.619 | 32.841 | 69.229 | 117.238 |
| | | | 0.6 | 3.158 | 8.151 | 27.622 | 57.361 | 95.751 |
| | | 0.8 | 0.8 | 3.239 | 9.586 | 33.109 | 69.444 | 116.798 |
| | | | 0.6 | 3.374 | 8.078 | 27.914 | 57.970 | 95.370 |
| | 0.50 | 1.0 | 0.8 | 2.926 | 9.710 | 30.711 | 61.768 | 95.623 |
| | | | 0.6 | 3.089 | 9.060 | 26.172 | 56.534 | 86.795 |
| | | 0.8 | 0.8 | 3.019 | 9.809 | 30.464 | 61.912 | 95.326 |
| | | | 0.6 | 3.251 | 9.211 | 25.893 | 56.578 | 86.647 |

**TABLE 5.17(d).** Beam elastically restrained at one end and guided at the other. Natural frequency coefficients $\lambda$. The beam at the left-hand end is free to rotate ($\phi' = \infty$) and elastically restrained in translation ($\kappa' = 10$)

| $\dfrac{I_1}{A_1 L_1^2}$ | $\dfrac{L_1}{L}$ | $\dfrac{b_2}{b_1}$ | $\dfrac{h_2}{h_1}$ | Mode | | | | |
|---|---|---|---|---|---|---|---|---|
| | | | | 1 | 2 | 3 | 4 | 5 |
| $10^{-7}$ | 0.25 | 1.0 | 0.8 | 1.752 | 7.943 | 24.772 | 61.662 | 117.334 |
| | | | 0.6 | 1.402 | 7.392 | 18.727 | 46.706 | 91.233 |
| | | 0.8 | 0.8 | 1.782 | 7.942 | 24.295 | 60.937 | 116.755 |
| | | | 0.6 | 1.409 | 7.422 | 18.400 | 45.932 | 90.284 |
| | 0.5 | 1.0 | 0.8 | 1.771 | 7.847 | 27.481 | 67.061 | 123.556 |
| | | | 0.6 | 1.417 | 7.277 | 22.282 | 59.390 | 103.078 |
| | | 0.8 | 0.8 | 1.775 | 7.772 | 27.123 | 67.560 | 122.939 |
| | | | 0.6 | 1.398 | 7.283 | 21.730 | 60.088 | 102.425 |
| 0.0036 | 0.25 | 1.0 | 0.8 | 1.751 | 7.931 | 24.535 | 60.219 | 112.322 |
| | | | 0.6 | 1.402 | 7.385 | 18.599 | 45.985 | 88.703 |
| | | 0.8 | 0.8 | 1.781 | 7.931 | 24.060 | 59.505 | 111.728 |
| | | | 0.6 | 1.408 | 7.415 | 18.269 | 45.213 | 87.771 |
| | 0.50 | 1.0 | 0.8 | 1.767 | 7.792 | 26.344 | 60.916 | 105.004 |
| | | | 0.6 | 1.417 | 7.234 | 21.602 | 54.957 | 91.209 |
| | | 0.8 | 0.8 | 1.771 | 7.718 | 25.984 | 61.337 | 104.473 |
| | | | 0.6 | 1.395 | 7.239 | 21.062 | 55.534 | 90.603 |
| 0.01 | 0.25 | 1.0 | 0.8 | 1.750 | 7.909 | 24.132 | 57.929 | 105.023 |
| | | | 0.6 | 1.400 | 7.371 | 18.379 | 44.794 | 84.754 |
| | | 0.8 | 0.8 | 1.779 | 7.910 | 23.662 | 57.230 | 104.403 |
| | | | 0.6 | 1.407 | 7.401 | 18.044 | 44.026 | 83.846 |
| | 0.50 | 1.0 | 0.8 | 1.760 | 7.700 | 24.680 | 53.626 | 86.999 |
| | | | 0.6 | 1.412 | 7.161 | 20.553 | 49.343 | 78.298 |
| | | 0.8 | 0.8 | 1.764 | 7.625 | 24.316 | 53.981 | 86.541 |
| | | | 0.6 | 1.390 | 7.162 | 20.028 | 49.796 | 77.754 |

## 5.4  TAPERED SIMPLY SUPPORTED BEAMS ON AN ELASTIC FOUNDATION

Consider a non-uniform simply supported beam resting on an elastic foundation; $K_F$ is the foundation modulus. Three types of linear tapers are presented in Fig. 5.14. They are referred as (a) breadth taper, (b) depth taper, and (c) diameter taper.

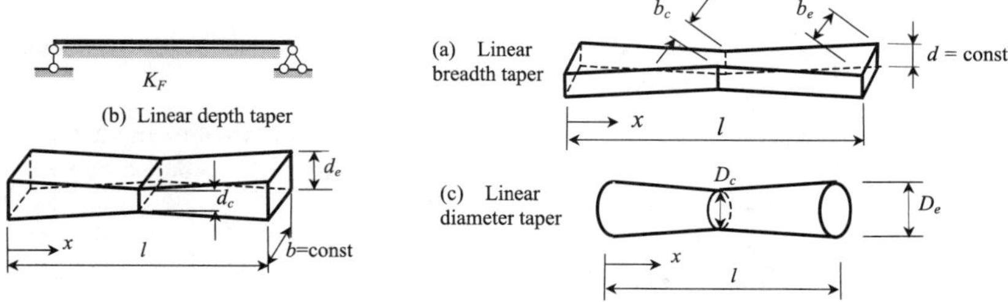


**FIGURE 5.14.**  Non-uniform simply supported beam on an elastic foundation. Types of linear taper.

The cross-sectional area $A(x)$ and the moment of inertia $I(x)$ may be expressed in common form

$$A(x) = A_c\left[1 - \beta + \frac{2x}{l}\beta\right]^{n_1}, \quad I(x) = I_c\left[1 - \beta + \frac{2x}{l}\beta\right]^{n_2}, \quad 0 \le x \le \frac{l}{2} \quad (5.23)$$

$$A(x) = A_c\left[1 + \beta - \frac{2x}{l}\beta\right]^{n_1}, \quad I(x) = I_c\left[1 + \beta - \frac{2x}{l}\beta\right]^{n_2}, \quad \frac{l}{2} \le x \le l \quad (5.24)$$

where $A_c$ and $I_c$ are the area of cross-section and the moment of inertia at the midpoint of the beam, respectively.

The dimensionless parameter of the foundation is $\gamma_F = \dfrac{K_F l^4}{\pi^4 E I_c}$, where $K_F$ is the foundation modulus. Parameter $\gamma_{FT}$ depends of the types of taper. The various types of linear taper and the corresponding power parameters $n_1$, $n_2$ as well as expressions for $\gamma_{FT}$

**TABLE 5.18.**  Expressions for $\gamma_{FT}$ for the first and second modes of vibration of tapered simply supported beams

| Type of taper | $\beta$ | $n_1$ | $n_2$ | Expression for $\gamma_{FT}$ |
|---|---|---|---|---|
| Breadth taper | $\dfrac{b_c - b_e}{b_c}$ | 1 | 1 | $-\dfrac{1}{\beta}(74.0220 - 59.0251\beta + 11.0070\beta^2)$ |
| Depth taper | $\dfrac{d_c - d_e}{d_c}$ | 1 | 3 | $-\dfrac{1}{\beta}(74.0220 - 135.0438\beta + 98.0387\beta^2 - 31.3730\beta^3 + 3.9184\beta^4)$ |
| Diameter taper | $\dfrac{D_c - D_e}{D_c}$ | 2 | 4 | $\dfrac{1}{\beta(0.1013 - 0.0380\beta)}(-3.75 + 9.8318\beta - 11.5007\beta^2 + 7.5146\beta^3 - 2.8629\beta^4 + 0.6049\beta^5 - 0.0567\beta^6)$ |

**TABLE 5.19.**   Parameters $\gamma_{FT}$ for the first and second mode of vibrations ($m = 1, 2$)

| Taper parameter $\beta$ | Breadth taper | Depth taper | Diameter taper |
|---|---|---|---|
| $-0.05$ | 1540.02 | 1620.5 | 827.77 |
| $-0.10$ | 800.34 | 885.39 | 462.03 |
| $-0.15$ | 554.16 | 643.95 | 343.33 |
| $-0.20$ | 431.34 | 526.05 | 286.56 |

for the first and second mode of vibration of simply supported beams are presented in Table 5.18 (Kanaka Raju, Venkateswara Rao, 1990).

If the foundation parameter $\gamma_F < \gamma_{FT}$ then the beam vibrates in the first mode ($m = 1$), and if $\gamma_F > \gamma_{FT}$ then the beam vibrates in the second mode ($m = 2$), with the values of $\lambda_f$ being the lowest, as listed in Table 5.20 later.

The subscripts $c$ and $e$ indicate the midpoint and ends for each of the beams. Several numerical results for parameters $\gamma_{FT}$ for different taper parameters $\beta$ and types of taper are presented in Table 5.19.

The non-dimensional frequency parameter is $\lambda^4 = \dfrac{\rho A_c \omega^2 l^4}{EI_c}$ and is presented for different types of taper in Table 5.20.

## 5.5   FREE–FREE SYMMETRIC PARABOLIC BEAM

A doubly symmetric parabolic beam is presented in Fig. 5.15.

The area of the cross-section and the moment of inertia are

$$A(x) = A_0(1 - cx^2) \tag{5.25}$$

$$I(x) = I_0(1 - bx^2) \tag{5.26}$$

where parameters $c$ and $b$ have dimensions (length)$^{-2}$.

The mode shape of vibration is

$$X_i(x) = \frac{\cosh k_i \cos k_i \dfrac{x}{l} + \cos k_i \cosh k_i \dfrac{x}{l}}{\sqrt{\cos^2 k_i + \cosh^2 k_i}} \tag{5.27}$$

where $k_i$ are the roots of the equation

$$\tan k_i + \tanh k_i = 0, \quad k_1 = 0, \quad k_1 = 2.3650$$

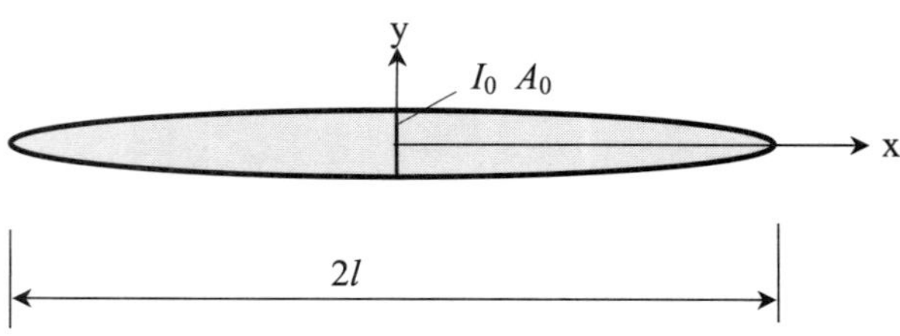


**FIGURE 5.15.**   Free–free doubly symmetric parabolic beam.

**TABLE 5.20.** Tapered simply supported beams on elastic foundation: Frequency parameters $\lambda^4$ for different types of taper

| Type of taper | Frequency parameter $\lambda^4 = a/b$ |
|---|---|

**Breadth taper**

$$a = \pi^4 \left\{ m^4 \left[ 1 + \left( -\frac{1}{2} + \frac{1}{m^2\pi^2} - \frac{\cos m\pi}{m^2\pi^2} \right)\beta \right] + \gamma_F \right\}, \quad b = 1 - \left( \frac{1}{2} + \frac{\cos m\pi}{m^2\pi^2} - \frac{1}{m^2\pi^2} \right)\beta$$

**Depth taper**

$$a = \pi^4 \left\{ m^4 \left[ 1 + \left( -\frac{3}{2} + \frac{3}{m^2\pi^2} - \frac{3\cos m\pi}{m^2\pi^2} \right)\beta + \left( 1 - \frac{6}{m^2\pi^2} \right)\beta^2 + \left( -\frac{1}{4} + \frac{3}{m^2\pi^2} + \frac{6\cos m\pi}{m^4\pi^4} - \frac{6}{m^4\pi^4} \right)\beta^3 \right] + \gamma_F \right\}$$

$$b = 1 + \left( -\frac{1}{2} + \frac{1}{m^2\pi^2} - \frac{\cos m\pi}{m^2\pi^2} \right)\beta$$

**Diameter taper**

$$a = \pi^4 \left\{ m^4 \left[ 1 + \left( -2 + \frac{2}{m^2\pi^2} - \frac{4\cos m\pi}{m^2\pi^2} \right)\beta + 2\left( 1 - \frac{6}{m^2\pi^2} \right)\beta^2 + \left( -1 + \frac{12}{m^2\pi^2} + \frac{24\cos m\pi}{m^4\pi^4} - \frac{24}{m^4\pi^4} \right)\beta^3 \right. \right.$$

$$\left. \left. + \left( \frac{1}{5} - \frac{4}{m^2\pi^2} + \frac{24}{m^4\pi^4} \right)\beta^4 \right] + \gamma_F \right\}$$

$$b = 1 + \left( -1 + \frac{2}{m^2\pi^2} - \frac{2\cos m\pi}{m^2\pi^2} \right)\beta + \left( \frac{1}{3} - \frac{2}{m^2\pi^2} \right)\beta^2$$

*Special cases*:
If a simply supported beam has a uniform cross section then the taper parameter $\beta = 0$ and the frequency parameter is $\lambda^4 = \pi^4(m^4 + \gamma_F)$.
If a simply supported beam has no elastic foundation then $\gamma_F = 0$ and $\lambda^4 = \pi^4 m^4$.

The first root $k = 0$ corresponds to the motion of the beam without the bending effect. Eigenfunctions $X_i(x)$ satisfy the boundary condition at free ends:

$$[X_i(x)]''_{x=l} = [X_i(x)]'''_{x=l} = 0 \tag{5.28}$$

The eigenfunctions that correspond to the found roots are

$$X_1(x) = \text{const}, \quad X_2(x) = \frac{k_2^2}{l^2} \frac{\cos k_2 \cosh k_2 \frac{x}{l} - \cosh k_2 \cos k_i \frac{x}{l}}{\sqrt{\cos^2 k_2 + \cosh^2 k_2}} \tag{5.29}$$

Let $X_1(x) = \dfrac{1}{\sqrt{2}}$. In this case, the Ritz method yields the following expression for the fundamental frequency of vibration (Morrow, 1905; Krasnoperov, 1916; Timoshenko and Gere, 1961)

$$\omega = \sqrt{\frac{\alpha}{\beta_{22}} \frac{1}{1 - \dfrac{\beta_{12}^2}{\beta_{11}\beta_{22}}}} \sqrt{\frac{I_0 E}{A_0 \rho}} \tag{5.30}$$

where

$$\alpha = \int\limits_{-l}^{+l} (1 - bx^2)(X_2'')^2 \, dx \quad \beta_{ij} = \int\limits_{-l}^{+l} (1 - cx^2)X_i X_j \, dx \tag{5.31}$$

Substitution of expression (5.29) into (5.31) leads to the following results

$$\alpha = \frac{31.28}{l^3}(1 - 0.087bl^2)$$

$$\beta_{11} = l(1 - 0.333cl^2); \quad \beta_{12} = 0.297cl^3; \quad \beta_{22} = l(1 - 0.481cl^2)$$

## REFERENCES


Balasubramanian, T.S., Subramanian, G. and Ramani, T.S. (1990) Significance and use of very high order derivatives as nodal degrees of freedom in stepped beam vibration analysis. *Journal of Sound and Vibration*, **137**(2), 353–356.

Brock, J.E. (1976) Dunkerley–Mikhlin estimates of gravest frequency of a vibrating system. *Journal of Applied Mechanics*, June, 345–348.

Conway, H.D., Becker, E.C.H. and Dubil, J.F. (1964) Vibration frequencies of tapered bars and circular plates. *ASME Journal of Applied Mechanics*, **33**, *Trans ASME*, **88**, Series E, 329–331.

Downs, B. (1977) Transverse vibrations of cantilever beams having unequal breadth and depth tapers. *ASME Journal of Applied Mechanics*, **44**, 737–742.

Filippov, A.P. (1970) *Vibration of Deformable Systems* (Moscow: Mashinostroenie) (in Russian).

Gaines, J.H. and Volterra, E. (1966) Transverse vibrations of cantilever bars of variable cross section. *The Journal of the Acoustical Society of America*, **39**(4), 674–679.

Gutierrez, R.H., Laura, P.A.A. and Rossi, R.E. (1990) Natural frequencies of a Timoshenko beam of non-uniform cross-section elastically restrained at one end and guided at the other. *Journal of Sound and Vibration*, **141**(1), 174–179.

Gutierrez, R.H., Laura, P.A.A. and Rossi, R.E. (1991) Fundamental frequency of vibrations of a Timoshenko beam of non-uniform thickness. *Journal of Sound and Vibration*, **145**(2), 341–344.

Jano, S.K. and Bert, C.W. (1989) Free vibration of stepped beams: exact and numerical solution. *Journal of Sound and Vibration*, **130**(2), 342–346.

Kanaka Raju, K. and Venkateswara, Rao G. (1990) Effect of elastic foundation on the mode shapes in stability and vibration problems of tapered columns/beams. *Journal of Sound and Vibration*, **136**(1), 171–175.

Karnovsky, I.A. and Lebed, O. (2003). Free vibrations of beams and frames. Eigenvalues and eigenfunctions. (New York: McGraw-Hill).

Laura, P.A.A., Paloto, J.C., Santos, R.D. and Carnicer, R. (1989) Vibrations of a non-uniform beam elastically restrained against rotation at one end and carrying a guided mass at the other. *Journal of Sound and Vibration*, **129**(3), 513–516.

Lau, J.H. (1984) Vibration frequencies of tapered bars with end mass. *ASME Journal of Applied Mechanics*, **51**, 179–181.

Lee, T.W. (1976) Transverse vibrations of a tapered beam carrying a concentrated mass. *ASME Journal of Applied Mechanics*, **43**, *Trans ASME*, **98**, Series E, 366–367.

Magrab, E.B. (1979) *Vibrations of Elastic Structural Members* (Alphen aan den Rijn, The Netherlands/Germantown, Maryland, USA: Sijthoff and Noordhoff).

Rossi, R.E., Laura, P.A.A. and Gutierrez, R.H. (1990) A note on transverse vibrations of a Timoshenko beam of non-uniform thickness clamped at one end and carrying a concentrated mass at the other. *Journal of Sound and Vibration*, **143**(3), 491–502.

Sankaran, G.V., Kanaka Raju, K. and Venkateswara, Rao G. (1975) Vibration frequencies of a tapered beam with one end spring-hinged and carrying a mass at the other free end. *ASME Journal of Applied Mechanics*, September, 740–741.

## FURTHER READING

Abramovitz, M. and Stegun, I.A. (1970) *Handbook of Mathematical Functions* (New York: Dover).

Avakian, A. and Beskos, D.E. (1976) Use of dynamic influence coefficients in vibration of nonuniform beams. *Journal of Sound and Vibration*, **47**(2), 292–295.

Bambill, E.A. and Laura, P.A.A. (1989) Application of the Rayleigh–Schmidt method when the boundary conditions contain the eigenvalues of the problem. *Journal of Sound and Vibration*, **130**(1), 167–170.

Banks, D.O. and Kurowski, G.J. (1977) The transverse vibration of a doubly tapered beam. *ASME Journal of Applied Mechanics*, March, 123–126.

Blevins, R.D. (1979) *Formulas for Natural Frequency and Mode Shape* (New York: Van Nostrand Reinhold).

Conn, J.F.C. (1944) Vibration of truncated wedge. *Aircraft Engineering*, **16**, 103–105.

Conway, H.D. (1946) The calculation of frequencies of vibration of a truncated cone. *Aircraft Engineering*, **18**, 235–236.

Conway, H.D. and Dubil, J.F. (1965) Vibration frequencies of truncated-cone and wedge beams. *ASME Journal of Applied Mechanics*, **32**, 932–934.

Cranch, E.T. and Adler, A.A. (1956) Bending vibrations of variable section beams. *Journal of Applied Mechanics*, **29**, *Trans ASME*, **84**, 103–108.

Dinnik, A.N. (1955) *Selected Transactions*, Vol. 2 (Kiev: AN Ukraine SSR), pp. 125–221 (in Russian).

Gast, R.G. and Sneck, H.J. (1991) Modal analysis of non-prismatic beams: uniform segments method. *Journal of Sound and Vibration*, **149**(3), 489–494.

Goel, R.P. (1976) Transverse vibrations of tapered beams. *Journal of Sound and Vibration*, **47**, 1–7.

Grossi, R.O. and Bhat, R.B. (1991) A note on vibrating tapered beams. *Journal of Sound and Vibration*, **147**(1), 174–178.

Gupta, A.K. (1985) Vibration of tapered beams. *Journal of Structural Engineering, American Society of Civil Engineers*, **111**, 19–36.

Gutierrez, R.H., Laura, P.A.A. and Rossi, R.E. (1991) Numerical experiments on vibrational characteristics of Timoshenko beams of non-uniform cross-section and clamped at both ends. *Journal of Sound and Vibration*, **150**(3), 501–504.

Housner, G.W. and Keightley, W.O. (1962) Vibrations of linearly tapered cantilever beams. *Journal Engineering Mechanics Division, Proceedings ASCE*, **88**, EM2, 95–123.

Kirchhoff, G.R. (1879) Uber die Transversalschwingungen eines Stabes von veranderlichen Querschnitt. *Akademie der Wissenschaften* (Berlin: Monatsberichte), S.815–828.

Klein, L. (1974) Transverse vibrations of nonuniform beams. *Journal of Sound and Vibration*, **37**(4), 491–505.

Krasnoperov, E.B. (1916) *Application of Ritz Method to the Free Vibration of a Beam* (Petrograd: Politechnical Institute), **25**, pp. 377–400.

Lau, J.H. (1984) Vibration frequencies for a non-uniform beam with end mass. *Journal of Sound and Vibration*, **97**, 513–521.

Lee, H.C. (1963) A generalized minimum principle and its application to the vibration of a wedge with rotary inertia and shear. *Journal of Applied Mechanics*, **30**, *Trans ASME*, **85**, *Series E*, 176–180.

Lee, S.Y. and Ke, H.Y. (1990) Free vibrations of a non-uniform beam with general elastically restrained boundary conditions. *Journal of Sound and Vibration*, **136**(3), 425–437.

Lee, Ho Chong and Bisshopp, K.E. (1964) Application of integral equations to the flexural vibration of a wedge with rotary inertia. *Journal of The Franklin Institute*, **277**, 327–336.

Levinson, M. (1976) Vibrations of stepped strings and beams. *Journal of Sound and Vibration*, **49**, 287–291.

Lindberg, G.M. (1963) Vibrations of nonuniform beams. *The Aeronautical Quarterly*, November, 387–395.

Mononobe, N. (1921) *Z. Angew. Math. Mech.*, **1**(6), 444–451.

Morrow, J. (1905) On the lateral vibration of bars of uniform and varying sectional area. *Philosophical Magazine and Journal of Science*, Series 6, **10**(55), 113–125.

Krishna Murty, A.V. and Prabhakaran, K.R. (1969) Vibrations of tapered cantilever beams and shafts. *The Aeronautical Quarterly*, May, 171–177.

Mabie, H.H. and Rogers, C.B. (1974) Transverse vibrations of double-tapered cantilever beams with end support and with end mass. *Journal of Acoustical Society of America*, **55**(5), 986–991.

Pfeiffer, F. (1934) Vibration of elastic systems (Moscow-Leningrad: ONTI) 154 p. Translated from the German-Mechanik Der Elastischen Korper, Handbuch Der Physik, Band IV (Berlin) 1928.

Sato, H. (1983) Free vibrations of beams with abrupt changes of cross-section. *Journal of Sound and Vibration*, **89**, 59–64.

Sekhniashvili, E.A. (1960) *Free Vibration of Elastic Systems* (Tbilisi: Sakartvelo).

Subramanian, G. and Balasubramanian, T.S. (1989) Effects of steps on the free vibrations characteristics of short beams. *Journal of the Aeronautical Society of India*, **41**, 71–74.

Subramanian, G. and Balasubramanian, T.S. (1987) Beneficial effects of steps on the free vibrations characteristics of beams. *Journal of Sound and Vibration*, **118**, 555–560.

Taleb, N.J. and Suppiger, E.W. (1961) Vibration of stepped beams. *Journal of the Aerospace Sciences*, **28**, 295–298.

Timoshenko, S.P. and Gere, J.M. (1961) *Theory of Elastic Stability*, 2nd ed. (New York: McGraw Hill).

Thomson, W.T. (1949) Vibrations of slender bars with discontinuities in stiffness. *Journal of Applied Mechanics*, **16**, 203–207.

Todhunter, I. and Pearson, K. (1960) *A History of the Theory of Elasticity and of the Strength of Materials* (New York: Dover). *Volume II. Saint-Venant to Lord Kelvin*. Part 1–762 p., part 2–546 p.

Volterra, E. and Zachmanoglou, E.C. (1965) *Dynamics of Vibrations* (Columbus, Ohio: Charles E. Merrill Books).

Wang, H.C. (1967) Generalized hypergeometric function solutions on transverse vibrations of a class of nonuniform beams. *Journal of Applied Mechanics*, **34**, *Trans ASME*, **89**, *Series E*, pp. 702–708.

Ward, P.F. (1913) Transverse vibration of a rod of varying cross section. *Philosophical Magazine*, **46**, 85–106.

Weaver, W., Timoshenko, S.P. and Young, D.H. (1990) *Vibration Problems in Engineering*, 5th edn (New York: Wiley).

Wrinch, D. (1922) On the lateral vibrations of bars of conical type. *Proceedings of the Royal Society, London, Series A*, **101**, 493–508.

Yang, K.Y. (1990) The natural frequencies of a non-uniform beam with a tip mass and with translation and rotational springs. *Journal of Sound and Vibration*, **137**(2), 339–341.

# CHAPTER 6

# OPTIMAL DESIGNED BEAMS

This Chapter is devoted to the optimal design of vibrating one-span beams. Two main problems are discussed.

1. The volume–frequency problem: find a configuration of the cross-sectional area $A(x)$ along the beam for the minimum (or maximum) frequency $\omega$ of a beam, if the volume of the beam $V_0$ is given.
2. The frequency–volume problem: find a configuration of the cross-sectional area $A(x)$ for the minimum (or maximum) volume $V$ of a beam, if frequency $\omega = \omega_0$ is given.

The Bernoulli–Euler and Timoshenko beam theories are applicable. Analytical and numerical results for a beam with classical boundary conditions are presented. The maximum principle of Pontryagin has been applied.

## 6.1  STATEMENT OF A PROBLEM

The objects under study are non-uniform beams with different boundary conditions. The problem is to find the cross-sectional area distribution along the beam for a *minimum volume* of the beam, if the frequency of vibration is given. The dual problem is to find the cross-sectional area distribution along the beam for a *minimum frequency* of vibration of the beam if the volume of the beam is given.

***Mathematical model.***   Differential equations of the transverse vibration presented in normal form are

$$\frac{\mathrm{d}y}{\mathrm{d}x} = \phi$$

$$\frac{\mathrm{d}\phi}{\mathrm{d}x} = -\frac{M}{EI}$$

$$\frac{\mathrm{d}M}{\mathrm{d}x} = Q \qquad\qquad (6.1)$$

$$\frac{\mathrm{d}Q}{\mathrm{d}x} = -\omega^2 \rho A y$$

125

where a vector of the state variables consists of the amplitude values of a lateral displacement $y$, slope $\phi$, shear force $Q$ and bending moment $M$ with corresponding boundary conditions.

Boundary conditions may be presented in a common form

$$a_1 y(0) + b_1 Q(0) = 0; \qquad a_2 \phi(0) + b_2 M(0) = 0$$
$$a_3 y(l) + b_3 Q(l) = 0; \qquad a_4 \phi(l) + b_4 M(l) = 0 \tag{6.2}$$

Coefficients $a_i$ and $b_i$ for different types of the supports are listed in Table 6.1.

**TABLE 6.1.**   Boundary condition coefficients

| | Pinned | Fixed | Free | Elastic support |
|---|---|---|---|---|
| Left end $(x = 0)$ | $a_1 = b_2 = 1$ $a_2 = b_1 = 0$ | $a_1 = a_2 = 1$ $b_1 = b_2 = 0$ | $a_1 = a_2 = 0$ $b_1 = b_2 = 1$ | $b_1 = b_2 = 1$ $a_1 = -k_{tr}$ $a_2 = k_{rot}$ |

***Variable parameters.***   The cross-sectional area distribution along the beam, $A(x)$ (configuration), is the controlled variable.

***Restrictions.***   The configuration $A(x)$ at any $x \in [0, l]$ must satisfy the condition $A_1(x) \le A(x) \le A_2(x)$, where $A_1(x)$ and $A_2(x)$ are given functions, and represent the lower and upper bounds of the cross-sectional area of the beam.

***Criteria optimality.***   Optimal configuration $A(x)$ is such that it leads to the minimal volume of the beam:

$$V = \int A(x)\,dx \to \min$$

***Problem*** $\omega \to V$.   Find the configuration of $A(x)$ for the minimum (or maximum) volume $V = \int A(x)\,dx$ of a beam, if the frequency $\omega = \omega_0$ is given.

This problem may be solved if $\omega \in [\omega^-, \omega^+]$, where $\omega^-$ and $\omega^+$ are the lower and upper bounds of the frequency of vibration according to the restriction.

***Problem*** $V \to \omega$.   Find the configuration $A(x)$ for the minimum (or maximum) frequency $\omega$ of a beam, if the volume of the beam, $V_0$, is given

$$V = \int A(x)\,dx = V_0 \tag{6.3}$$

This problem may be solved if $V \in [V^-, V^+]$, where $V^-$ and $V^+$ are the lower and upper limits of the volume of the beam according to the restriction.

The *Hamiltonian* is defined in terms of the vector of state variables and the criteria of optimality as follows

$$H = k_1 \left( \frac{M^2}{EI} + \omega^2 \rho A y^2 \right) - k_2 A \tag{6.4}$$

where $k_1$ and $k_2$ are the Lagrange multipliers.

The optimal configuration $A(x)$ will be expressed in terms of state variables from the condition of the maximum Hamiltonian (Pontryagin *et al.*, 1962).

The relation between a moment of inertia and cross-sectional area is conveniently expressed by

$$I = \gamma A^n \tag{6.5}$$

where $\gamma$ is the proportional coefficient (Table 6.2). Parameters $n = 1$, $n = 2$ and $n = 3$ correspond to the variable width, homothetic cross-sections, and variable height, respectively.

**TABLE 6.2.** Presentation of moment of inertia for different cross-sections according to formula 6.5

| Cross-section | Variable parameter | Constant parameter | Moment of inertia $I$ vs area cross-section $A$ |
|---|---|---|---|
| (rectangle, $h$, $b$) | $b$ | $h$ | $I = \dfrac{h^2}{12} A$ |
| | $h$ | $b$ | $I = \dfrac{1}{12b^2} A^3$ |
| (circle, $2r$) | $r$ | — | $I = \dfrac{1}{4\pi} A^2$ |
| (ellipse, $\alpha$, $\beta$) | $\beta$ | $\alpha$ | $I = \gamma A$ |
| | $\alpha$ | $\beta$ | $I = \gamma A^3$ |
| (ellipse, $\eta\alpha$, $\eta\beta$) | $\eta$ | $\alpha, \beta$ | $I = \gamma A^2$ |

## 6.2 COMMON PROPERTIES OF $\omega \to V$ AND $V \to \omega$ PROBLEMS

The fundamental properties of the optimal designed beams are distinctly and completely characterized by using the characteristic curve, which is presented in Fig. 6.1 (Grinev and Filippov, 1979).

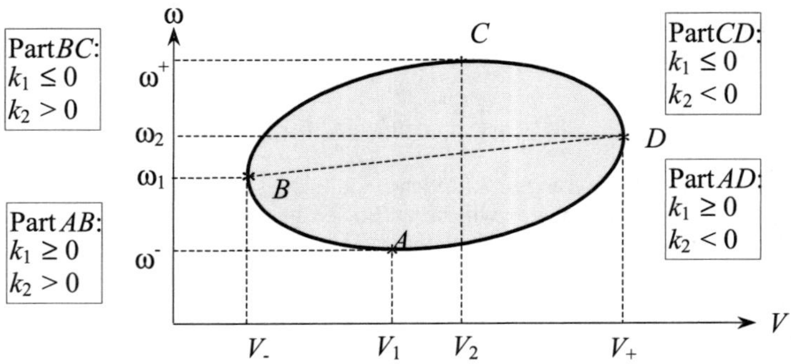



**FIGURE 6.1.**   Characteristic curve. Notation and signs of Lagrange's multipliers.

The frequencies of vibrations $\omega_1$ and $\omega_2$ correspond to configurations

$$A(x) \equiv A_1(x) \quad \text{and} \quad A(x) \equiv A_2(x), \text{ respectively}$$

The minimum and maximum volumes of the beam, according to restrictions on the cross-sectional area are

$$V_- = \int\limits_0^l A_1(x)\, dx; \quad V_+ = \int\limits_0^l A_2(x) dx \tag{6.6}$$

If volume $V_-$ is given, then configuration $A_1(x)$ is the solution of the problem $V \to \omega$ and its frequency vibration is $\omega_1$. If volume $V_+$ is given, then configuration $A_2(x)$ is the solution of the problem $V \to \omega$ and its frequency vibration is $\omega_2$.

The frequencies of vibrations $\omega^-$, $\omega^+$ present the minimal and maximal frequencies for the $V \to \omega$ problem, if we need only minimal and maximal frequencies without the condition $V = $ constant. This means that, in the expression of the Hamilton function, the Lagrange multiplier $k_2$ equals zero.

***Properties of the characteristic curve.***    The points $A$ and $C$ correspond to volumes $V_1$ and $V_2$ and eigenvalues $\omega^-$ and $\omega^+$. The points $B$ and $D$ correspond to eigenvalues $\omega_1$ and $\omega_2$ and volumes $V_-$ and $V_+$. The line $BD$ corresponds to a uniform beam.

***Problem $\omega \to V$.***    Part $ABC$ corresponds to minimal volumes.
Part $CDA$ corresponds to maximal volumes.

***Problem $V \to \omega$.***    Part $BCD$ corresponds to maximal eigenvalues.
Part $BAD$ corresponds to minimal eigenvalues.

***Types of solution.***    The problem of maximal eigenvalues, if the volume of the beam is given, has a continuous solution (see Figs. 6.2–6.5).
The problem of minimal eigenvalues, if the volume of the beam is given, has a discontinuous solution (see Figs. 6.2–6.5).

## 6.3 ANALYTICAL SOLUTION $\omega \to V$ AND $V \to \omega$ PROBLEMS

The optimal configuration $A(x)$ may be expressed in terms of state variables.

***Continuous solution, $k_1 < 0$.***  This condition corresponds to the part $BCD$ of the characteristic curve

$$
A(x) = \begin{cases}
A_2, & b \leq 0 \\[2mm]
A_2, & A \geq A_2, \quad b > 0 \\[2mm]
\left(\dfrac{n|k_1|M^2}{\gamma bE}\right)^{1/1+n}, & A_1 \leq A \leq A_2, \quad b > 0 \\[2mm]
A_1, & A \leq A_1, \quad b > 0
\end{cases}
\tag{6.7}
$$

where $b = |k_1|\omega^2\rho y^2 + k_2$; parameter $n$ is listed in Table 6.2.

***Discontinuous solution, $k_1 \geq 0$.***  This condition corresponds to the part $BAD$ of the characteristic curve

$$
A(x) = \begin{cases}
A_1, & k_1 M^2 \geq \gamma E\xi(b - 2k_2) \\[2mm]
A_2, & k_1 M^2 < \gamma E\xi(b - 2k_2)
\end{cases}
\tag{6.8}
$$

where

$$
\xi = A_1^n A_2^n \frac{A_2 - A_1}{A_2^n - A_1^n}
$$

The numerical procedures for finding the optimal configuration and the location of the points of the switch were developed and compehensively discussed by Grinev and Filippov (1979).

The properties of the solution are presented in Table 6.3.

**TABLE 6.3.**  Properties of continuous and discontinuous solutions for $\omega \to V$ and $V \to \omega$ problems

| Continuous solution<br>$(k_1 < 0)$ | Discontinuous solution<br>$(k_1 \geq 0)$ |
| --- | --- |
| 1. Maximum $\omega$, if $V = $ constant | 1. Minimum $\omega$, if $V = $ constant |
| 2. Minimum $V$, if $\omega \in [\omega_1 - \omega^+]$ | 2. Minimum $V$, if $\omega \in [\omega^- - \omega_1]$ |
| 3. Maximum $V$, if $\omega \in [\omega_2 - \omega^+]$ | 3. Maximum $V$, if $\omega \in [\omega^- - \omega_2]$ |

## 6.4  NUMERICAL RESULTS

The numerical results of optimal designed beams with different boundary conditions are presented in Figs. 6.2–6.4, (Grinev and Filippov, 1979). There are characteristic curves for fundamental frequencies of vibration in the dimensionless coordinates $V/V_+$ and $\omega/\omega_1$ and the corresponding optimal configuration $A/A_2$ in terms of $x/l$ for different volumes (lines 1, 2, 3) of the beam. The continuous solution corresponds to the problem $V \to \omega_{max}$ and the discontinious one corresponds to the problem $V \to \omega_{min}$.

The numerical results for fundamental and second modes of vibration (clamped–free beam) are presented in Fig. 6.5. Lines I and II on the characteristic curve correspond to the Bernoulli–Euler and Timoshenko beam theories, respectively.

The numerical procedures developed and vast numerical results for longitudinal, bending and torsional vibrations were obtained by Grinev and Filippov (1971–1979).

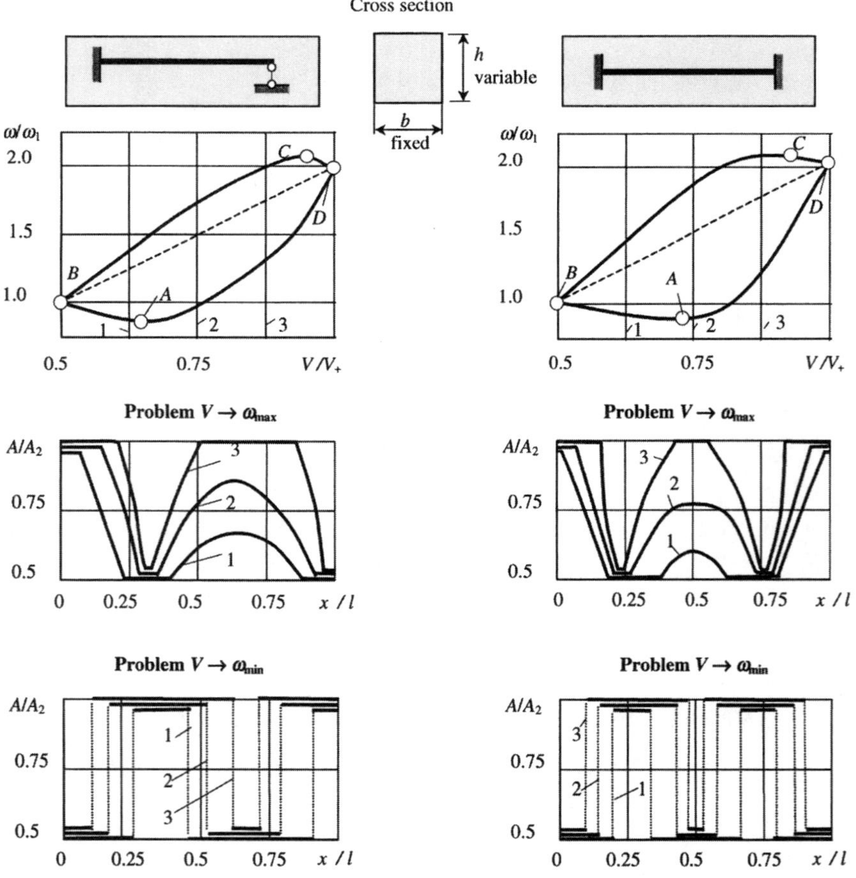


**FIGURE 6.2(a).**  Optimal designed cantilever and simply-supported beams. Cross-section: rectangle; width $b = 0.02\,$m; height $h$ is variable (parameter of cross-section $n = 3$). Length of a beam $l = 1.2\,$m. Restriction: $A_1 = 4 \times 10^{-4}\,$m$^2$, $A_2 = 2A_1$. Material: $E = 1.96 \times 10^{11}\,$N/m$^2$, $\rho = 7.8 \times 10^3\,$kg/m$^3$.

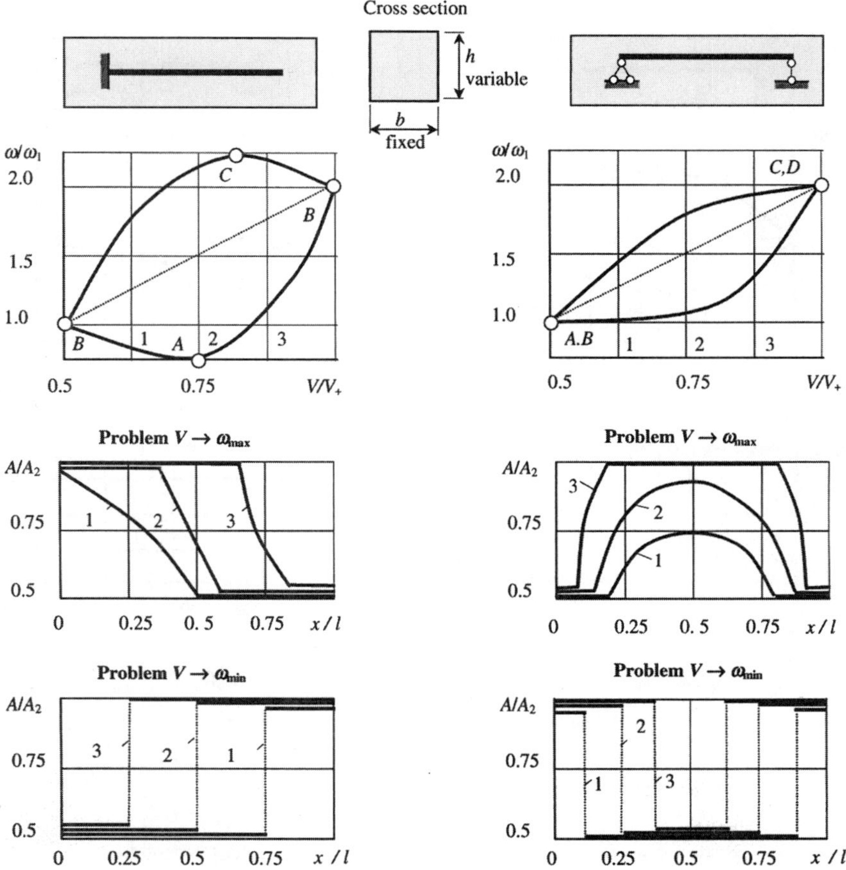


**FIGURE 6.2(b).** Optimal designed clamped–pinned and clamped–clamped beams. Cross-section: rectangle; width $b = 0.02$ m; height $h$ is variable (parameter of cross-section $n = 3$). Length of a beam $l = 1.2$ m. Restriction: $A_1 = 4 \times 10^{-4}$ m$^2$, $A_2 = 2A_1$. Material: $E = 1.96 \times 10^{11}$ N/m$^2$, $\rho = 7.8 \times 10^3$ kg/m$^3$.

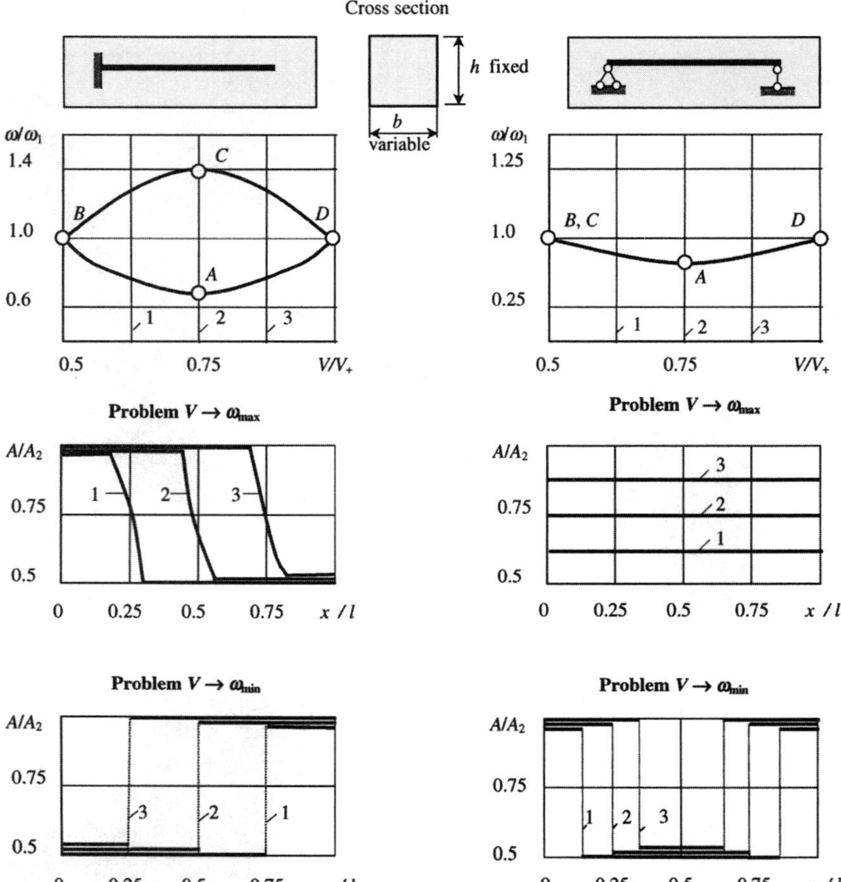


**FIGURE 6.3(a).** Optimal designed cantilever and simply-supported beams. Cross-section: rectangle; height $h = 0.02$ m; width $b$ is variable (parameter of cross-section $n = 1$). Length of a beam $l = 1.2$ m. Restriction: $A_1 = 4 \times 10^{-4}$ m$^2$, $A_2 = 2A_1$. Material: $E = 1.96 \times 10^{11}$ N/m$^2$, $\rho = 7.8 \times 10^3$ kg/m$^3$.

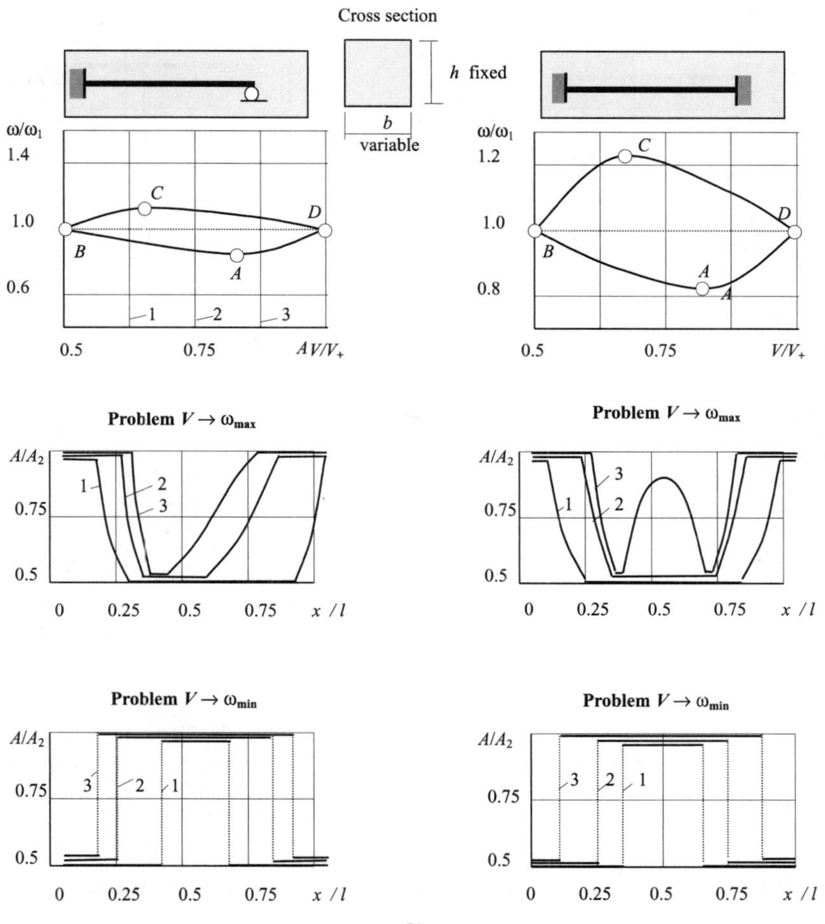



**FIGURE 6.3(b).** Optimal designed clamped–pinned and clamped–clamped beams. Cross-section: rectangle; height $h = 0.02$ m; width $b$ is variable (parameter of cross-section $n = 1$). Length of a beam $l = 1.2$ m. Restriction: $A_1 = 4 \times 10^{-4}$ m$^2$, $A_2 = 2A_1$. Material: $E = 1.96 \times 10^{11}$ N/m$^2$, $\rho = 7.8 \times 10^3$ kg/m$^3$.

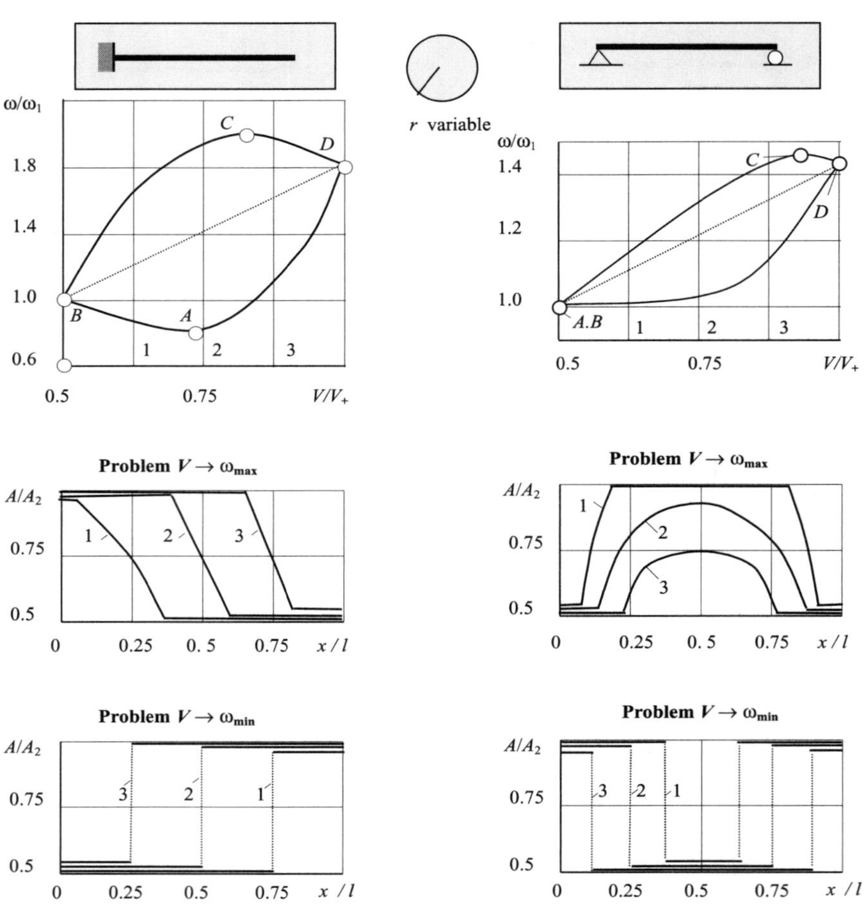


**FIGURE 6.4(a).** Optimal designed cantilever and simply-supported beams. Cross-section: circle; radius $r$ is variable ($n = 2$). Restriction: $A_1 = 4 \times 10^{-4}\,\text{m}^2$, $A_2 = 2A_1$. Length of a beam $l = 1.2\,\text{m}$. Material: $E = 1.96 \times 10^{11}\,\text{N/m}^2$, $\rho = 7.8 \times 10^3\,\text{kg/m}^3$.

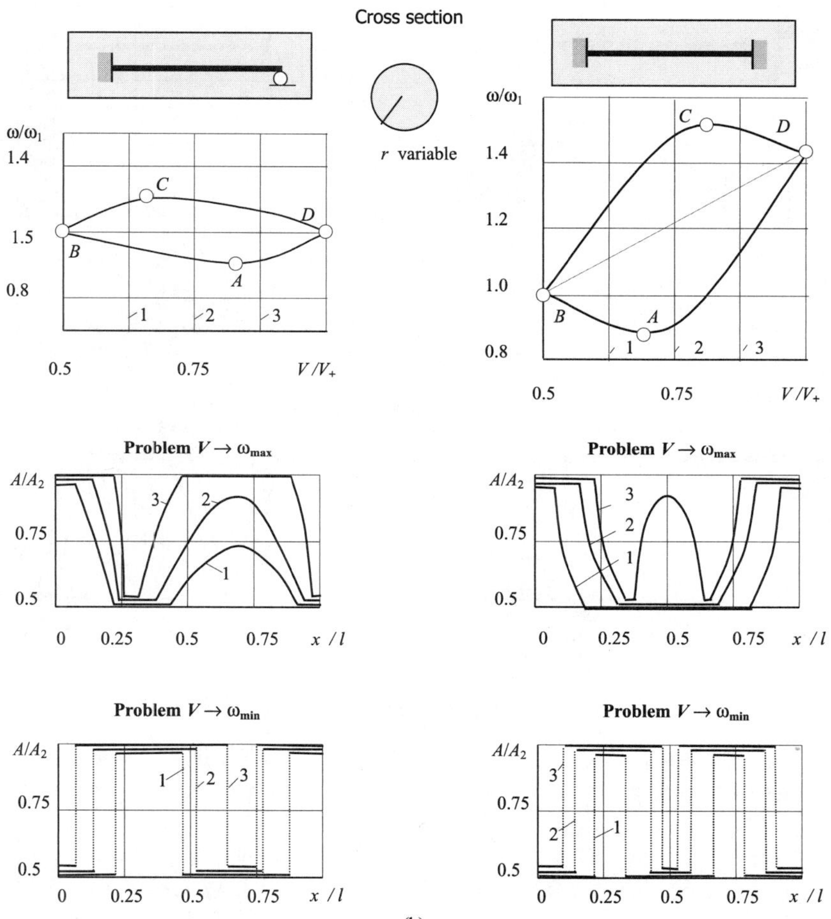

FIGURE 6.4(b). Optimal designed clamped–pinned and clamped–clamped beams. Cross-section: circle; radius $r$ is variable ($n = 2$). Length of a beam $l = 1.2$ m. Restriction: $A_1 = 4 \times 10^{-4}$ m$^2$, $A_2 = 2A_1$. Material: $E = 1.96 \times 10^{11}$ N/m$^2$, $\rho = 7.8 \times 10^3$ kg/m$^3$.

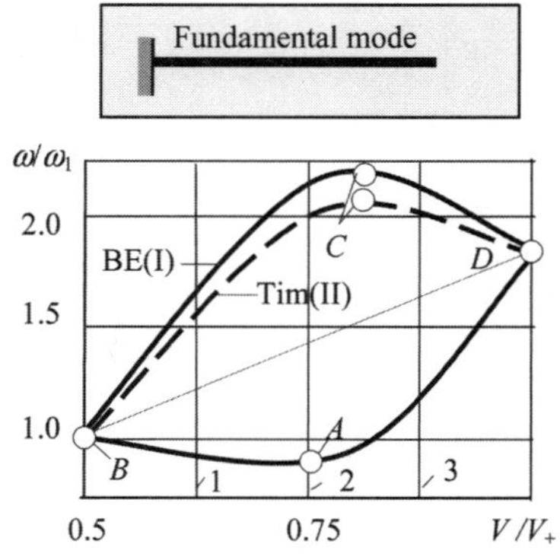


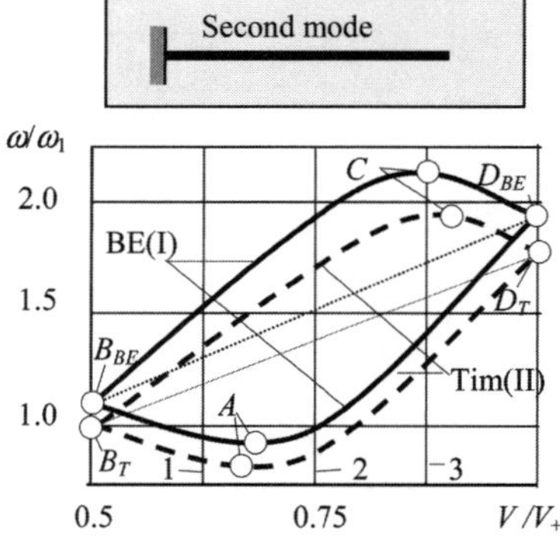


### Timoshenko theory (Tim)

**Problem $V \to \omega_{max}$**



**Problem $V \to \omega_{max}$**

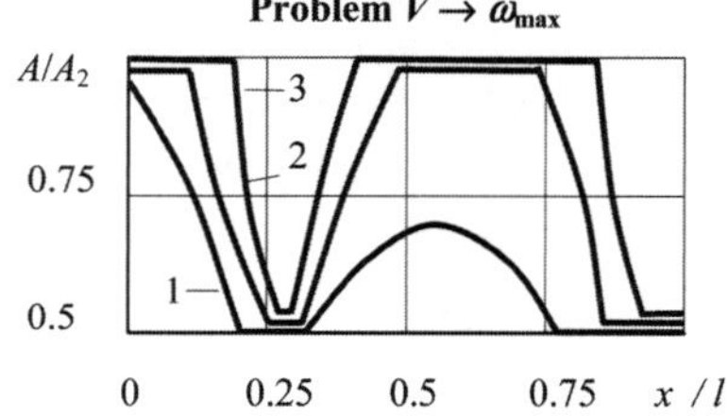


### Bernoulli-Euler theory (BE)

**Problem $V \to \omega_{max}$**

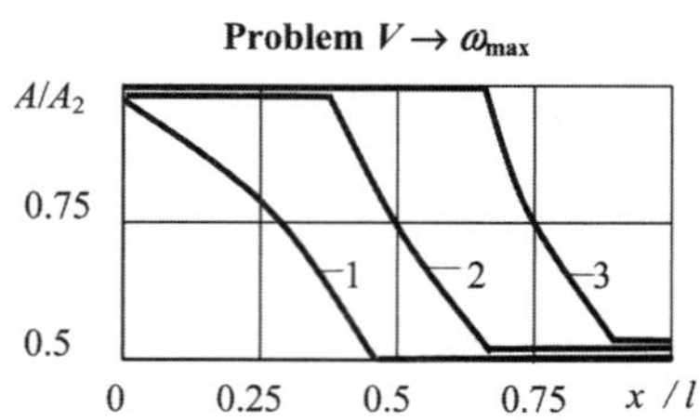


**Problem $V \to \omega_{max}$**

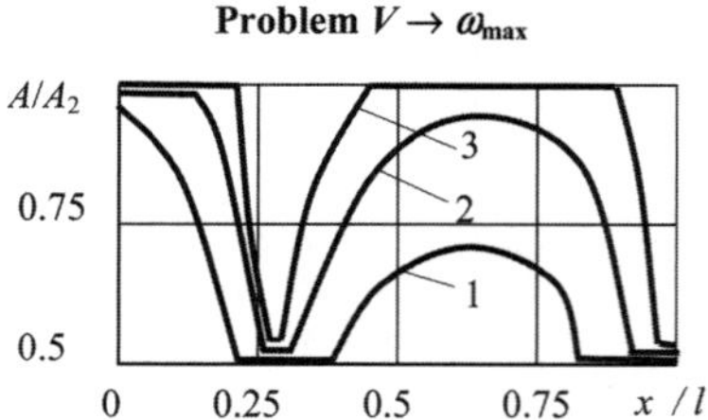


### Bernoulli-Euler and Timoshenko theories

**Problem $V \to \omega_{min}$**

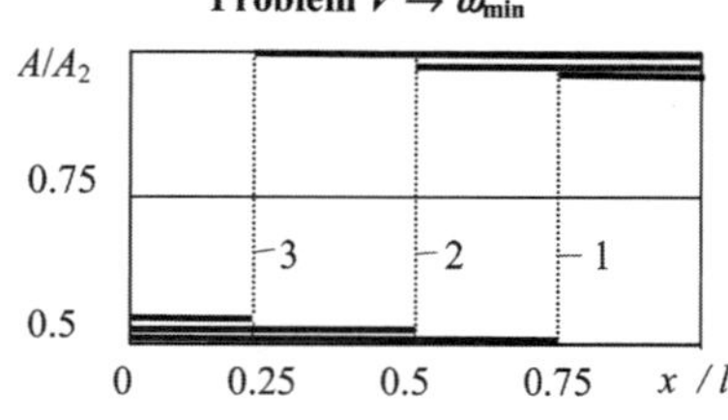


**Problem $V \to \omega_{min}$**

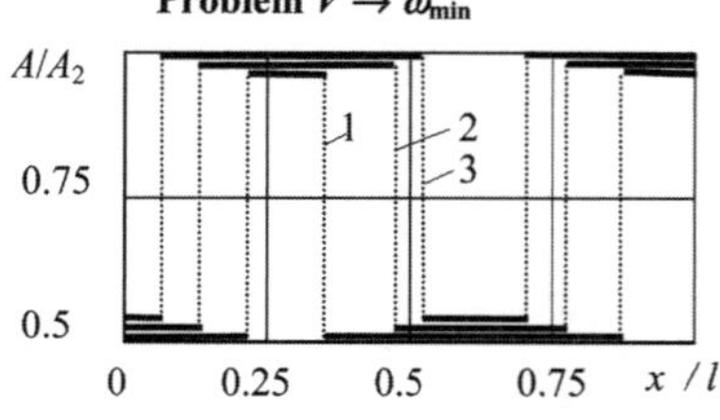


**FIGURE 6.5.** Optimal designed cantilever beam. Length of a beam $l = 1.2\,$m. Cross-section: rectangle; $b = 0.02\,$m; $h$ is variable ($n = 3$). Material: $E = 1.96 \times 10^{11}\,$N/m$^2$, $\rho = 7.8 \times 10^3\,$kg/m$^3$. Restriction: $A_1 = 4 \times 10^{-4}\,$m$^2$, $A_2 = 2A_1$.

## REFERENCES

Grinev, V.B. and Filippov, A.P. (1979) *Optimization of Rods by Eigenvalues* (Kiev: Naukova Dumka) (in Russian).

Karihaloo, B.L. and Niordson, F.I. (1973) Optimum design of vibrating cantilevers. *Journal of Optimization Theory and Application*, **11**, 638–654.

Olhoff, N. (1977) Maximizing higher order eigenfrequencies of beams with constraints on the design geometry. *Journal of Structural Mechanics*, **5**(2), 107–134.

Olhoff, N. (1980) Optimal design with respect to structural eigenvalues. *Proceedings of the XVth International Congress of Theoretical and Applied Mechanics*, Toronto.

Pontryagin, L.S., Boltyanskii, V.G., Gamkrelidze, R.V. and Mishchenko, E.F. (1962) *The Mathematical Theory of Optimal Processes* (New York: Pergamon).

## FURTHER READING

Banichuk, N.V. and Karihaloo, B. L. (1976) Minimum-weight design of multipurpose cylindrical bars. *International Journal of Solids and Structures*, **12**(4).

Brauch, R. (1973) Optimized design: characteristic vibration shapes and resonators. *Journal of Acoustic Society of America*, **53**(1).

Bryson, A.E., Jr. and Ho, Yu-Chi. (1969) *Applied Optimal Control* (Waltham, Massachusetts: Toronto).

Collatz, L. (1963) *Eigenwertaufgaben mit technischen Anwendungen* (Leipzig: Geest and Portig).

Elwany, M.H.S. and Barr, A.D.S. (1983) Optimal design of beams under flexural vibration. *Journal of Sound and Vibration*, **88**, 175–195.

Haug, E.J. and Arora, J.S. (1979) *Applied Optimal Design. Mechanical and Structural Systems* (New York: Wiley).

Johnson, M.R. (1968) Optimum frequency design of structural elements. Ph.D. Dissertation, Department of Engineering Mechanics, The University of Iowa.

Karnovsky, I.A. (1989) Optimal vibration protection of deformable systems with distributed parameters. Doctor of Science Thesis, Georgian Polytechnical University, (in Russian).

Liao, Y.S. (1993) A generalized method for the optimal design of beams under flexural vibration. *Journal of Sound and Vibration*, **167**(2), 193–202.

Miele, A. (Ed) (1965) *Theory of Optimum Aerodynamic Shapes* (New York: Academic Press).

Niordson, F.I. (1965) On the optimal design of vibrating beam. *Quart. Appl. Math.*, **23**, 47–53.

Olhoff, N. (1976) Optimization of vibrating beams with respect to higher order natural frequencies. *Journal of Structural Mechanics*, **4**(1), 87–122.

Sippel, D.L. (1970) Minimum-mass design of structural elements and multi-element systems with specified natural frequencies. Ph.D. Dissertation, September, University of Minnesota.

Tadjbakhsh, I. and Keller, J. (1962) Strongest columns and isoperimetric inequalities for eigenvalues. *Journal of Applied Mechanics*, **9**, 159–164.

Taylor, J.E. (1967) Minimum mass bar for axial vibrations at specified natural frequencies. *AIAA Journal*, **5**(10).

Turner, M.J. (1967) Design of minimum mass structures with specified natural frequencies. *AIAA Journal*, **5**(3), 406–412.

Troitskii, V.A. (1975) On some optimum problems of vibration theory. *Journal of Optimization Theory and Application*, **15**(6), 615–632.

Troitskii, V.A. and Petukhov L.V. (1982) Optimization of form of elastic bodies. (Moscow: Nauka) (in Russian).

Vepa, K. (1973–1974) On the existence of solutions to optimization problems with eigenvalue constraints. *Quart. Appl. Math.*, **31**, 329–341.

Weisshaar, T.A. (1972) Optimization of simple structures with higher mode frequency constraints. *AIAA Journal*, **10**, 691–693.

# NONLINEAR TRANSVERSE VIBRATIONS

This chapter is devoted to nonlinear transverse vibrations of beams. Static, physical and geometrical nonlinearities are discussed. In many cases, a method of reduction of nonlinear problems to linear ones with modified parameters is applied. This method is developed and presented by Bondar (1971). Some of the examples from the above mentioned book, are presented in this chapter. These are beams in magnetic fields, beams on nonlinear foundations, pipelines under moving liquids and internal pressure, etc. The frequency equations and the fundamental modes of vibrations are presented.

## 7.1  ONE-SPAN PRISMATIC BEAMS WITH DIFFERENT TYPES OF NONLINEARITY

### 7.1.1  Static nonlinearity

The uniform beam rests on two inmovable end supports, so displacements in the longitudinal direction of the beam at the support points are impossible, and the axial force $N$, e.g. thrust, is the response due to a vibration (Fig. 7.1). (Bondar', 1971; Lou and Sikarskie, 1975; Filin, 1981).

*Fundamental relationships.*  Axial deformation and strain

$$\Delta l = \int_0^l \varepsilon(x)\mathrm{d}x = \frac{1}{2}\int_0^l \left(\frac{\partial y}{\partial x}\right)^2 \mathrm{d}x, \quad \varepsilon_{\mathrm{avr}} = \frac{\Delta l}{l} = \frac{1}{2l}\int_0^l \left(\frac{\partial y}{\partial x}\right)^2 \mathrm{d}x \tag{7.1}$$

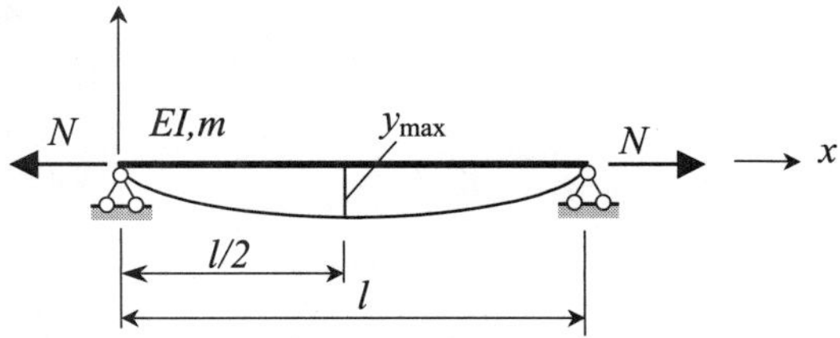


**FIGURE 7.1.**  Beam on two immovable end supports.

The length of the curve, axial force and bending moment are

$$S = \int_0^l \sqrt{1 + \left(\frac{dy}{dx}\right)^2}\, dx \approx \int_0^l \left[1 + \left(\frac{dy}{dx}\right)^2\right] dx, \tag{7.2}$$

$$N = EA\varepsilon_{avr} = \frac{EA}{2l}\int_0^l \left(\frac{\partial y}{\partial x}\right)^2 dx, \quad M = M_0 - Ny \tag{7.3}$$

where $M_0$ is the bending moment due to the transverse inertial force only.

***Differential equation of transverse vibration.***   The governing differential equation is

$$EI\frac{\partial^4 y}{\partial x^4} - \frac{EA}{2l}\frac{\partial^2 y}{\partial x^2}\left[\int_0^l \left(\frac{\partial y}{\partial x}\right)^2 dx\right] + \rho A\frac{\partial^2 y}{\partial t^2} = 0 \tag{7.4}$$

The integral term in (7.4) represents the axial tension induced by the deflection and is the source of the nonlinearity in the problem.

***Approximate solution***
We assume the solution in the form

$$y(x, t) = X(x)\cos \phi(t), \quad \phi(t) = \omega t + \psi \tag{7.5}$$

where $X(x)$ = fundamental functions
$\quad \phi(t)$ = phase functions
$\quad \omega$ = frequency of vibration
$\quad \psi$ = initial phase angle

***Equation for normal functions***
After a series of assumptions the equation for normal function may be presented as

$$X^{IV} - \frac{3}{4}aX''\int_0^l (X')^2\, dx - b\omega^2 X = 0, \quad a = \frac{A}{2Il}, \quad b = \frac{m}{EI} \tag{7.6}$$

The solution of equation (7.6) may be presented as

$$X(x) = y_{max}\sin\frac{\pi}{l}x$$

where $y_{max}$ is the transverse displacement of a beam at $x = 0.5l$.

***Fundamental frequency of nonlinear vibration***

$$\omega = \left(\frac{\pi}{l}\right)^2 \sqrt{\frac{1}{b}\left[1 + \left(\frac{l}{\pi}\right)^2 q_*\right]} \approx \left(\frac{\pi}{l}\right)^2\sqrt{\frac{1}{b}}\left[1 + \frac{1}{2}\left(\frac{l}{\pi}\right)^2 q_*\right] = \frac{\pi^2}{l^2}\sqrt{\frac{EI}{m}}\left[1 + \frac{3}{8}\left(\frac{y_{max}}{2r}\right)^2\right],$$

$$\tag{7.7}$$

where

$$q_* = \frac{3}{8}\frac{a}{l}\pi^2 y_{max}^2, \quad r = \sqrt{\frac{I}{A}}$$

This result may be obtained in another way. Let the transverse displacement be

$$y(x, t) = y_{max} T(t) \sin \frac{\pi x}{l}$$

The Bubnov–Galerkin procedure is

$$\int_0^l L(x, t) \sin \frac{\pi x}{l} \, dx = 0$$

where $L$ is the left part of the differential equation of transverse vibration; this algorithm yields to Duffing's equation (Hayashi, 1964)

$$\ddot{T} + \frac{\pi^4 EI}{ml^4} \left[ 1 + \left( \frac{y_{max}}{2r} \right)^2 T^2 \right] T = 0 \tag{7.8}$$

The fundamental frequency of vibration $\omega$ is as in equation 7.7.

If a free vibration is a response of a transverse shock on the beam, then the nonlinear mode shape and frequency vibration are

$$X(x) = -\frac{V_{max}}{\omega_0} \sin \frac{\pi}{l} x$$

$$\omega = \omega_0 \left[ 1 + \frac{3}{8} \left( \frac{V_{max}}{2r\omega_0} \right)^2 \right], \qquad \omega_0 = \frac{\pi^2}{l^2} \sqrt{\frac{EI}{m}}$$

where $V_{max}$ is the initial velocity in the middle of the beam due to the transverse shock.

*Note.* The static nonlinearity does not have an influence on the mode shape of vibration.

### 7.1.2  Physical nonlinearity

Hook's law cannot be applied to the material of the beam.

*Hardening nonlinearity.* The 'stress–strain' relationship for the material of the beam is

$$\sigma = E\varepsilon + \beta \varepsilon^3 \tag{7.9}$$

where $\sigma$, $\varepsilon$ are the stress and strain of the beam's material and $\beta$ is the nonlinear parameter of the beam's material.

Bending moment

$$M = -y''[EI_2 + \beta I_4(y'')^2]$$

The moment of inertia of the cross-sectional area of the order $n$ is

$$I_n = \int_{(A)} z^n \, dA \tag{7.10}$$

where $z$ is the distance from a neutral axis.

For a rectangular cross section, $b \times h$:

$$I_2 = \frac{bh^3}{12}, \quad I_4 = \frac{bh^5}{80}$$

for a circular section of diameter $d$

$$I_2 = \frac{\pi d^4}{64}, \quad I_4 = \frac{\pi d^6}{512}$$

for a pipe with inner and outer diameters $d$ and $D$, respectively

$$I_2 = \frac{\pi}{64}(D^4 - d^4), \quad I_4 = \frac{\pi}{512}(D^6 - d^6)$$

**Differential equation**

$$EI_2 \frac{\partial^4 y}{\partial x^4} + 6\beta I_4 \frac{\partial^2 y}{\partial x^2}\left(\frac{\partial^3 y}{\partial x^3}\right)^2 + 3\beta I_4 \left(\frac{\partial^2 y}{\partial x^2}\right)^2 \frac{\partial^4 y}{\partial x^4} + m\frac{\partial^2 y}{\partial t^2} = 0 \qquad (7.11)$$

**Approximate solution**

$$y(x, t) = X(x)\cos\phi(t), \quad \phi(t) = \omega t + \psi \qquad (7.12)$$

**Equation for normal functions**

$$X^{IV}\left[EI_2 + \frac{9}{4}\beta I_4(X'')^2\right] + \frac{9}{2}\beta I_4 X''(X''')^2 - m\omega^2 X = 0 \qquad (7.13)$$

**Fundamental frequency of nonlinear vibration**

$$\omega = \frac{\pi^2}{l^2\sqrt{b}}\frac{1}{\left(1 - \frac{3}{16}\lambda\frac{\pi^4}{l^4}y_{max}^2\right)^2} \approx \frac{\pi^2}{l^2\sqrt{b}}\left(1 + \frac{6}{16}\lambda\frac{\pi^4}{l^4}y_{max}^2\right) = \frac{\pi^2}{l^2}\sqrt{\frac{EI_2}{m}}\left(1 + \frac{27}{32}\frac{\pi^4\beta I_4}{l^4 EI_2}y_{max}^2\right)$$

$$(7.14)$$

The fundamental mode of vibration may be presented as

$$X = \frac{y_{max}(1-s)\sin\frac{\pi}{l}x}{1 - s\left(1 + \cos^2\frac{\pi}{l}x\right)} \approx y_{max}(1-s)\left[1 + s\left(1 + \cos^2\frac{\pi}{l}x\right)\right]\sin\frac{\pi}{l}x \qquad (7.14a)$$

where

$$s = \frac{\lambda}{8}\left(\frac{\pi}{l}\right)^4 y_{max}^2, \quad \lambda = \frac{9}{4}\frac{\beta I_4}{EI_2}, \quad b = \frac{m}{EI_2}$$

**Softening nonlinearity.**   The stress–strain relationship for the beam material is (Kauderer, 1958; Khachian and Ambartsumyan, 1981)

$$\sigma = E\varepsilon - \beta_1 E^3 \varepsilon^3 \qquad (7.14b)$$

Displacement: $y(\xi, \tau) = X(\xi)Y(\tau)$, $\quad \xi = \pi \dfrac{x}{l}$, $\quad \tau = \omega t$.

The differential equation for the time function $Y(\tau)$

$$\frac{\partial^2 Y}{\partial \tau^2} + \frac{v^2 a^2}{\omega^2} Y \left(1 - \frac{1}{3}\lambda b_2 Y^2\right) = 0 \tag{7.15}$$

where

$$a^2 = \frac{\pi^4 E I_2}{ml^4}, \quad \lambda = \frac{3\pi^4 \beta_1 E^2}{l^4}\frac{I_4}{I_2}, \quad v^2 = \frac{1}{b_0}\int_0^\pi X''^2 \, d\xi,$$

$$b_0 = \int_0^\pi X^2 \, d\xi, \quad b_2 = \frac{1}{v^2 b_0}\int_0^\pi X''^4 \, d\xi$$

***Period of nonlinear vibration***

$$T = \frac{1}{va}\left[1 + \frac{3}{8}\frac{\lambda b_2}{2}y_{\max}^2 + \frac{57}{256}\left(\frac{\lambda b_2}{3}\right)^2 y_{\max}^4 + \frac{315}{2048}\left(\frac{\lambda b_2}{3}\right)^6 y_{\max}^6 + \cdots\right] \tag{7.16}$$

where $y_{\max}$ is the fixed initial maximum lateral displacement of a beam.

***Notes***

**1.** Physical nonlinearity has an influence on the mode shape of vibration.

**2.** A hardening nonlinearity increases the frequency of vibration.

**3.** A softening nonlinearity decreases the frequency of vibration.

***Special cases.*** The shape mode and period of the nonlinear vibration for beams with different boundary conditions may be calculated by the following formulas. (Khachian and Ambartsumyan, 1981)

***Simply supported beam.*** Normal function

$$X(\xi) = \sin \xi, \quad b_0 = 0.5\pi, \quad v^2 = 1, \quad b_2 = 0.75$$

Nonlinear vibration period

$$T = T_0[1 + 0.09375\lambda y_{\max}^2 + 0.013916\lambda^2 y_{\max}^4 + 0.002403\lambda^3 y_{\max}^6 + \cdots]$$

where $T_0 = 2\pi/a$ is the period of linear vibration.

***Clamped–clamped beam.*** Fundamental function

$$X(\xi) = \frac{1}{1.61643}\left[\sin\frac{k}{\pi}\xi - 1.0178\cos\frac{k}{\pi}\xi - \sinh\frac{k}{\pi}\xi + 1.0178\cosh\frac{k}{\pi}\xi\right],$$

$$X_{\max} = 1, \quad k = 4.73$$

therefore $b_0 = 1.24542$, $v^2 = 5.1384$, $b_2 = 3.77213$.

Nonlinear vibration period

$$T = T_0[1 + 0.47152\lambda y_{\max}^2 + 0.35203\lambda^2 y_{\max}^4 + 0.30577\lambda^3 y_{\max}^6 + \cdots]$$

where $T_0 = 2\pi/2.267a$.

***Pinned–clamped beam.***   Normal function is

$$X(\xi) = \frac{1}{1.06676}\left[\sin\frac{k}{\pi}\xi + 0.027875\sinh\frac{k}{\pi}\xi\right], \quad k = 3.927$$

where $X_{\max} = 1$ at $x = 0.421l$ from the pinned support.

For the above-mentioned expression $X$ parameters $b_0$, $v^2$, $b_2$ and period of nonlinear vibration are

$$b_0 = 1.37911, \quad v^2 = 2.4423, \quad b_2 = 1.8116$$

$$T = T_0[1 + 0.22645\lambda y_{\max}^2 + 0.081193\lambda^2 y_{\max}^4 + 0.03387\lambda^3 y_{\max}^6 + \cdots]$$

where the period of linear vibration is

$$T_0 = 2\pi/1.563a$$

***Cantilever beam***

$$X(\xi) = \frac{1}{2.7242}\left[\sin\frac{k}{\pi}\xi - 1.3622\cos\frac{k}{\pi}\xi - \sinh\frac{k}{\pi}\xi + 1.3622\cosh\frac{k}{\pi}\xi\right], \quad k = 1.875$$

$$b_0 = 0.78536, \quad v^2 = 0.1269, \quad b_2 = 0.074514$$

$$T = T_0[1 + 0.009314\lambda y_{\max}^2 + 0.000137\lambda^2 y_{\max}^4 + 0.0000023\lambda^3 y_{\max}^6 + \cdots]$$

where $T_0 = 2\pi/0.362a$.

***Notes***

1. A hardening nonlinearity increases the frequency of vibration and has an influence on the fundamental shape of the mode of vibration.

2. A softening nonlinearity decreases the frequency of vibration and does not have an influence on the fundamental shape of the mode of vibration.

***Example.***   The clamped–clamped beam has the following parameters: $l = 300\,\text{cm}$, $b = h = 40\,\text{cm}$. The beam material is concrete; a maximum strain of a concrete equals 0.003. Calculate the period of nonlinear vibration.

***Solution.***   Expression (7.14) may be used if

$$\frac{d\sigma}{d\varepsilon} = E(1 - 3\beta_1 E^2 \varepsilon^2) > 0$$

which leads to a maximum value of the nonlinear coefficient

$$\beta_1 = \frac{1}{3E^2\varepsilon^2}$$

In this case, the coefficient

$$\lambda = \frac{3\pi^4 \beta_1 E^2}{l^4}\frac{I_4}{I_2} = \frac{\pi^4}{l^4 \varepsilon^2}\frac{I_4}{I_2} = \frac{\pi^4}{l^4 \varepsilon^2}\frac{12h^2}{80}$$

If the maximum strain is 0.003 then the corresponding parameter

$$\lambda = 0.32 \ (1/\mathrm{cm}^2)$$

If $y_{\max} = l/200 \cong 1.5 \,\mathrm{cm}$, then the period of nonlinear vibration

$$T = T_0[1 + 0.47152\lambda y_{\max}^2 + 0.35203\lambda^2 y_{\max}^4 + 0.30577\lambda^3 y_{\max}^6 + \cdots] = 1.6T_0$$

### 7.1.3.  Geometrical nonlinearity (large amplitude vibration)

Geometrical nonlinearity occurs at large beam deflections.

***Uniform beams.***   The differential equation of the free vibration of the geometrical nonlinear beam with different boundary conditions is given by Bondar (1971). The notation of the pinned–pinned beam with large transversal displacements is presented in Fig. 7.2.

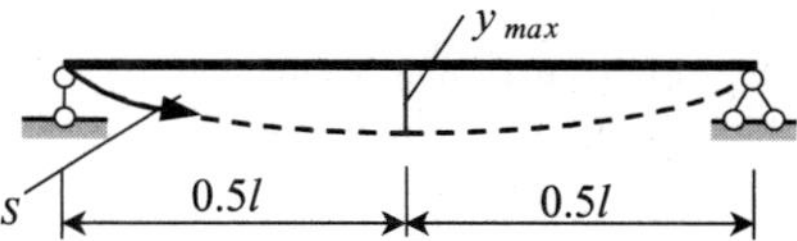


**FIGURE 7.2.**   Simply supported beam with large transversal displacements.

The approximate fundamental frequency of vibration for a beam with various boundary conditions may be calculated by the formula

$$\omega \cong \frac{\lambda^2}{l^2}\sqrt{\frac{EI}{m}}\left[1 + \frac{1}{4l}\int_0^1 (X')^2 \,\mathrm{d}s\right] \tag{7.17}$$

where $\lambda$ is a frequency parameter, which depends from boundary condition (Karnovsky and Lebed, 2003), and $X$ is a mode shape of vibration for the linear problem.

***Period of nonlinear vibration***

$$T = \frac{2\pi}{\omega} \cong T_0\left[1 - \frac{1}{4l}\int_0^l (X')^2 \,\mathrm{d}s\right] \tag{7.18}$$

For calculation of the integral $\int_0^l (X')^2 \,\mathrm{d}s$ for beams with different boundary conditions Table 4.8, (Karnovsky and Lebed, 2003) may be used.

***Special cases.***   The period of nonlinear vibration for beams with different boundary conditions may be presented in the following form.

*Pinned–pinned beam.*   The fundamental mode of vibration is

$$X(s) = y_{\max} \sin\frac{\pi s}{l}, \qquad \text{so} \quad \int_0^l (X')^2 \, ds = \frac{\pi^2}{2l} y_{\max}$$

and the period of nonlinear vibration in terms of initial maximum displacement is

$$T = T_0 \left[ 1 - \frac{\pi^2}{8} \left(\frac{y_{\max}}{l}\right)^2 \right]$$

*Cantilever beam.*   The fundamental mode of vibration is

$$X(s) = \frac{1}{2} y_{\max} \left[ \cosh\frac{\alpha s}{l} - \cos\frac{\alpha s}{l} - 0.734 \left( \sinh\frac{\alpha s}{l} - \sin\frac{\alpha s}{l} \right) \right],$$

$$\alpha = 1.875, \qquad \text{so} \quad \int_0^l (X')^2 \, ds = \frac{\pi^2}{2l} y_{\max}$$

and the period of nonlinear vibration

$$T = T_0 \left[ 1 - \frac{1}{8} \left(\frac{y_{\max}}{l}\right)^2 \right]$$

*Free–free beam.*   The period of nonlinear vibrations is

$$T = T_0 \left[ 1 - 1.55 \left(\frac{y_{\max}}{l}\right)^2 \right]$$

where $T_0$ is the period of vibration of the linear problem; and $y_{\max}$ is the maximum lateral displacement of the middle point for a pinned–pinned beam and free–free beam and at the free end for a cantilever beam.

***Note.***   Geometrical nonlinearity decreases the period of vibration and does have an influence on the fundamental shape of the mode of vibration.

***Tapered beams.***   *Free–free tapered beam.*   A free–free tapered beam with large displacements is presented in Fig. 7.3, where $l$ is a length of the beam and $\alpha$ is a slope at the tip. Table 7.1 present the types of linear tapered beams: breadth taper, depth taper, and diameter taper and their characteristics.

The mass per unit length $m$ and the moment of inertia $I$ of the tapered beams

$$m = m_R \left[ 1 - \varepsilon + \frac{\eta}{l}\varepsilon \right]^{n_1}$$

$$I = I_R \left[ 1 - \varepsilon + \frac{\eta}{l}\varepsilon \right]^{n_2} \tag{7.19}$$

where $\eta = s/l$.

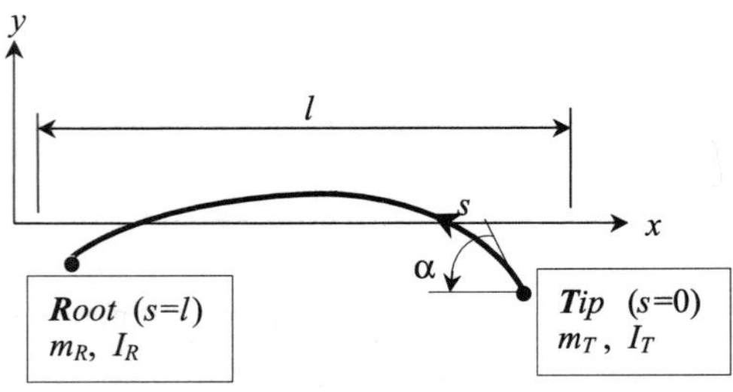


**FIGURE 7.3.**   Free–free tapered beam with large displacements.

**TABLE 7.1.** Types of linear tapered beams

| Type of taper | Geometry of tapered beams | Taper parameter $\varepsilon$ | $n_1$ | $n_2$ |
|---|---|---|---|---|
| Linear breadth taper | | $1 - \dfrac{B_T}{B_R}$ | 1 | 1 |
| Linear depth taper | | $1 - \dfrac{D_T}{D_R}$ | 1 | 3 |
| Linear diameter taper | | $1 - \dfrac{D_T}{D_R}$ | 2 | 4 |

***Numerical results.*** The governing equations for the geometry of the nonlinear vibration of the beam take into account both the axial and transverse inertia terms. The numerical method had been applied (Nageswara Rao and Venkateswara Rao, 1990).

Subscripts $T$ and $R$ refer to the quantities at the tip and at the root of a beam, respectively. The frequency of vibration may be calculated by

$$\omega = \frac{\lambda}{l^2}\sqrt{\frac{EI_R}{m_R}} \tag{7.20}$$

Fundamental natural frequency parameters, $\lambda$, for different types of taper are presented in Table 7.2. Here, $\varepsilon$ is a tapered parameter and $\alpha^\circ$ is the slope at the tip.

*Cantilever tapered beam* (Nageswara Rao and Venkateswara Rao, 1988). A cantilever tapered beam with large displacements is presented in Fig. 7.4, where $l$ is the length of the beam and $\alpha$ is the slope at the tip.

The frequency of vibration may be calculated by formula (7.20). Amplitude vibrations $x_T, y_T$ and fundamental parameter $\lambda$ for different types of taper are presented in Table 7.3; $\varepsilon$ is a tapered parameter and $\alpha^\circ$ is the slope at the tip.

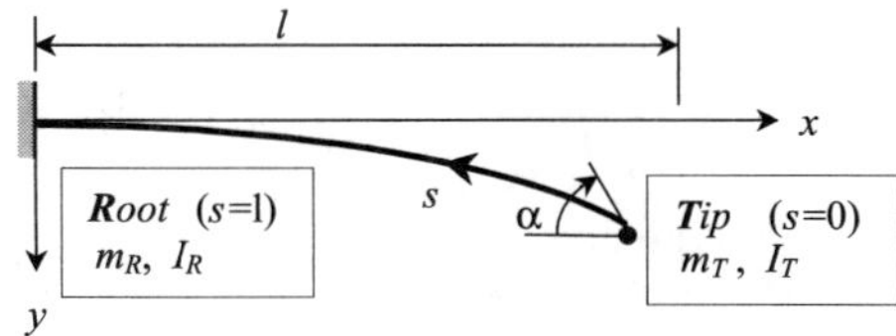


**FIGURE 7.4.** Cantilever tapered beam with large displacements.

**TABLE 7.2.**   Free–free linear tapered beams: fundamental frequency parameter $\lambda$

| Type of taper | Geometry of tapered beams | $\varepsilon$ | $n_1$ | $n_2$ | $\alpha(°)$ | $\varepsilon$ | | | |
|---|---|---|---|---|---|---|---|---|---|
| | | | | | | **0.0**** | 0.2 | 0.4 | 0.6 |
| Linear breadth taper | | $1 - \dfrac{B_T}{B_R}$ | 1 | 1 | **0.01*** | 22.373 | 22.407 | 22.552 | 22.940 |
| | | | | | 10 | 22.295 | 22.334 | 22.486 | 22.885 |
| | | | | | 20 | 22.069 | 22.123 | 22.296 | 22.726 |
| | | | | | 30 | 21.722 | 21.797 | 22.001 | 22.477 |
| | | | | | 40 | 21.292 | 21.390 | 21.629 | 22.160 |
| | | | | | 50 | 20.822 | 20.941 | 21.215 | 21.801 |
| | | | | | 60 | 20.355 | 20.490 | 20.792 | 21.429 |
| Linear depth taper | | $1 - \dfrac{D_T}{D_R}$ | 1 | 3 | **0.01*** | 22.373 | 20.127 | 17.863 | 15.586 |
| | | | | | 10 | 22.295 | 20.068 | 17.822 | 15.565 |
| | | | | | 20 | 22.069 | 19.896 | 17.706 | 15.503 |
| | | | | | 30 | 21.722 | 19.629 | 17.523 | 15.405 |
| | | | | | 40 | 21.292 | 19.294 | 17.290 | 15.279 |
| | | | | | 50 | 20.822 | 18.919 | 17.025 | 15.133 |
| | | | | | 60 | 20.355 | 18.537 | 16.748 | 14.978 |
| Linear diameter taper | | $1 - \dfrac{D_T}{D_R}$ | 2 | 4 | **0.01*** | 22.373 | 20.171 | 18.066 | 16.173 |
| | | | | | 10 | 22.295 | 20.115 | 18.035 | 16.162 |
| | | | | | 20 | 22.069 | 19.957 | 17.943 | 16.133 |
| | | | | | 30 | 21.722 | 19.710 | 17.800 | 16.087 |
| | | | | | 40 | 21.292 | 19.397 | 17.615 | 16.028 |
| | | | | | 50 | 20.822 | 19.045 | 17.402 | 15.960 |
| | | | | | 60 | 20.355 | 18.683 | 17.178 | 15.891 |

*Linear vibration.
** Uniform free–free beam.

## 7.2   BEAMS IN A MAGNETIC FIELD

### 7.2.1   Physical nonlinear beam in a nonlinear magnetic field

A uniform simply-supported beam in a magnetic field is presented in Fig. 7.5.

*Types of nonlinearity*

**1.** Physical nonlinearity: stress–strain relation for a beam material

$$\sigma = E\varepsilon + \beta\varepsilon^3 \tag{7.21}$$

**2.** Nonlinearity of the magnetic field: attractive force of the magnetic field

$$q_M = k\frac{ay}{(a^2 - y^2)^2} \approx k\frac{y}{a^3}\left(1 + 2\frac{y^2}{a^2}\right), \quad (y \ll a) \tag{7.22}$$

where  $a$ = distance between beam and magnet
      $y$ = transverse displacement of a beam
      $k$ = proportional coefficient ($k$ = constant)

**TABLE 7.3.** Amplitude vibrations $x_T$, $y_T$ and fundamental parameter $\lambda$ for a cantilever tapered beam

| Type of taper | $\alpha(°)$ | $\varepsilon = 0.2$ | | | $\varepsilon = 0.4$ | | | $\varepsilon = 0.6$ | | | $\varepsilon = 0.8$ | | |
|---|---|---|---|---|---|---|---|---|---|---|---|---|---|
| | | $x_T$ | $y_T$ | $\lambda$ | $x_T$ | $y_T$ | $\lambda$ | $x_T$ | $y_T$ | $\lambda$ | $x_T$ | $y_T$ | $\lambda$ |
| Linear breadth taper | **0.01*** | 1.000 | 0.000 | 3.763 | 1.000 | 0.000 | 4.097 | 1.000 | 0.000 | 4.585 | 1.000 | 0.000 | 5.398 |
| | 10 | 0.991 | 0.125 | 3.770 | 0.991 | 0.124 | 4.105 | 0.991 | 0.122 | 4.594 | 0.991 | 0.121 | 5.407 |
| | 20 | 0.963 | 0.248 | 3.792 | 0.964 | 0.245 | 4.128 | 0.965 | 0.242 | 4.619 | 0.965 | 0.239 | 5.436 |
| | 30 | 0.918 | 0.365 | 3.829 | 0.920 | 0.361 | 4.168 | 0.921 | 0.356 | 4.663 | 0.923 | 0.352 | 5.485 |
| | 40 | 0.857 | 0.474 | 3.883 | 0.859 | 0.469 | 4.225 | 0.862 | 0.464 | 4.725 | 0.864 | 0.459 | 5.555 |
| | 50 | 0.779 | 0.574 | 3.954 | 0.783 | 0.568 | 4.301 | 0.787 | 0.562 | 4.808 | 0.790 | 0.556 | 5.649 |
| | 60 | 0.688 | 0.661 | 4.045 | 0.693 | 0.655 | 4.399 | 0.698 | 0.648 | 4.914 | 0.703 | 0.642 | 5.768 |
| Linear depth taper | **0.01*** | 1.000 | 0.000 | 3.608[1] | 1.000 | 0.000 | 3.737[1] | 1.000 | 0.000 | 3.934[1] | 1.000 | 0.000 | 4.292[1] |
| | 10 | 0.991 | 0.122 | 3.615 | 0.992 | 0.116 | 3.744 | 0.993 | 0.107 | 3.941 | 0.994 | 0.095 | 4.299 |
| | 20 | 0.965 | 0.241 | 3.636 | 0.967 | 0.229 | 3.764 | 0.971 | 0.213 | 3.961 | 0.976 | 0.188 | 4.317 |
| | 30 | 0.922 | 0.355 | 3.671 | 0.927 | 0.338 | 3.799 | 0.935 | 0.315 | 3.994 | 0.946 | 0.279 | 4.348 |
| | 40 | 0.862 | 0.462 | 3.721 | 0.872 | 0.441 | 3.848 | 0.885 | 0.412 | 4.042 | 0.904 | 0.366 | 4.392 |
| | 50 | 0.788 | 0.560 | 3.788 | 0.802 | 0.535 | 3.914 | 0.821 | 0.501 | 4.106 | 0.850 | 0.448 | 4.452 |
| | 60 | 0.699 | 0.646 | 3.873 | 0.719 | 0.619 | 3.998 | 0.746 | 0.582 | 4.188 | 0.786 | 0.523 | 4.528 |
| Linear diameter taper | **0.01*** | 1.000 | 0.000 | 3.855[2] | 1.000 | 0.000 | 4.319[2] | 1.000 | 0.000 | 5.009[2] | 1.000 | 0.000 | 6.196[2] |
| | 10 | 0.991 | 0.120 | 3.862 | 0.992 | 0.112 | 4.326 | 0.993 | 0.102 | 5.017 | 0.995 | 0.086 | 6.203 |
| | 20 | 0.965 | 0.238 | 3.884 | 0.969 | 0.223 | 4.349 | 0.973 | 0.202 | 5.040 | 0.979 | 0.171 | 6.225 |
| | 30 | 0.923 | 0.351 | 3.921 | 0.930 | 0.329 | 4.387 | 0.939 | 0.299 | 5.079 | 0.952 | 0.253 | 0.261 |
| | 40 | 0.864 | 0.457 | 3.974 | 0.877 | 0.429 | 4.442 | 0.893 | 0.391 | 5.135 | 0.915 | 0.333 | 6.314 |
| | 50 | 0.791 | 0.554 | 4.044 | 0.810 | 0.522 | 4.515 | 0.834 | 0.477 | 5.210 | 0.868 | 0.410 | 6.385 |
| | 60 | 0.704 | 0.640 | 4.134 | 0.730 | 0.605 | 4.609 | 0.763 | 0.555 | 5.306 | 0.811 | 0.481 | 6.475 |

* Linear problem.
[1] Table 5.2 for cases $\chi = 1 - \varepsilon$.
[2] Table 5.5 for cases $\delta = 1 - \varepsilon$.

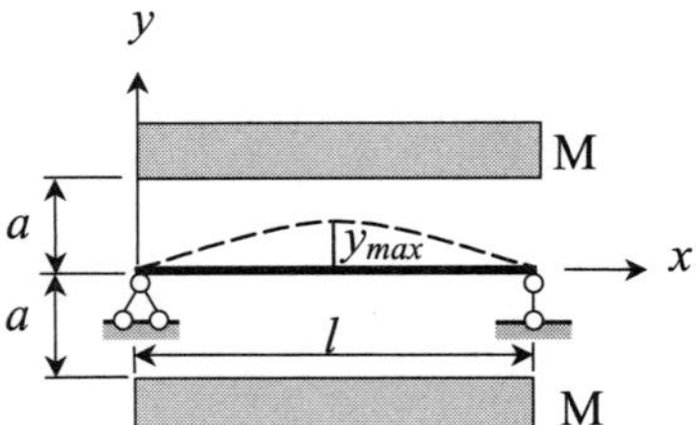


**FIGURE 7.5.** Simply-supported beam in the magnetic field.

### Differential equation of the transverse vibration

$$EI_2 \frac{\partial^4 y}{\partial x^4} + 6\beta I_4 \frac{\partial^2 y}{\partial x^2}\left(\frac{\partial^3 y}{\partial x^3}\right)^2 + 3\beta I_4 \left(\frac{\partial^2 y}{\partial x^2}\right)^2 \frac{\partial^4 y}{\partial x^4} + m\frac{\partial^2 y}{\partial t^2} - k\frac{y}{a^3}\left(1 + 2\frac{y^2}{a^2}\right) = 0 \qquad (7.23)$$

This equation takes into account a physical nonlinearity of the beam material (the terms with coefficient $\beta$) and the nonlinearity of the magnetic field (term $ky^3/a^5$) (Bondar', 1971).

Bending moment and deflection are related as follows:

$$M = -y''[EI_2 + \beta I_4(y'')^2]$$

Approximate solution

$$y(x, t) = X(x)\cos\phi(t), \quad \phi = \omega t + \psi$$

where $X(x)$ = fundamental mode of vibration
$\quad\ \phi(t)$ = phase functions
$\quad\quad \omega$ = frequency of vibration
$\quad\quad \psi$ = initial phase angle

### Equation for fundamental normal function

$$X^{IV}\left[EI_2 + \frac{9}{4}\beta I_4(X'')^2\right] + \frac{9}{2}\beta I_4 X''(X''')^2 - \left(m\omega^2 + \frac{k}{a^3}\right)X - \frac{3k}{2a^5}X^3 = 0 \qquad (7.24)$$

### Frequency of vibration

$$\omega \approx \sqrt{\frac{1}{m}\left\{\left(\frac{\pi}{l}\right)^4 EI_2\left[1 + \frac{3}{8}\frac{y_{max}^2}{EI_2}\left(\frac{9}{2}\frac{\pi^4\beta I_4}{l^4} - \frac{kl^4}{\pi^4 a^5}\right)\right] - \frac{k}{a^3}\right\}} \qquad (7.25)$$

Vibration is unstable, if

$$\left(\frac{\pi}{l}\right)^4 EI_2\left[1 + \frac{3}{8}\frac{y_{max}^2}{EI_2}\left(\frac{9}{2}\frac{\pi^4\beta I_4}{l^4} - \frac{3k^4}{\pi^4 a^5}\right)\right] \leq \frac{k}{a^3}$$

Fundamental mode shape of vibration

$$X \approx y_{\max}\left[1 + \frac{\lambda}{8}\left(\frac{\pi}{l}\right)^4 y_{\max}^2\left(1 + \cos^2\frac{\pi x}{l}\right) - \frac{k^*}{8}\left(\frac{l}{\pi}\right)^4 y_{\max}^2 \sin^2\frac{\pi x}{l}\right]\sin\frac{\pi x}{l} \qquad (7.26)$$

where parameters $\lambda$ and $k^*$ are as follows

$$\lambda = \frac{9\beta I_4}{4EI_2}, \quad k^* = \frac{3k}{2a^5 EI_2}$$

Different limiting cases are presented in Table 7.4.

**TABLE 7.4.** Physical nonlinear beam in nonlinear magnetic field and its limiting cases

| No. | Description of problem | Conditions |
|---|---|---|
| 1 | Nonlinear physical beam in a linear magnetic field | $k/a^5 = 0$ |
| 2 | Linear physical beam in a nonlinear magnetic field | $\beta = 0$ |
| 3 | Linear physical beam in a linear magnetic field | $k/a^5 = 0$ and $\beta = 0$ |
| 4 | Linear physical beam without a magnetic field | $k = 0$ and $\beta = 0$ (Table 4.3)* |

*Karnovsky, Lebed (2003)

### 7.2.2  Geometrical nonlinear beam in a nonlinear magnetic field

Consider a simply supported beam with a large displacement placed in a nonlinear magnetic field (Fig. 7.5). The attractive force of the magnetic field ($k = \text{const}$)

$$q_M = k\frac{ay}{(a^2 - y^2)^2} \approx k\frac{y}{a^3}\left(1 + 2\frac{y^2}{a^2}\right), \quad (y \ll a)$$

For moderately large displacements, the differential equation for the fundamental mode of vibration is

$$EI_2 X^{IV} - \left(m\omega^2 + \frac{k}{a^2}\right)X - \frac{3k}{2a^5}X^3 = 0 \qquad (7.27)$$

The frequency of vibration

$$\omega = \sqrt{\frac{1}{m\psi}\left\{\left(\frac{\pi}{l}\right)^4 EI\left(1 - \frac{3}{8}\frac{k\psi l^4}{EI\pi^4 a^5}y_{\max}^2\right) - \frac{k\psi}{a^3}\right\}}, \quad \psi = 1 - \left(\frac{\pi}{2l}y_{\max}\right)^2 \qquad (7.28)$$

where $y_{\max}$ is the fixed initial maximum lateral displacement of the beam.

Vibration is unstable, if

$$\left(\frac{\pi}{l}\right)^4 EI\left(1 - \frac{3}{8}\frac{k\psi l^4}{EI\pi^4 a^5}y_{\max}^2\right) \leq \frac{k\psi}{a^3}$$

For the cantilever beam, coefficient $\psi = 1$.

**Special cases**

**1.** Geometrical nonlinear beam in a linear magnetic field: $k/a^5 = 0$. In this case, the frequency of vibration

$$\omega = \sqrt{\frac{1}{m\psi}\left\{\left(\frac{\pi}{l}\right)^4 EI - \frac{k\psi}{a^3}\right\}}$$

**2.** Geometrical nonlinear beam without a magnetic field: $k = 0$ (Section 7.1.3). In this case, the frequency of vibration

$$\omega = \frac{\pi^2}{l^2}\sqrt{\frac{EI}{m\psi}}$$

## 7.3  BEAMS ON AN ELASTIC FOUNDATION

### 7.3.1  Physical nonlinear beams on massless foundations

Consider simply supported physically nonlinear beams resting on a massless foundation.

**Types of nonlinearity**

**1.** Physical nonlinearity: the stress–strain relation for a beam material is presented by (7.21).

**2.** Nonlinearity of foundation: reaction of the foundation

$$q_F = -k_F y(1 + \beta_F y^2) \tag{7.29}$$

where  $y$ = transverse displacement of a beam
$k_F$ = stiffness of a foundation
$\beta_F$ = nonlinearity parameter of the foundation

**Differential equation of the transverse vibration of the beam**

$$EI_2 \frac{\partial^4 y}{\partial x^4} + 6\beta I_4 \frac{\partial^2 y}{\partial x^2}\left(\frac{\partial^3 y}{\partial x^3}\right)^2 + 3\beta I_4 \left(\frac{\partial^2 y}{\partial x^2}\right)^2 \frac{\partial^4 y}{\partial x^4} + m\frac{\partial^2 y}{\partial t^2} + k_F(1 + \beta_F y^2) = 0 \tag{7.30}$$

Moment of inertia of the order $n$ may be calculated by formula (7.10).

**Frequency of vibration**

$$\omega = \sqrt{\frac{1}{m}\left\{\left(\frac{\pi}{l}\right)^4 EI_2\left[1 - \frac{3}{8}\frac{y_{\max}^2}{EI_2}\left(\frac{9\pi^4\beta I_4}{l^4} + \frac{k_F\beta_F l^4}{\pi^4}\right)\right] + k_F\right\}} \tag{7.31}$$

Different limiting cases are presented in Table 7.5.

**TABLE 7.5.** Physical nonlinear beam on massless foundation and its limiting cases

| No. | Description of problem | Conditions |
|---|---|---|
| 1 | Physical nonlinear beam without foundation | $k_F = 0$: (Section 7.1.2) |
| 2 | Linear elastic beam on a nonlinear foundation | $\beta = 0$ |
| 3 | Nonlinear beam on a linear foundation | $\beta_F = 0$ |
| 4 | Linear elastic beam on a linear foundation | $\beta = 0$ and $\beta_F = 0$ (Section 2.2.2) |
| 5 | Simply supported linear elastic beam without a foundation | $\beta = 0$ and $\beta_F = 0$, $k_F = 0$ |

### 7.3.2 Physical nonlinear beam on a nonlinear inertial foundation

The physical nonlinear beam length $l$ and mass $m$ per unit length rest on a nonlinear elastic foundation. The stress–strain relation for a beam material is presented in formula (7.21). A nonlinear inertial foundation is a two-way communication one. A reaction to the foundation equals

$$q_F = -k_F y(1 + \beta_F y^2)$$

where   $y =$ transverse displacement of a beam
      $k_F =$ stiffness of a foundation
      $\beta_F =$ nonlinearity parameter of the foundation

The model of the foundation is represented as separate rods with the following parameters: modulus $E_F$, cross-sectional area $A_F = b_b \times 1$, and density of material $\rho_F$; the length of the rods are $l_0$ and they are shown in Fig. 7.6.

     Reaction of the rods

$$q_0 = -E_F A_F \left\{ \frac{\partial u}{\partial z} \left[ 1 + \frac{\beta_F}{E_F} \left( \frac{\partial u}{\partial z} \right)^2 \right] \right\}_{z=l_0} \tag{7.32}$$

where $u$ is the longitudinal displacement of the rod.

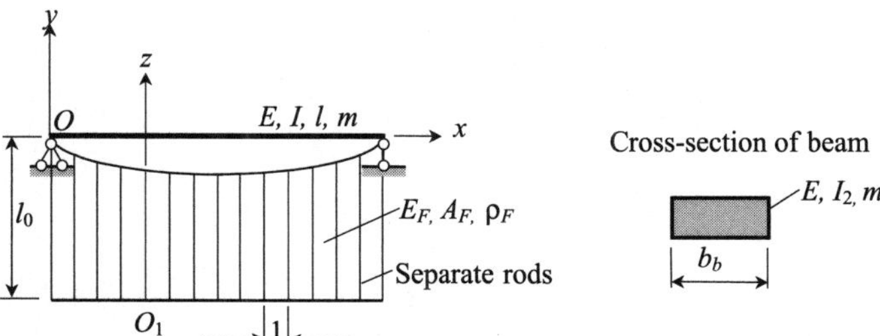


**FIGURE 7.6.** Mechanical model of nonlinear elastic foundation. Systems coordinates: for beam $xOy$; for rods $O_1 z$.

*Differential equations*

**1.** Longitudinal nonlinear vibration of the rods (Kauderer, 1958)

$$\frac{\partial^2 u}{\partial t^2} = a^2 \frac{\partial^2 u}{\partial z^2}\left[1 + \lambda\left(\frac{\partial u}{\partial z}\right)^2\right] \tag{7.33}$$

where

$$a^2 = \frac{E_F A_F}{m_F} = \frac{E_F}{\rho_F}, \quad \lambda = 3\frac{\beta_F}{E_F}$$

**2.** Transverse vibration of the beam

$$EI_2 \frac{\partial^4 y}{\partial x^4} + 6\beta I_4 \frac{\partial^2 y}{\partial x^2}\left(\frac{\partial^3 y}{\partial x^3}\right)^2 + 3\beta I_4 \left(\frac{\partial^2 y}{\partial x^2}\right)^2 \frac{\partial^4 y}{\partial x^4}$$
$$+ m\frac{\partial^2 y}{\partial t^2} + \left\{E_F A_F \frac{\partial u}{\partial z}\left[1 + \frac{\beta_F}{E_F}\left(\frac{\partial u}{\partial z}\right)^2\right]\right\}_{z=l_0} = 0 \tag{7.34}$$

where moments of inertia of the order $n(I_2$ and $I_4)$ may be calculated by formula (7.10).

*Approximate solution.* Transverse displacement $y(x, t)$ of a beam and longitudinal displacement of rods $u(x, t)$ may be presented as

$$\begin{aligned} y(x, t) &= X(x)\cos\phi(t)\\ u(z, x, t) &= Z(z, x)\cos\phi(t) \end{aligned} \tag{7.35}$$

where $\phi = \omega t + \psi_0$, $\psi_0$ is initial phase.

The normal function for a pinned–pinned beam

$$X(x) = y_{\max}\sin\frac{\pi}{l}x$$

*Approximate equations for normal functions.* For longitudinal linear vibration of the rods after averaging (Bondar', 1971)

$$\frac{\mathrm{d}^2 Z}{\mathrm{d}z^2} + \left(\frac{\omega}{\alpha}\psi_z\right)^2 Z = 0 \tag{7.36}$$

where parameter

$$\psi_z = \sqrt{1 - 3\lambda\left(\frac{\omega y_{\max}}{4a}\right)^2 \delta} \approx 1 - \frac{3}{2}\lambda\left(\frac{\omega y_{\max}}{4a}\right)^2\delta, \quad \delta = \frac{1 + \dfrac{a}{3\omega l_0}\sin\dfrac{2\omega}{a}l_0}{\left(\sin\dfrac{\omega}{a}l_0\right)^2}$$

For transverse nonlinear vibrations of the beam

$$X^{IV}\left[EI_2 + \frac{9}{4}\beta I_4 (X'')^2\right] + \frac{9}{2}\beta I_4 X''(X''')^2$$

$$+ \left(\frac{\omega}{a}E_F A_F \psi_z \cot\frac{\omega}{a}\psi_z l_0 - m\omega^2\right)X - \frac{3}{4}\beta_F A_F \psi_x^3 X^3 = 0 \qquad (7.37)$$

where parameter

$$\psi_x = \frac{\omega}{\alpha}\psi_z \cot\frac{\omega}{\alpha}\psi_z l_0$$

***Nonlinear frequency equation***

$$\omega^2 b^2 - \frac{\omega}{\alpha}C\cot\frac{\omega}{\alpha}\psi_z l_0 = \left(\frac{\pi}{l}\right)^4\left[1 + \frac{3}{16}\frac{y_{\max}^2}{EI_2}\left(\frac{9\pi^4\beta I_4}{l^4} - \frac{l^4\beta_F A_F}{\pi^4}\psi_x^3\right)\right] \qquad (7.38)$$

where parameter

$$C = \frac{E_F A_F}{EI_2} = \frac{E_F b_b}{EI_2}, \quad b^2 = \frac{m}{EJ_2}$$

***Special case***
Linear vibration: $y_{\max} = 0$ (Section 2.3.3). In this case a frequency equation becomes

$$\omega^2 b^2 - \frac{\omega}{\alpha}C\cot\frac{\omega}{\alpha}l_0 = \left(\frac{\pi}{l}\right)^4$$

1. *Beam without an elastic foundation* ($E_F = 0 \rightarrow C/a = 0$). *The frequency equation of the beam is*

$$\left(\frac{\pi}{l}\right)^4 - b^2\omega^2 = 0$$

and the fundamental frequency of vibration of the beam is $\omega = \dfrac{\pi^2}{l^2}\sqrt{\dfrac{EI_2}{m}}$.

2. *Elastic foundation without a beam* ($EI_2 = 0 \rightarrow b = \infty,\ C = \infty$). *The* frequency equation of the longitudinal vibration of the rod is

$$\tan\gamma = \infty, \quad \gamma = \pi/2$$

and the fundamental frequency of the longitudinal vibration is

$$\omega = \frac{\pi}{2l_0}\sqrt{\frac{E_F}{\rho_F}}$$

This case corresponds to a longitudinal vibration of the clamped–free rods.

**3.** *The beam is absolutely rigid* $(EI_2 = 0 \to b = 0,\ C = 0)$. The frequency equation becomes

$$\tan \gamma = 0, \quad \gamma = \pi$$

and the fundamental frequency of the longitudinal vibration is

$$\omega = \frac{\pi}{l_0}\sqrt{\frac{E_F}{\rho_F}}$$

This case corresponds to longitudinal vibration of the clamped–clamped rods.

## 7.4  PINNED–PINNED BEAM UNDER MOVING LIQUID

Consider a vertical pipeline, or a horizontal one, without initial deflections, carrying a moving liquid; the velocity of the liquid, $V$, and stiffness of the beam, $EI$, are constant. The quazi-static regime is discussed.

### 7.4.1  Static nonlinearity

The beam rests on two inmovable supports (Fig. 7.7).
   The distributed load on the beam is

$$w = -m\frac{\partial^2 y}{\partial t^2} - m_L\left(\frac{\partial^2 y}{\partial t^2} + V^2\frac{\partial^2 y}{\partial x^2}\right) \tag{7.39}$$

where $m$ and $m_L$ are the mass per unit length of the beam and the mass of the moving liquid, respectively. The first term in equation (7.39) describes the inertial force of the beam. The first and second terms in the brackets take into account the relative and transfer forces of inertia of a moving liquid, respectively.

***Differential equation of transverse vibration of the beam***

$$\frac{\partial^4 y}{\partial x^4} + \frac{m + m_L}{EI}\frac{\partial^2 y}{\partial t^2} + \frac{\partial^2 y}{\partial x^2}\left[\frac{m_L V^2}{EI} - \frac{1}{2lr^2}\int_0^l\left(\frac{\partial y}{\partial x}\right)^2 dx\right] = 0 \tag{7.40}$$

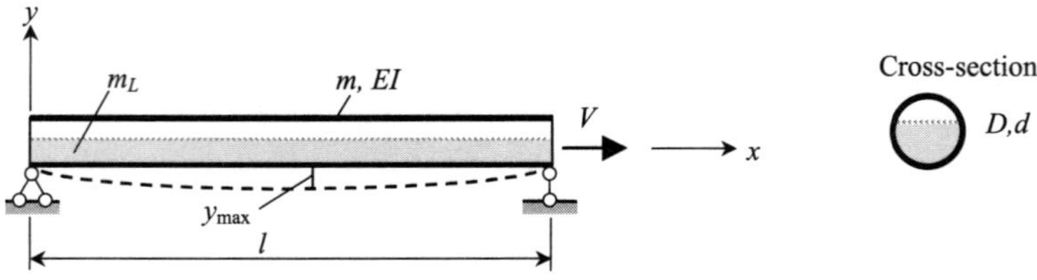


**FIGURE 7.7.**    Pipeline under infinite moving liquid. Quasi-static regime.

where $y$ is a transverse displacement of a beam; and $r^2 = I/A$, a square of the radius of gyration of a cross-section area.

This equation is approximate because it does not take into account the Coriolis inertia force. The solution of equation (7.40) may be presented in a form

$$y(x, t) = X(x) \cos \phi(t)$$

where $\phi = \omega t + \varphi_0$, $\varphi_0$ is initial phase.

***Equation for normal function***

$$X^{IV} - \frac{m + m_L}{EI} \omega^2 X + X'' \left[ \frac{m_L V^2}{EI} - \frac{3}{4} \cdot a \int_0^l (X')^2 \, dx = 0 \right], \quad a = \frac{1}{2lr^2} \tag{7.41}$$

The second term in the brackets takes into account the nonlinear effect.

The expression for the normal function

$$X(x) = y_{max} \sin \frac{\pi x}{l}$$

leads to a fundamental nonlinear frequency of vibration

$$\omega_1 = \frac{\pi^2}{l^2} \sqrt{\frac{EI}{m + m_L}} \sqrt{1 + \frac{l^2}{\pi^2} \left[ \frac{3}{4} \left( \frac{\pi y_{max}}{2lr} \right)^2 - \frac{m_L V^2}{EI} \right]} \tag{7.42}$$

Condition of stability loss

$$\frac{l^2}{\pi^2} \left[ \frac{3}{4} \left( \frac{\pi y_{max}}{2lr} \right)^2 - \frac{m_L V^2}{EI} \right] = -1$$

First critical velocity

$$V_{1cr} = \frac{\pi}{l} \sqrt{\frac{EI}{m_L}} \sqrt{1 + 3 \left( \frac{y_{max}}{4r} \right)^2} \approx \frac{\pi}{l} \sqrt{\frac{EI}{mL}} \left[ 1 + \frac{3}{2} \left( \frac{y_{max}}{4r} \right)^2 \right]$$

***Second frequency of vibration and critical velocity.*** These characteristics may be obtained by using the Bubnov–Galerkin method.

The mode shape

$$X(x) = y_{max} \sin \frac{2\pi x}{l}, \quad \text{where } y_{max} = y(0.25l)$$

which corresponds to the second frequency of vibration, and leads to the following expressions for the natural frequency of vibration and critical velocity

$$\omega_2 = \frac{4\pi^2}{l^2}\sqrt{\frac{EI}{m+m_L}}\sqrt{1+\frac{l^2}{4\pi^2}\left[\frac{3}{4}\left(\frac{\pi y_{\max}}{lr}\right)^2 - \frac{m_L V^2}{EI}\right]}$$

$$V_{2\mathrm{cr}} = \frac{2\pi}{l}\sqrt{\frac{EI}{m_L}}\sqrt{1+3\left(\frac{y_{\max}}{2r}\right)} \approx \frac{2\pi}{l}\sqrt{\frac{EI}{m_L}}\left[1+\frac{3}{2}\left(\frac{y_{\max}}{2r}\right)^2\right]$$

For the linear problem $V_{2\mathrm{cr}} = 2V_{1\mathrm{cr}}$.

### Special cases

1. Linear problem, if $y_{\max} = 0$.
2. If the Coriolis inertia force is taken into account, then the distributed load is

$$w = -m\frac{\partial^2 y}{\partial t^2} - m_L\left(\frac{\partial^2 y}{\partial t^2} + 2V\frac{\partial^2 y}{\partial t\,\partial x} + V^2\frac{\partial^2 y}{\partial x^2}\right)$$

and the differential equation of the transverse vibration of the beam becomes

$$\frac{\partial^4 y}{\partial x^4} + \frac{m+m_L}{EI}\frac{\partial^2 y}{\partial t^2} + \frac{\partial^2 y}{\partial x^2}\left[\frac{m_L V^2}{EI} - \frac{A}{2lI}\int_0^l\left(\frac{\partial y}{\partial x}\right)^2 dx\right] + \frac{2m_L V}{EI}\frac{\partial^2 y}{\partial x\,\partial t} = 0$$

In this case, the frequency of vibration is very close to the results that were obtained by using the expression for $\omega_1$ (Karnovsky, 1970).


### 7.4.2  Physical nonlinearity

Consider a simply supported beam carrying the moving load. The velocity of the liquid, $V$, and stiffness of the beam, $EI$, are constant (Fig. 7.7). The stress–strain relationship for the material of the beam $\sigma = E\varepsilon + \beta\varepsilon^3$. The initial deflection of the beam under self-weight is ignored. The quasi-static regime is discussed.

The bending moment is

$$M = -y''\left[EI_2 + \beta I_4(y'')^2\right]$$

The second term in brackets describes the effect of physical nonlinearity.

### *Differential equation of the transverse vibration*

$$EI_2\frac{\partial^4 y}{\partial x^4} + 6\beta I_4\frac{\partial^2 y}{\partial x^2}\left(\frac{\partial^3 y}{\partial x^3}\right)^2 + 3\beta I_4\left(\frac{\partial^2 y}{\partial x^2}\right)^2\frac{\partial^4 y}{\partial x^4} + (m+m_L)\frac{\partial^2 y}{\partial t^2} + m_L V^2\frac{\partial^2 y}{\partial x^2} = 0 \qquad (7.43)$$

where $m$ and $m_L$ are the mass per unit length of the beam and the mass of the moving liquid, respectively. Expressions for the cross-sectional area moments of inertia $I_2$ and $I_4$ are presented in Section 7.1.2.

The expression for transverse displacement in the form

$$y(x, t) = X(x) \cos \omega(t)$$

leads to the following equation for a normal function

$$X^{IV}\left[EI_2 + \frac{9}{4}\beta I_4(X'')^2\right] + X''\left[m_L V^2 + \frac{9}{2}\beta I_4(X''')^2\right] - m\omega^2 X = 0 \qquad (7.44)$$

The approximate presentation of a normal function

$$X(x) = y_{\max} \sin \frac{\pi x}{l}$$

leads to the following frequency of vibration of the beam under the quasi-static regime

$$\omega = \frac{\pi^2}{l^2}\sqrt{\frac{EI_2}{m + m_L}}\sqrt{1 + \frac{l^2}{\pi^2}\left(\frac{27}{16}\frac{\beta I_4 \pi^6}{EI_2 l^6}y_{\max}^2 - \frac{m_L V^2}{EI_2}\right)} \qquad (7.45)$$

The first critical velocity

$$V_{\mathrm{cr}} = \frac{\pi}{l}\sqrt{\frac{EI_2}{m_L}}\sqrt{1 + \frac{27}{16}\frac{\beta I_4 \pi^4}{EI_2 l^4}y_{\max}^2} \approx \frac{\pi}{l}\sqrt{\frac{EI_2}{m_L}}\left(1 + \frac{27}{32}\frac{\beta I_4 \pi^4}{EI_2 l^4}y_{\max}^2\right)$$

***Special case.*** Consider a physically linear beam under the quasi-static regime. In this case, the parameter nonlinearity $\beta = 0$ and the expression for the linear frequency of vibration and critical velocity are

$$\omega = \frac{\pi^2}{l^2}\sqrt{\frac{EI_2}{m + m_L}}\sqrt{1 - \frac{l^2 m_L V^2}{\pi^2 EI_2}}$$

$$V_{\mathrm{cr}} = \frac{\pi}{l}\sqrt{\frac{EI_2}{m_L}}$$

## 7.5  PIPELINE UNDER MOVING LOAD AND INTERNAL PRESSURE

A vertical pipeline or a horizontal one without initial deflections is kept under moving liquid and internal pressure $P$; the velocity of the liquid $V$ and stiffness of the beam $EI$ are constant. The quasi-static regime is discussed.

Distributed load on a beam

$$w = -m\frac{\partial^2 y}{\partial t^2} - m_L\left(\frac{\partial^2 y}{\partial t^2} + 2V\frac{\partial^2 y}{\partial t \partial x}\right) - (m_L V^2 + PA_0)\frac{\partial^2 y}{\partial x^2} \qquad (7.46)$$

where $m$ and $m_L$ are the mass per unit length of the beam and the mass of the moving liquid, respectively, and $A_0$ is the open cross-section.

The first term describes the inertial force of the beam. The first terms in the first and second brackets take into account relative and transfer forces of inertia of a moving liquid, respectively. The second terms in the first and second brackets take into account Coriolis inertia force of a moving liquid and internal pressure.

### 7.5.1  Static nonlinearity

The beam rests on two immovable supports (Fig. 7.7).

***Frequency of vibration***

$$\omega = \frac{\pi^2}{l^2}\sqrt{\frac{EI}{m+m_L}}\sqrt{1+\frac{l^2}{\pi^2}\left[\frac{3}{4}\left(\frac{\pi y_{\max}}{2lr}\right)^2-\frac{1}{EI}(m_L V^2 + PA_0)\right]} \qquad (7.47)$$

Critical velocity

$$V_{\mathrm{cr}} = \sqrt{\frac{1}{m_L}\left\{EI\,\frac{\pi^2}{l^2}\left[1+\frac{3}{4}\left(\frac{y_{\max}}{2r}\right)^2\right]-PA_0\right\}}, \qquad r=\sqrt{\frac{T}{A}}$$

***Special case.***   Consider a statically linear beam under an infinite moving load and internal pressure. In this case $y_{\max}=0$ and the expressions for the linear frequency of vibration and critical velocity are

$$\omega = \frac{\pi^2}{l^2}\sqrt{\frac{EI}{m+m_L}}\sqrt{1-\frac{l^2}{\pi^2 EI}(m_L V^2+PA_0)}$$

$$V_{\mathrm{cr}_0}=\sqrt{\frac{1}{m_L}\left(EI\,\frac{\pi^2}{l^2}-PA_0\right)}$$

From these equations, one may easily obtain formulas for frequency of vibration and critical velocity if internal pressure $P=0$, or the velocity of the liquid $V=0$.

### 7.5.2  Physical nonlinearity

***Frequency of vibration***

$$\omega = \frac{\pi^2}{l^2}\sqrt{\frac{EI_2}{m+m_l}}\cdot\sqrt{1+\frac{l^2}{\pi^2}\left[\frac{27\beta I_4\pi^6}{16EI_2 l^6}y_{\max}^2-\frac{1}{EI_2}(m_L V^2+PA_0)\right]}$$

Critical velocity

$$V_{cr}=\sqrt{\frac{1}{m_L}\left[\frac{EI_2\pi^2}{l^2}\left(1+\frac{27\beta I_4\pi^6}{16EI_2 l^6}y_{\max}^2\right)-PA_0\right]}$$

The case $\beta=0$ corresponds to linear problem.

## 7.6 HORIZONTAL PIPELINE UNDER A MOVING LIQUID AND INTERNAL PRESSURE

Consider a horizontal pipeline under self-weight, moving liquid and internal pressure $P$; the velocity of the liquid, $V$, and the stiffness of the beam, $EI$, are constant (Fig. 7.8). The initial deflection is taken into account. The quasi-static regime is discussed.

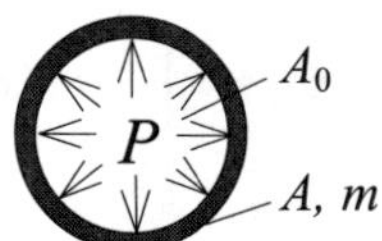



**FIGURE 7.8.**

Distributed load on the beam

$$w = w_0 - (m + m_L)\frac{\partial^2 y}{\partial t^2} - (m_L V^2 + PA_0)\frac{\partial^2 y}{\partial x^2} \tag{7.48}$$

where $w_0$ is the uniformly distributed force due to self-weight.

### 7.6.1 Static nonlinearity

The beam rests on two immovable supports (Fig. 7.9).
    Total deflection of a beam

$$y(x, t) = y_1(x) + y_2(x, t) \tag{7.49}$$

where $y_1$ and $y_2$ are deflections that correspond to the quasi-static regime only and the process of vibration, respectively.

***Differential equation of the beam under static nonlinearity***

$$\frac{\partial^4 y}{\partial x^4} + b\frac{\partial^2 y}{\partial t^2} + \frac{\partial^2 y}{\partial x^2}\left(C^* - a\int_0^l \left(\frac{\partial y}{\partial x}\right)^2 dx\right) = \frac{w_0}{EJ} \tag{7.50}$$

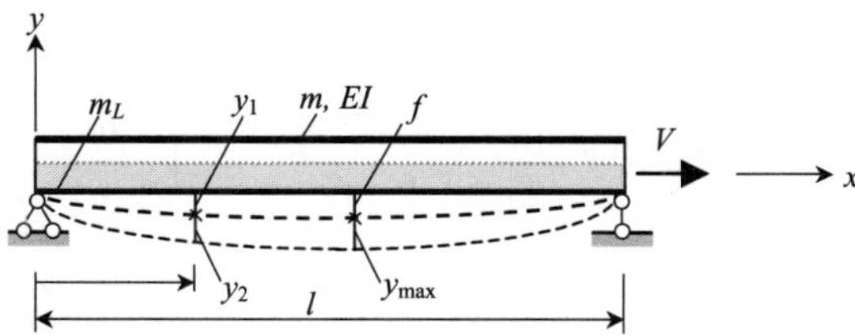



**FIGURE 7.9.**   Pipeline under moving liquid. The initial deflection is taken into account.

where

$$b = \frac{m + m_L}{EI}; \quad a = \frac{1}{2lr^2} = \frac{A}{2lI}; \quad C^* = \frac{m_L V^2 + PA_0}{EI}$$

This equation is non-homogeneous one.

***Approximate equation for normal functions.***   Quasi-static regime

$$y_1^{IV} + y_1'' \left[ C^* - a \int_0^l (y')^2 \, dx \right] = \frac{w_0}{EI}$$

Dynamic regime

$$\frac{\partial^4 y_2}{\partial x^4} + b \frac{\partial^2 y_2}{\partial t^2} - 2a \frac{\partial^2 y_1}{\partial x^2} \int_0^l \frac{\partial y_1}{\partial x} \frac{\partial y_2}{\partial x} \, dx + \frac{\partial^2 y_2}{\partial x^2} \left( C^* - a \int_0^l \left[ \left( \frac{\partial y_1}{\partial x} \right)^2 + \left( \frac{\partial y_2}{\partial x} \right)^2 \right] dx \right) = 0$$

If $y_2(x, t) = X(x) \cos \omega(t)$, then the equation for the normal function, $X$, becomes

$$X^{IV} - b\omega^2 X + X'' \left\{ C^* - a \left[ \int_0^l (y_1')^2 \, dx + \frac{3}{4} \int_0^l (X')^2 \, dx \right] \right\} - 2y_1'' a \int_0^l y_1' X' \, dx = 0$$

The deflection of the beam in the quasi-static regime is

$$y_1(x) = f \sin \frac{\pi x}{l}$$

where the maximum deflection of the linear problem is

$$f = \frac{4w_0 l^4}{\pi^5 EI} \frac{1}{1 - \dfrac{l^2}{\pi^2 EI} (m_L V^2 + PA_0)}$$

Normal function $X(x) = y_{\max} \sin \dfrac{\pi x}{l}$.

***Frequency of vibration***

$$\omega = \frac{\pi^2}{l^2} \sqrt{\frac{EI}{m + m_L}} \sqrt{1 + \frac{l^2}{\pi^2} \left[ \frac{3}{4} \pi^2 \left( \frac{f + y_{\max}}{2lr} \right)^2 - \frac{1}{EI} (m_L V^2 + PA_0) \right]} \qquad (7.51)$$

Deflection $f$ of the beam in the quasi-static regime increases the frequency of vibration. The equation for nonlinear critical velocity

$$1 + \frac{l^2}{\pi^2} \left[ 3 \left( \frac{w_0 l^3}{\pi^4 rEI} \right)^2 \frac{1}{S^2} - \frac{m_L V_{cr}^2 + PA_0}{EI} \right] = 0$$

where

$$S = 1 - \frac{l^2}{\pi^2 EI} (m_L V_{cr}^2 + PA_0)$$

### 7.6.2 Physical nonlinearity

If the material of a simply-supported beam has a hardening characteristic of nonlinearity (7.9), then the frequency of free vibration may be calculated as follows

$$
\omega = \frac{\pi^2}{l^2}\sqrt{\frac{EI_2}{m+m_L}}\sqrt{1+\frac{l^2}{\pi^2}\left[\frac{27}{16}\frac{\beta I_4 \pi^6}{EI_2 l^6}(f+y_{max})^2 - \frac{1}{EI_2}(m_L V^2 + PA_0)\right]}
$$

The equation for critical velocity

$$
1+\frac{l^2}{\pi^2}\left[\frac{3\beta I_4}{EI_2}\left(\frac{3w_0 l}{\pi^2 EI_2}\right)^2\frac{1}{S^2} - \frac{m_L V_{cr}^2 + PA_0}{EI_2}\right] = 0
$$

where

$$
S = 1 - \frac{l^2}{\pi^2 EI_2}(m_L V_{cr}^2 + PA_0)
$$

*Note*

1. Hardening nonlinearity ($\beta > 0$): the critical velocity for a physically nonlinear problem is more than for a linear one, $V_{cr.} > V_{cr.lin}$.
2. Softening nonlinearity ($\beta < 0$): the critical velocity for a physically nonlinear problem is less then for a linear one, $V_{cr.} < V_{cr.lin}$.

## *REFERENCES*


Bondar', N.G. (1971) *Non-Linear Problems of Elastic Systems* (Kiev: Budivel'nik) (in Russian).

Filin, A.P. (1981) *Applied Mechanics of a Solid Deformable Body*, vol. 3, (Moscow: Nauka) (in Russian).

Karnovsky, I.A. and Cherevatsky, B.P. (1970) Linearization of nonlinear oscillatory systems with an arbitrary number of degrees of freedom, New York. *Soviet Applied Mechanics*. **6**(9), 1018–1020.

Karnovsky, I.A. and Lebed, O. (2003) Free vibrations of beams and frames: Eigenvalues and eigenfunctions. (New York: McGraw-Hill).

Kauderer, H. (1961) *Nonlinear Mechanics*, (Izd. Inostr. Lit. Moscow) translated from *Nichtlineare Mechanik* (Berlin, 1958).

Khachian, E.E. and Ambartsumyan, V.A. (1981) *Dynamical Models of Structures in the Seismic Stability Theory*, Moscow, Nauka, 204 pp.

Lou, C.L. and Sikarskie, D.L. (1975) Nonlinear vibration of beams using a form-function approximation. *ASME Journal of Applied Mechanics*, pp. 209–214.

Nageswara Rao, B. and Venkateswara Rao, G. (1988) Large-amplitude vibrations of a tapered cantilever beam. *Journal of Sound and Vibration*, **127**(1), 173–178.

Nageswara Rao, B. and Venkateswara Rao, G. (1990) Large-amplitude vibrations of free–free tapered beams. *Journal of Sound and Vibration*, **141**(3), 511–515.

## *FURTHER READING*


Blekhman, I.I. (Ed) (1979) *Vibration of Nonlinear Mechanical Systems*, vol. 2. In *Handbook: Vibration in Tecnnik*, vol. 1–6 (Moscow: Mashinostroenie) (in Russian).

Evensen, D.A. (1968) Nonlinear vibrations of beams with various boundary conditions. *American Institute of Aeronautics and Astronautics Journal*, **6**, 370–372.

Hayashi, C. (1964) *Nonlinear Oscillations in Physical Systems* (New York: McGraw Hill).

Ho, C.H., Scott, R.A. and Eisley, J.G. (1976) Non-planar, non-linear oscillations of a beam, *Journal of Sound and Vibration*, **47**, 333–339.

Holmes, P.J. (1979) A nonlinear oscillator with a strange attractor. *Philosophical Transactions of the Royal Society*, London, **292** (1394), 419–448.

Inman, D.J. (1996) *Engineering Vibration*, (Prentice-Hall).

Karnovsky, I.A. (1970) Vibrations of plates and shells carrying a moving load. Ph.D. Thesis, Dnepropetrovsk (in Russian).

Masri, S.F., Mariamy, Y.A. and Anderson, J.C. (1981) Dynamic response of a beam with a geometric nonlinearity. *ASME Journal of Applied Mechanics*, **48**, 404–410.

Nageswara Rao, B. and Venkateswara Rao, G. (1990a) On the non-linear vibrations of a free–free beam of circular cross-section with linear diameter taper. *Journal of Sound and Vibration*, **141**(3), 521–523.

Nageswara Rao, B. and Venkateswara Rao, G. (1990b) Large amplitude vibrations of clamped–free and free–free uniform beams. *Journal of Sound and Vibration*, **134**, 353–358.

Nageswara Rao, B. and Venkateswara Rao, G. (1988) Large amplitude vibrations of a tapered cantilever beam. *Journal of Sound and Vibration*, **127**, 173–178.

Nayfeh, A.H. and Mook, D.T. (1979) *Nonlinear Oscillations* (New York: Wiley).

Ray, J.D. and Bert, C.W. Nonlinear vibrations of a beam with pinned ends. *Transactions of the American Society of Mechanical Engineers, Journal of Engineering for Industry*, **91**, 997–1004.

Sathyamoorthy, M. (1982) Nonlinear analysis of beams, part I: a survey of recent advances. *Shock and Vibration Digest*, **14**(17), 19–35.

Singh, G., Sharma, A.K. and Venkateswara Rao, G. (1990) Large-amplitude free vibrations of beams – a discussion on various formulations and assumptions. *Journal of Sound and Vibration*, **142**(1), 77–85.

Singh, G., Venkateswara Rao, G. and Iyengar, N.G.R. (1990) Re-investigation of large-amplitude free vibrations of beams using finite elements. *Journal of Sound and Vibration*, **143**(2), 351–355.

Wagner, H. (1965) Large amplitude free vibrations of a beam. *Transactions of the American Society of Mechanical Engineers, Journal of Applied Mechanics*, **32**, 887–892.

# ARCHES

Chapter 8 considers the vibration of arches. Fundamental relationships for uniform and non-uniform arches are presented – they are the differential equations of vibrations, strain and kinetic energy, as well as the governing functional. Eigenvalues for arches with different equations of the neutral line, different boundary conditions, uniform, continuously and discontinuously varying cross-sections are presented.

## 8.1 FUNDAMENTAL RELATIONSHIPS

This section presents the geometric parameters for arches with various equations of the neutral line (i.e. different shapes of arches) as well as differential equations of vibrations of arches.

### 8.1.1 Equation of the neutral line in terms of span *l* and arch rise *h*

Different shapes of arches and notations are presented in Fig. 8.1 (Bezukhov, 1969; Lee and Wilson, 1989).

### 8.1.2 Equation of the neutral line in terms of radius of curvature $R_0$ and slope $\alpha$

Figure 8.2 presents arch notation in the polar coordinate system. The radius of curvature is given by the functional relation (Romanelli and Laura, 1972; Laura *et al.*, 1988; Rossi *et al.*, 1989):

$$R(\alpha) = R_0 \cos^n \alpha \qquad (8.1)$$

where $R_0$ is the radius of curvature at $\alpha = 0$ and $n$ is an integer specified for a typical line (Table 8.1).

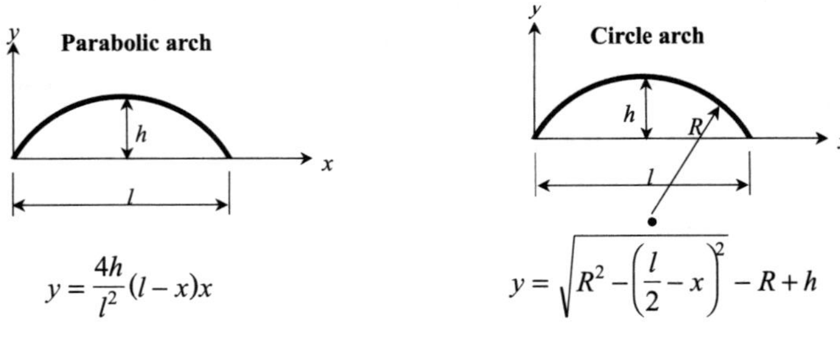

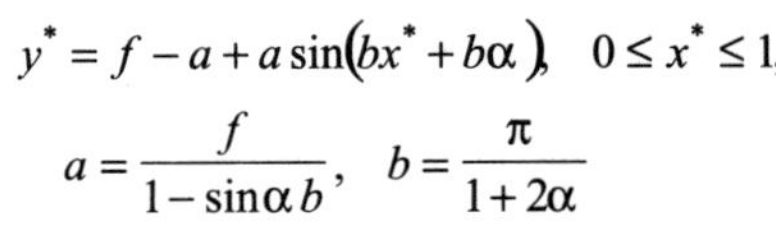

$$y = \frac{4h}{l^2}(l-x)x$$

$$y = \sqrt{R^2 - \left(\frac{l}{2}-x\right)^2} - R + h$$

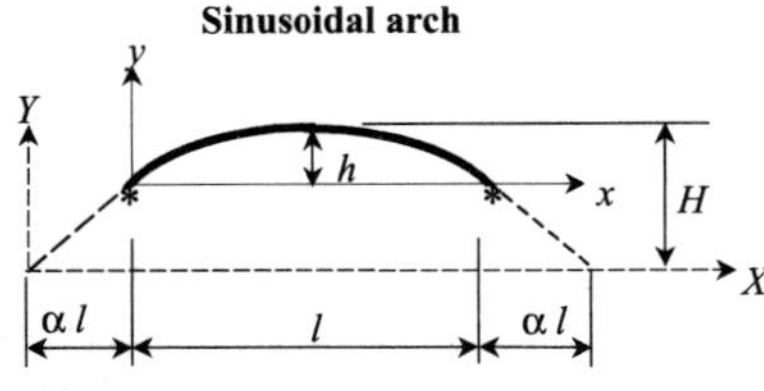

$$X = \alpha\,l + x, \quad Y = H - h + y$$

$$y^* = f - a + a\sin(bx^* + b\alpha), \quad 0 \le x^* \le 1,$$

$$a = \frac{f}{1 - \sin\alpha\, b}, \quad b = \frac{\pi}{1 + 2\alpha}$$

$$f = \frac{h}{l}, \quad x^* = \frac{x}{l}, \quad y^* = \frac{y}{l}$$

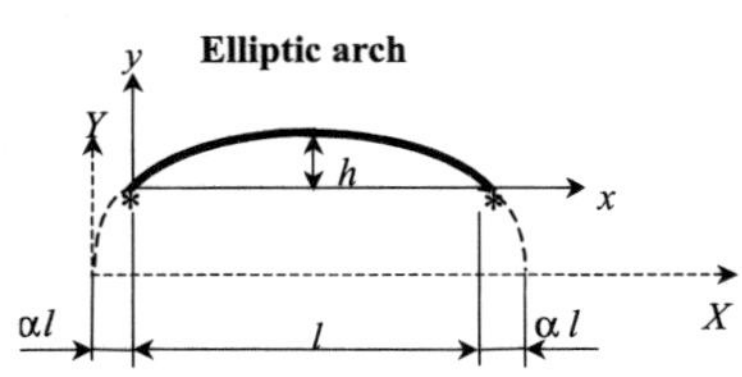

$$y^* = f - c + \frac{c}{e}\cdot\sqrt{e^2 - \left(x^* - \frac{1}{2}\right)^2}, \quad 0 \le x^* \le 1$$

$$c = \frac{ef}{e - \sqrt{\alpha + \alpha^2}}, \quad e = \frac{1 + 2\alpha}{2}$$

$$f = \frac{h}{l}, \quad x^* = \frac{x}{l}, \quad y^* = \frac{y}{l}$$

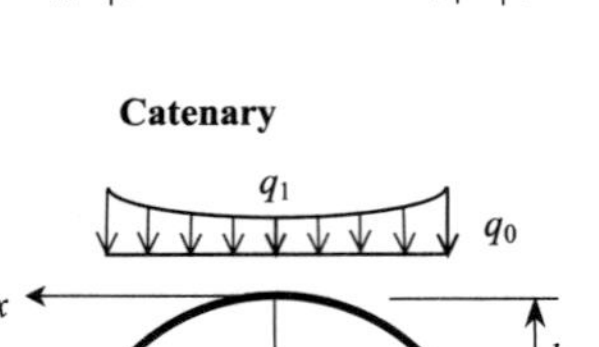

$$y = \frac{h}{n_1 - 1}(\cosh\xi\,k - 1),$$

$$n_1 = \frac{q_0}{q_1}, \quad k = \ln\!\left(n_1 + \sqrt{n_1^2 - 1}\right) \quad \xi = \frac{2x}{l}$$

**FIGURE 8.1.** Types of arches and equations of the neutral line.

For an elliptic arch, the equations of neutral line in standard form and the radius of curvature are as follows:

$$\frac{x^2}{a^2} + \frac{y^2}{b^2} = 1 \qquad R = a^2 b^2 \left(\frac{1 + \tan^2\alpha}{a^2\tan^2\alpha + b^2}\right)^{3/2}$$

The instantaneous radius of curvature, $R$, of an axis of any type of arch is expressed in terms of polar coordinates $\rho$ and $\alpha$ as follows

$$R = \frac{[\rho^2 + (\rho')^2]^{3/2}}{\rho^2 - \rho\rho'' + 2(\rho')^2}, \quad \rho' = \frac{d\rho}{d\alpha}, \quad \rho'' = \frac{d^2\rho}{d\alpha^2}$$

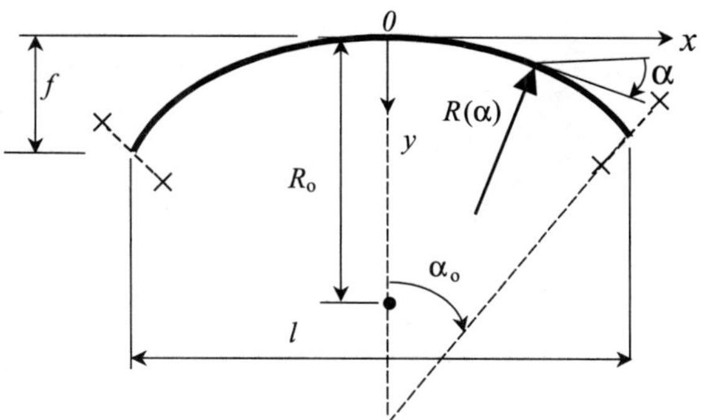

**FIGURE 8.2.** Arch geometry. Boundary conditions are not shown.

**TABLE 8.1.** Geometry relationships of the arches with different equations of the neutral line

| Curve | Parameter $n$ | Equation of the neutral line | $l/R_0$ | $f/R_0$ |
|---|---|---|---|---|
| Parabola | $-3$ | $y = x^2/2R_0$ | $2\tan\alpha_0$ | $\dfrac{1}{2}\tan^2\alpha_0$ |
| Catenary | $-2$ | $y = R_0(\cosh x/R_0 - 1)$ | $2\,\mathrm{arc\,sinh}\,\tan\alpha_0$ | $1/\cos\alpha_0 - 1$ |
| Spiral | $-1$ | $y = -R_0 \ln\cos x/R_0$ | $2\alpha_0$ | $-\ln\cos\alpha_0$ |
| Circle | $0$ | $y = R_0 - \sqrt{R_0^2 - x^2}$ | $2\sin\alpha_0$ | $1 - \cos\alpha_0$ |
| Cycloid | $1$ | $x = \dfrac{R_0}{4}\arccos\left(\dfrac{4y}{R_0} - 1\right) \pm \sqrt{y\left(\dfrac{R_0}{2} - y\right)} - \dfrac{R_0\pi}{4}$ | $\alpha_0 + \dfrac{1}{2}\sin 2\alpha_0$ | $1 - \cos 2\alpha_0$ |

### 8.1.3 Differential equations of in-plane vibrations of arches

This section considers the fundamental relationship for an arch of non-uniform cross-section with a variable radius of curvature.

The geometry of a uniform, symmetric arch with a variable radius of curvature $R(\alpha)$ is defined in Fig. 8.3. Its span length, rise, shape of middle surface, and inclination with the $x$-axis are $l, h, y(x)$ and $\alpha$, respectively. The positive radial, $w$, and tangential, $v$, displacements, positive rotation $\psi$ of cross-section, as well as all internal forces – axial forces $N$, bending moments $M$, shear forces $V$, and rotary inertia couple $T_{in}$ – are shown (Rzhanitsun, 1982; Lee and Wilson, 1989; Borg and Gennaro, 1959 convention is used). Radial and tangential displacement, and angle of rotation, are related by the following formulas

$$w = \frac{\partial v}{\partial \alpha}, \ \psi = \frac{1}{R}(w' - v), \text{ where } (') = \mathrm{d}/\mathrm{d}\alpha \tag{8.2}$$

***Assumptions (Navier's hypothesis)***

(1) A plane section before bending remains plane after bending.

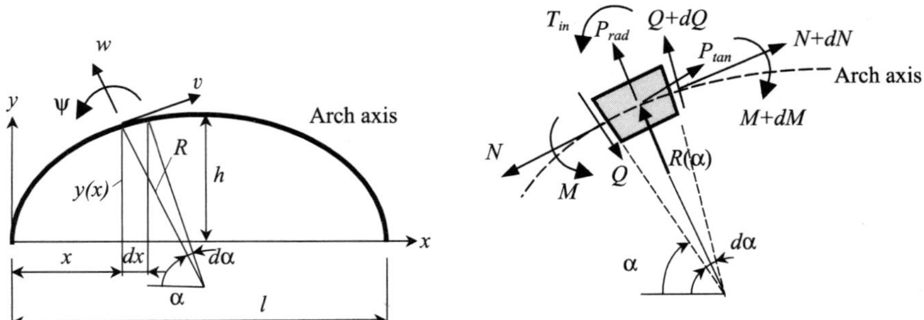

**FIGURE 8.3.**   Arch geometry and loads on the arch element.

(2) The cross-section of an arch is symmetric with respect to the loading plane.

(3) Hook's law applies.

(4) Deformations are small enough so that the values of the stresses and moments are not substantially affected by these deformations.

With the specified assumptions, the fundamental equations for the 'dynamic equilibrium' of an element are

$$\frac{dN}{d\alpha} + Q + RP_{\text{tan}} = 0$$

$$\frac{dQ}{d\alpha} - N + RP_{\text{rad}} = 0 \tag{8.3}$$

$$\frac{dM}{d\alpha} - RQ - RT_{\text{in}} = 0$$

The inertial forces and couple are

$$P_{\text{rad}} = m\omega^2 w$$

$$P_{\text{tan}} = m\omega^2 v \tag{8.4}$$

$$T_{\text{in}} = m\omega^2 r^2 \psi = \frac{m\omega^2 r^2}{R}(\omega' - v)$$

where $m$ = mass per unit length

$\quad\quad r$ = radius of gyration of a cross-section

$\quad\quad \omega$ = frequency of vibration

Internal forces in terms of tangential and radial displacements, and radius of curvature are

$$N = \frac{EA}{R}\left[v' + w + \frac{r^2}{R^2}(w'' + w)\right]$$

$$M = -EA\frac{r^2}{R^2}(w'' + w) \tag{8.5}$$

$$Q = \frac{1}{R}\cdot\frac{dM}{d\alpha} - ST = -EAr^2\left[\frac{1}{R^3}(w''' + w') - 2\frac{R'}{R^4}(w'' + w)\right] - S\frac{m\omega^2 r^2}{R}(w' - v)$$

where $S$ is a switch function. The switch function $S = 1$ if the rotatory inertia couple $T_{\text{in}}$ is included, and $S = 0$ if $T_{\text{in}}$ is excluded.

Differential equations for $N$, $V$ and $M$ are as follows

$$
\begin{aligned}
\frac{\mathrm{d}N}{\mathrm{d}\alpha} &= EA\left[\frac{1}{R}(v'' + w') + \frac{r^2}{R^3}(w''' + w') - \frac{R'}{R^2}(v' + w) - \frac{3r^2 R'}{R^4}(w'' + w)\right] \\
\frac{\mathrm{d}M}{\mathrm{d}\alpha} &= -EAr^2\left[\frac{1}{R^2}(w''' + w') - \frac{2R'}{R^3}(w'' + w)\right] \\
\frac{\mathrm{d}Q}{\mathrm{d}\alpha} &= -EAr^2\left[\frac{1}{R^3}(w'''' + w') - \frac{5R'}{R^4}(w''' + w') + 2\left(\frac{4R'^2}{R^5} - \frac{R''}{R^4}\right)\cdot(w'' + w)\right] \\
&\quad - Sm\omega^2 r^2\left[\frac{1}{R}(w'' - v') - \frac{R'}{R^2}(w' - v)\right]
\end{aligned}
\tag{8.6}
$$

**Special case.**   Consider arches with constant curvature ($R = R_0 = \text{constant}$). In this case, the in-plane vibration of a thin curve element is described by the following differential equations, which are obtained from (8.3) ($\mathrm{d}s = R\mathrm{d}\alpha$):

$$
\begin{aligned}
\frac{\mathrm{d}N}{\mathrm{d}s} + \frac{Q}{R} + P_{\text{tan}} &= 0 \\
\frac{\mathrm{d}Q}{\mathrm{d}s} - \frac{N}{R} + P_{\text{rad}} &= 0 \\
\frac{\mathrm{d}M}{\mathrm{d}s} - Q - T_{\text{in}} &= 0
\end{aligned}
\tag{8.3a}
$$

If $R = \infty$, then the equations break down into two independent equations, which describe the longitudinal and transverse vibrations of a beam.

**Strain and kinetic energy of an arch.**   Potential energy $U$ and kinetic energy $T$ of a non-uniform arch may be presented in the following forms.

**Form 1.** This form uses a curvilinear coordinate $s$ along the arch

$$
U = U_1 + U_2 \tag{8.7}
$$

$$
U_1 = \frac{1}{2}\int_0^l EI\left[\frac{\partial^2 w}{\partial s^2} + \frac{\partial}{\partial s}\left(\frac{v}{R}\right)\right]^2 \mathrm{d}s \tag{8.8}
$$

$$
U_2 = \frac{1}{2}\int_0^l EA\left(\frac{\partial v}{\partial s} - \frac{w}{R}\right)^2 \mathrm{d}s \tag{8.9}
$$

$$
T = \frac{1}{2}\int_0^l \rho A\left[\left(\frac{\partial v}{\partial t}\right)^2 + \left(\frac{\partial w}{\partial t}\right)^2\right]\mathrm{d}s \tag{8.10}
$$

**Form 2.** This form uses angular coordinate $\alpha$: $ds = R\,d\alpha$

$$U_1 = \frac{1}{2}\int_0^\alpha \frac{EI(\alpha)}{R^3(\alpha)}\left(\frac{\partial^2 w}{\partial\alpha^2} - \frac{1}{R(\alpha)}\frac{\partial R}{\partial\alpha}\frac{\partial w}{\partial\alpha} + w - \frac{1}{R(\alpha)}\frac{\partial R}{\partial\alpha}v\right)^2 d\alpha \qquad (8.11)$$

$$T = \frac{1}{2}\int_0^\alpha \rho A\left[\left(\frac{\partial v}{\partial t}\right)^2 + \left(\frac{\partial w}{\partial t}\right)^2\right]R\,d\alpha \qquad (8.12)$$

Expression (8.11) may be presented in a form that contains only tangential displacement

$$U_1 = \frac{1}{2}\int_0^\alpha \frac{EI(\alpha)}{R^3(\alpha)}\left(\frac{\partial^3 v}{\partial\alpha^3} + \frac{\partial v}{\partial\alpha} - \frac{1}{R}\frac{\partial R}{\partial\alpha}\left(\frac{\partial^2 v}{\partial\alpha^2} + v\right)\right)^2 d\alpha \qquad (8.13)$$

***Governing functional.***   Tangential and radial displacements are

$$\begin{aligned}
v(a, t) &= V_0(a)\exp(i\omega t)\\
w(a, t) &= W_0(a)\exp(i\omega t)
\end{aligned} \qquad (8.14)$$

where $V_0$ and $W_0$ are normal modes that are related by the formula $W_0 = dV_0/da$.

The maximum kinetic energy of lumped mass $M_0$, which is attached at $\alpha = \alpha_0$, is determined by the formula

$$T = \frac{1}{2}M_0\omega^2(V_0^2 + W_0^2)\bigg|_{\alpha=\alpha_0} \qquad (8.15)$$

where $\omega$ is the frequency of vibration.

Ritz's classical method requires that the functional

$$J[V_0] = U_{\max} - T_{\max} \qquad (8.16)$$

be a minimum. This leads to the governing functional

$$J[V_0] = \frac{E}{2}\int_{(\alpha)} \frac{I(\alpha)}{R^3(\alpha)}\left[V_0''' + V_0' - \frac{R'(\alpha)}{R(\alpha)}(V_0'' + V_0)\right]^2 d\alpha$$

$$- \frac{\rho\omega^2}{2}\int_{(\alpha)} A(\alpha)R(\alpha)(V_0'^2 + V_0^2)d\alpha - \frac{M}{2}\omega^2(V_0'^2 + V_0^2)\bigg|_{\alpha=0} \qquad (8.17)$$

where $V_0$ = tangential displacement amplitude
  $R(\alpha)$ = radius of curvature at any arbitrary point
  $A(\alpha)$ = cross-sectional area
  $I(\alpha)$ = second moment of inertia of cross-sectional area

The last term in the equation (8.17) takes into account a lumped mass $M_0$ at $\alpha = \alpha_0$.

## 8.2  ELASTICALLY CLAMPED UNIFORM CIRCULAR ARCHES

This section provides eigenvalues for uniform circular arches with elastic supports.

A circular arch with constant cross-section and elastic supports is shown in Fig. 8.4. Here $m$ = constant distributed mass

$R$ = radius of the arch

$2\alpha$ = angle of opening

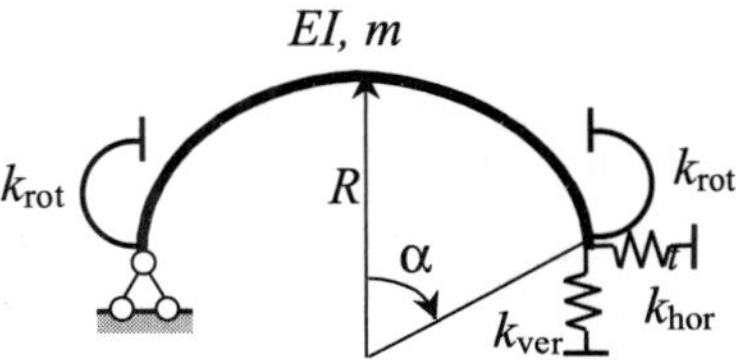


**FIGURE 8.4.**  Clamped uniform circular arch with various types of elastic supports.

*Assumptions:*

(1) Axis of arch is inextensible.

(2) Shear effects and rotary inertia can be neglected.

(3) Cross-section is small in comparison with the radius of an arch.

The non-dimensional rotational stiffness parameter and the vertical and horizontal stiffnesses are as follows

$$\beta_r = \frac{k_{\mathrm{rot}}R}{EI}$$

$$\beta_v = \frac{k_{\mathrm{vert}}R^3}{EI}$$

$$\beta_h = \frac{k_{\mathrm{hor}}R^3}{EI}$$

The mathematical model, numerical procedure and results (Table 8.2) are obtained and discussed by De Rosa (1991). The square of frequency vibration is

$$\omega^2 = \frac{\lambda EI}{mR^4} \tag{8.18}$$

Table 8.2(a)–(e) present the frequency parameter $\lambda$ for different types of elastic supports.

*Type 1. Circle.*   $(\beta_{\text{vert}} = \infty,\ \beta_{\text{hor}} = \infty)$

$$\beta_r = \frac{k_{\text{rot}} R}{EI}$$



*Type 2. Circle.*   $(\beta_{\text{rot}} = \infty,\ \beta_{\text{vert}} = \infty)$

$$\beta_h = \frac{k_{\text{hor}} R^3}{EI}$$



*Type 3. Circle.*   $(\beta_{\text{rot}} = \infty,\ \beta_{\text{vert}} = 0)$

$$\beta_h = \frac{k_{\text{hor}} R^3}{EI}$$



*Type 4. Circle.*   $(\beta_{\text{rot}} = \infty,\ \beta_{\text{hor}} = \infty)$

$$\beta_v = \frac{k_{\text{ver}} R^3}{EI}$$



*Type 5. Circle.*   $(\beta_{\text{rot}} = \infty,\ \beta_{\text{hor}} = 0)$

$$\beta_v = \frac{k_{\text{vert}} R^3}{EI}$$



Difference between types 2 and 3 is: type 3 allows vertical displacement on the right support, while in type 2 such displacement is impossible.

**TABLE 8.2(a).**   Type 1. Non-dimensional frequency parameter $\lambda$ for different values of rotational stiffness

| $\alpha° \backslash \beta_r$ | 0 | 6 | 12 | 18 | 24 | 100 | $10^7$ |
|---|---|---|---|---|---|---|---|
| 10 | 102930 | 122100 | 136981 | 148839 | 158496 | 209038 | 250560 |
| 20 | 6146.0 | 8225.1 | 9540.7 | 10443 | 11100 | 13710 | 15188 |
| 30 | 1126.3 | 1659.8 | 1944.7 | 2120.7 | 2240.0 | 2654.1 | 2854.0 |
| 40 | 321.55 | 515.89 | 605.75 | 657.17 | 690.37 | 796.68 | 843.63 |
| 50 | 115.75 | 201.40 | 236.37 | 255.23 | 267.00 | 302.69 | 317.58 |
| 60 | 47.842 | 90.388 | 105.95 | 113.95 | 118.82 | 133.03 | 138.73 |
| 70 | 21.599 | 44.541 | 52.139 | 55.908 | 58.144 | 64.500 | 66.973 |
| 80 | 10.332 | 23.468 | 27.441 | 29.345 | 30.462 | 33.555 | 34.731 |
| 90 | 5.1286 | 13.004 | 15.192 | 16.213 | 16.803 | 18.407 | 19.005 |

**TABLE 8.2(b).**  Type 2. Non-dimensional frequency parameter $\lambda$ for different values of horizontal axial stiffness

| $\beta_h\backslash\alpha°$ | 10 | 20 | 30 | 40 | 50 | 60 | 70 | 80 | 90 |
|---|---|---|---|---|---|---|---|---|---|
| 0 | 33001 | 1968.9 | 361.79 | 104.47 | 38.617 | 16.738 | 8.1345 | 4.3191 | 2.4639 |
| 1 | 33001 | 1969.4 | 362.39 | 105.18 | 39.379 | 17.512 | 8.8915 | 5.0391 | 3.1342 |
| 5 | 33002 | 1971.3 | 364.81 | 108.02 | 42.422 | 20.581 | 11.844 | 7.7586 | 5.5262 |
| 10 | 33003 | 1973.4 | 367.84 | 111.57 | 46.207 | 24.352 | 15.363 | 10.803 | 7.9228 |
| 50 | 33012 | 1991.2 | 392.03 | 139.75 | 75.655 | 51.650 | 36.615 | 24.260 | 15.219 |
| 100 | 33024 | 2013.4 | 422.21 | 174.55 | 110.18 | 77.918 | 49.682 | 29.225 | 17.050 |
| 500 | 33119 | 2191.1 | 661.30 | 430.66 | 264.48 | 127.91 | 63.808 | 33.658 | 18.613 |
| 1000 | 33237 | 2412.9 | 953.71 | 658.16 | 297.14 | 133.76 | 65.434 | 34.199 | 18.809 |
| 5000 | 34182 | 4170.7 | 2613.1 | 828.90 | 314.36 | 137.80 | 66.673 | 34.626 | 18.966 |
| $10^4$ | 35362 | 6322.6 | 2801.8 | 837.01 | 316.02 | 138.27 | 66.824 | 34.678 | 18.986 |
| $10^5$ | 56501 | 15149 | 2850.7 | 843.02 | 317.43 | 138.68 | 66.958 | 34.726 | 19.003 |
| $10^{10}$ | 250565 | 15188 | 2854.0 | 843.63 | 317.58 | 138.73 | 66.973 | 34.731 | 19.005 |

**TABLE 8.2(c).**  Type 3. Non-dimensional frequency parameter $\lambda$ for different values of horizontal axial stiffness

| $\beta_h\backslash\alpha°$ | 10 | 20 | 30 | 40 | 50 | 60 | 70 | 80 | 90 |
|---|---|---|---|---|---|---|---|---|---|
| 0 | 2047.0 | 129.57 | 26.142 | 8.5196 | 3.6246 | 1.8308 | 1.0436 | 0.6514 | 0.4365 |
| 1 | 2047.0 | 129.57 | 26.143 | 8.5214 | 3.6269 | 1.8336 | 1.0469 | 0.6551 | 0.4404 |
| 5 | 2047.0 | 129.58 | 26.148 | 8.5284 | 3.6354 | 1.8427 | 1.0556 | 0.6625 | 0.4464 |
| 10 | 2047.0 | 129.58 | 26.155 | 8.5367 | 3.6442 | 1.8506 | 1.0614 | 0.6634 | 0.4487 |
| 50 | 2047.1 | 129.62 | 26.204 | 8.5875 | 3.6826 | 1.8730 | 1.0729 | 0.6719 | 0.4515 |
| 100 | 2047.1 | 129.66 | 26.256 | 8.6273 | 3.7013 | 1.8800 | 1.0754 | 0.6729 | 0.4519 |
| 500 | 2047.3 | 129.97 | 26.503 | 8.7222 | 3.7280 | 1.8875 | 1.0778 | 0.6738 | 0.4523 |
| 1000 | 2047.5 | 130.29 | 26.636 | 8.7473 | 3.7328 | 1.8886 | 1.0781 | 0.6739 | 0.4523 |
| 5000 | 2049.1 | 131.62 | 26.843 | 8.7723 | 3.7368 | 1.8895 | 1.0784 | 0.6740 | 0.4524 |
| $10^5$ | 2072.3 | 133.24 | 26.920 | 8.7790 | 3.7378 | 1.8897 | 1.0785 | 0.6740 | 0.4524 |
| $10^8$ | 2107.0 | 133.40 | 26.924 | 8.7794 | 3.7379 | 1.8897 | 1.0785 | 0.6740 | 0.4524 |

**TABLE 8.2(d).**  Type 4. Non-dimensional frequency parameter $\lambda$ for different values of vertical axial stiffness

| $\beta_v\backslash\alpha°$ | 10 | 20 | 30 | 40 | 50 | 60 | 70 | 80 | 90 |
|---|---|---|---|---|---|---|---|---|---|
| 0 | 2107.0 | 133.40 | 26.924 | 8.7795 | 3.7379 | 1.8897 | 1.0785 | 0.6740 | 0.452 |
| 1 | 2114.4 | 137.09 | 29.363 | 10.586 | 5.1593 | 3.0506 | 2.0496 | 1.4996 | 1.161 |
| 5 | 2144.1 | 151.85 | 39.113 | 17.804 | 10.839 | 7.6888 | 5.9306 | 4.7969 | 3.981 |
| 10 | 2181.2 | 170.30 | 51.290 | 26.811 | 17.923 | 13.474 | 10.774 | 8.9062 | 7.457 |
| 50 | 2478.0 | 317.62 | 148.24 | 98.224 | 73.843 | 59.015 | 48.878 | 34.628 | 18.768 |
| 100 | 2848.8 | 501.16 | 268.20 | 185.55 | 140.80 | 109.78 | 66.495 | 34.716 | 18.927 |
| 500 | 5808.3 | 1942.4 | 1158.7 | 683.79 | 307.12 | 137.91 | 66.936 | 34.729 | 18.993 |
| 1000 | 9490.2 | 3664.9 | 1972.1 | 795.31 | 313.59 | 138.37 | 66.957 | 34.730 | 18.999 |
| 5000 | 38177 | 12043 | 2763.6 | 836.98 | 316.92 | 138.66 | 66.970 | 34.731 | 19.004 |
| $10^5$ | 233294 | 15097 | 2859.2 | 843.32 | 317.55 | 138.72 | 66.973 | 34.731 | 19.005 |

**TABLE 8.2(e).**    Type 5. Non-dimensional frequency parameter $\lambda$ for different values of vertical axial stiffness

| $\beta_v \backslash \alpha^\circ$ | 10 | 20 | 30 | 40 | 50 | 60 | 70 | 80 | 90 |
|---|---|---|---|---|---|---|---|---|---|
| 0 | 2047.0 | 129.57 | 26.142 | 8.5196 | 3.6246 | 1.8308 | 1.0436 | 0.6514 | 0.4365 |
| 1 | 2054.0 | 133.05 | 28.431 | 10.206 | 4.9411 | 2.8927 | 1.9163 | 1.3747 | 1.0352 |
| 5 | 2082.0 | 146.89 | 37.454 | 16.716 | 9.8353 | 6.5877 | 4.6029 | 3.1567 | 2.0823 |
| 10 | 2116.9 | 164.05 | 48.425 | 24.305 | 15.070 | 9.8706 | 6.2487 | 3.7871 | 2.2949 |
| 50 | 2395.0 | 295.81 | 123.89 | 64.293 | 31.795 | 15.406 | 7.8178 | 4.2291 | 2.4341 |
| 100 | 2739.1 | 446.74 | 190.03 | 82.912 | 35.358 | 16.104 | 7.9817 | 4.2752 | 2.4492 |
| 500 | 5352.1 | 1197.1 | 320.22 | 100.38 | 38.005 | 16.616 | 8.1048 | 4.3105 | 2.4610 |
| 1000 | 8277.0 | 1537.7 | 341.29 | 102.46 | 38.313 | 16.677 | 8.1197 | 4.3148 | 2.4624 |
| 5000 | 21438 | 1881.5 | 357.77 | 104.08 | 38.556 | 16.726 | 8.1315 | 4.3182 | 2.4636 |
| $10^4$ | 32355 | 1964.6 | 361.59 | 104.45 | 38.614 | 16.737 | 8.1343 | 4.3190 | 2.4638 |

*Type 6. Elastically cantilevered circular arch*

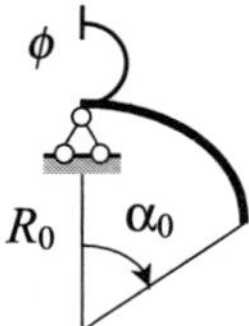


The parameter of the flexibility, $\phi$, of the elastic support defines the angle of rotation in accordance with the relation $\theta = \phi M$. The fundamental frequency of vibration may be calculated by the formula

$$\omega = \frac{\lambda_1}{(R_0\alpha_0)^2} \sqrt{\frac{EI_0}{\rho A_0}} \tag{8.19}$$

Upper and lower bounds for the fundamental frequency coefficient, $\lambda_1$, for different angles $\alpha_0^\circ$ and dimensionless parameter, $\phi^* = \dfrac{\phi EI_0}{R_0\alpha_0}$ (where $\alpha_0$ is in radians), are presented in Table 8.3 (Laura *et al.*, 1987).

Upper and lower bounds for the fundamental frequency coefficients are determined by using the Rayleigh–Ritz and modified Dunkerley's methods, respectively.

## 8.3  TWO-HINGED UNIFORM ARCHES

### 8.3.1  Circular arch. In-plane vibrations

A two-hinged circular uniform arch with central angle $2\alpha_0$ is presented in Fig. 8.5(a).

**TABLE 8.3.**   Type 6. Elastically cantilevered uniform circular arch: upper and lower bounds of fundamental frequency parameter

| $\alpha_0^\circ$ | Bound | $\phi^* = 0.10$ | 1.0 | 5.0 | 10.0 | 50.0 |
|---|---|---|---|---|---|---|
| 10 | Upper | 2.97 | 1.55 | 0.75 | 0.54 | 0.24 |
|    | Lower | 2.93 | 1.55 | 0.75 | 0.54 | 0.24 |
| 30 | Upper | 2.98 | 1.56 | 0.76 | 0.54 | 0.24 |
|    | Lower | 2.64 | 1.56 | 0.76 | 0.54 | 0.24 |
| 60 | Upper | 3.04 | 1.60 | 0.77 | 0.55 | 0.25 |
|    | Lower | 3.00 | 1.59 | 0.77 | 0.55 | 0.25 |
| 90 | Upper | 3.14 | 1.66 | 0.80 | 0.57 | 0.26 |
|    | Lower | 3.09 | 1.64 | 0.80 | 0.57 | 0.26 |

The frequency of vibration of an arch is calculated by the formula

$$\omega = \frac{\beta(\beta^2 - 1)}{\sqrt{\lambda(\beta^2 + 3)}} \tag{8.20}$$

where $\lambda = \dfrac{mR^4}{EI}$, $\beta = \dfrac{\pi}{\alpha_0}$.

The frequency of vibration, $\omega$, of an arch in terms of the frequency of vibration, $\omega_0$, of the simply supported 'reference' beam (Bezukhov *et al.*, 1969) is

$$\omega = \omega_0 \sqrt{\frac{1 - n^2}{1 + 3n^2}}, \quad n = \frac{\alpha_0}{\pi} \tag{8.21}$$

The frequencies of vibration, which correspond to fundamental (antisymmetric) and second (symmetric) modes, may be calculated as follows

$$\omega = \frac{\lambda}{4R^2\alpha_0^2} \sqrt{\frac{EI}{\rho A}} \tag{8.22}$$

where $\alpha_0$ is in radians.

These results are obtained under the following assumptions: rotation of the cross sections of the arch are neglected and radius $R$ remains constant.

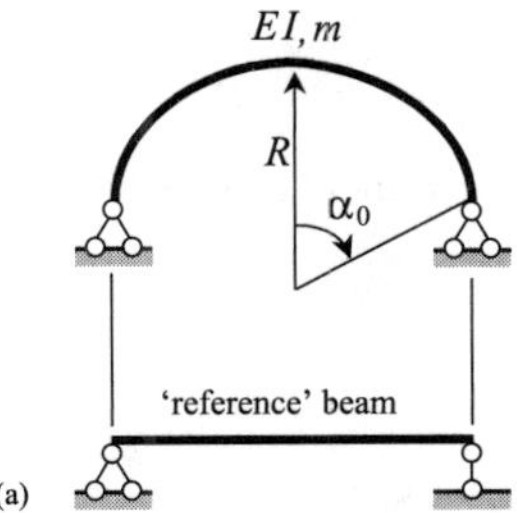


**FIGURE 8.5.**   (a) Two-hinged uniform circular arch and end-supported 'reference' beam.

**TABLE 8.4.** Two-hinged uniform circular arch: frequency parameters for first and second modes of vibration

| $2\alpha_0$ (degrees) | Antisymmetric mode $\lambda_1$ | Symmetric mode $\lambda_2$ |
|---|---|---|
| 10 | 39.40 | 84.25 |
| 20 | 39.18 | 84.09 |
| 30 | 38.80 | 83.81 |
| 40 | 38.29 | 83.43 |

The corresponding frequency coefficients $\lambda_1$ and $\lambda_2$ are presented in Table 8.4 (Gutierrez *et al.*, 1989).

The fundamental frequency of vibration corresponds to the first antisymmetric mode of vibration. Frequency coefficients $\lambda_1$ and $\lambda_2$, which are obtained using different numerical methods, are compared in Gutierrez *et al.* (1989).

### 8.3.2  Circular arch with constant radial load. Vibration in the plane

A circular arch of uniform cross-section under constant hydrostatic pressure $X$ is presented in Fig. 8.5(b).

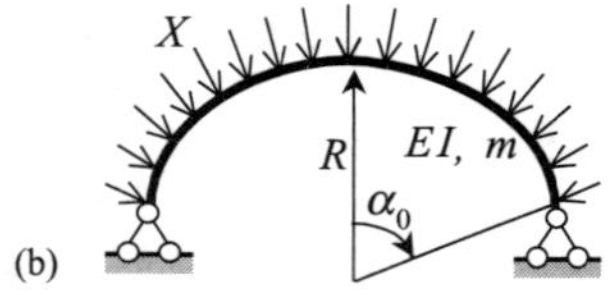


**FIGURE 8.5** (b) Two-hinged uniform circular arch under constant hydrostatic pressure.

The fundamental frequency of vibration of an arch is

$$\omega = \beta \sqrt{\frac{(\beta^2 - 1)(\beta^2 - 1 - c^2)}{\lambda(\beta^2 + 3)}} \tag{8.23}$$

where the non-dimensional parameters are

$$\lambda = \frac{mR^4}{EI}, \quad c^2 = \frac{XR^3}{EI}, \quad \beta = \frac{\pi}{\alpha_0}$$

The inertial force due to the angle of rotation is neglected (Bezukhov *et al.*, 1969).

Critical pressure

$$X \le \frac{EI}{R^3}(\beta^2 - 1)$$

These results are based on the assumptions, which are adopted in Section 8.3.1.

***Special case.***  Hydrostatic pressure $X = 0$. In this case expression (8.23) transforms to expression (8.20).

### 8.3.3  Shallow parabolic arch. In-plane vibration

A shallow parabolic arch carrying a uniformly distributed load $q$ including self-weight of the arch is presented in Fig. 8.6.

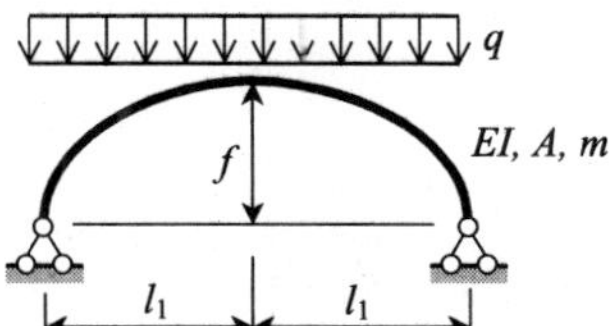


**FIGURE 8.6.**   Shallow parabolic arch carrying a uniformly distributed load.

The fundamental frequency of vibration of an arch

$$\omega^2 = \frac{\pi^4(1 - 0.5653k^2)}{1 + 4.1277k^2 + 1.6910k^4}\frac{EIg}{ql_1^4} \tag{8.24}$$

where $k = f/l_1$.

This expression has been obtained by the Ritz method (Federhofer, 1934) and yields good results if $f \leq 0.3l$, $l = 2l_1$ (Morgaevsky, 1940; Bezukhov *et al.*, 1969).

### 8.3.4  Non-circular arches

Figure 8.7 shows a design diagram of a symmetric arch with different equations of the neutral line. Geometric relationships for different curves are presented in Table 8.1.

The fundamental frequency of vibration for arches corresponds to the first antisymmetric mode of vibration.

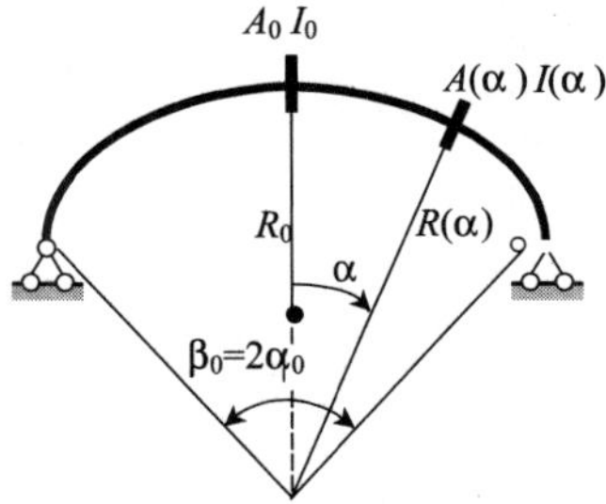


**FIGURE 8.7.**   Two-hinged non-circular arch.

The frequency of vibration of uniform symmetric arches with different equations of the neutral line may be presented by the formula

$$\omega = \frac{\lambda}{(R_0\beta_0)^2}\sqrt{\frac{EI_0}{\rho A_0}}$$ (8.25)

where $\beta_0 = 2\alpha_0 = $ the angle of opening of arch
$R_0 = $ radius at the axis of symmetry, which corresponds to $\alpha = 0$
$A_0 = $ cross-section area at $\alpha = 0$
$I_0 = $ second moment of inertia of the cross-section area at $\alpha = 0$

The first and second frequency parameters, $\lambda_1$ and $\lambda_2$, as a function of $\beta_0$ for different arch shapes, are tabulated in Table 8.5. These parameters correspond to fundamental (antisymmetric) and lowest symmetric modes, respectively. For their determination the governing functional (8.17, $M = 0$) has been used (Gutierrez *et al.*, 1989).

**TABLE 8.5.**  Two-hinged uniform arches with different shapes of neutral line: first and second frequency parameters

| Arch's shape | $\beta_0$ (degrees) | Antisymmetric mode $\lambda_1$ | Symmetric mode $\lambda_2$ |
|---|---|---|---|
| Parabola | 10 | 39.10 | 83.68 |
|  | 20 | 37.98 | 81.81 |
|  | 30 | 36.17 | 78.72 |
|  | 40 | 33.71 | 74.45 |
| Catenary | 10 | 39.20 | 83.87 |
|  | 20 | 38.38 | 82.57 |
|  | 30 | 37.03 | 80.40 |
|  | 40 | 35.19 | 77.39 |
| Spiral | 10 | 39.30 | 84.06 |
|  | 20 | 38.77 | 83.33 |
|  | 30 | 37.91 | 82.10 |
|  | 40 | 36.72 | 80.39 |
| Circle | 10 | 39.40 | 84.26 |
|  | 20 | 39.17 | 84.10 |
|  | 30 | 38.80 | 83.82 |
|  | 40 | 38.28 | 83.44 |
| Cycloid | 10 | 39.50 | 84.38 |
|  | 20 | 39.57 | 84.79 |
|  | 30 | 39.70 | 85.48 |
|  | 40 | 39.88 | 86.46 |

### 8.3.5  Circular arch. Out-of-plane vibration of arch

The $a-a$ axis of hinge supports, is horizontal and located in the plane of the arch (Fig. 8.8). The bending stiffness $EI$ and mass $m$ per unit length are constant.

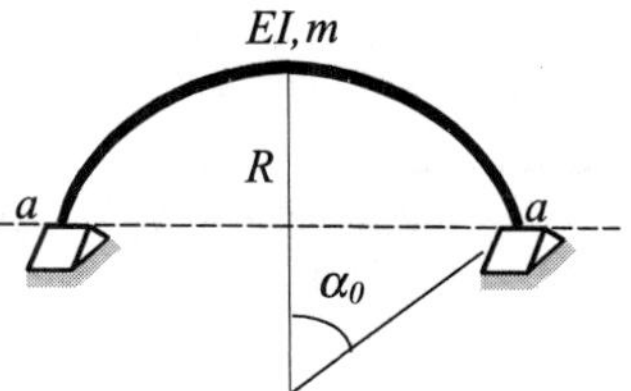

**FIGURE 8.8.**   Design diagram of uniform circular arch for out-of-plane vibration.

The antisymmetric frequency of vibration is determined by the following formula (Bezukhov *et al.*, 1969)

$$\omega = \frac{k\pi(\alpha_0^2 - k^2\pi^2)}{\alpha_0^2 R^2} \sqrt{\frac{EI}{m(k^2\pi^2 + \alpha_0^2\chi)}} \qquad (8.26)$$

The symmetric *k*th frequency of vibration

$$\omega_k = \frac{(2k-1)\pi[4\alpha_0^2 - (2k-1)^2\pi^2]}{4\alpha_0^2 R^2} \sqrt{\frac{EI}{m}\frac{1}{(2k-1)^2\pi^2 + 4\alpha_0^2\chi}} \qquad (8.27)$$

where $\chi = \dfrac{EI}{GJ}$, $I$ and $J$ are the axial and polar moments of inertia of the cross-sectional area; and $k = 1, 2, 3, \ldots$

## 8.4   *HINGELESS UNIFORM ARCHES*

### 8.4.1   Circular arch

The natural frequency of vibration of a circular uniform arch with clamped ends may be calculated by the formula

$$\omega = \frac{\lambda}{(R\beta_0)^2} \sqrt{\frac{EI}{\rho A}}$$

where $\beta_0$ is the angle of opening (radians).

Coefficients $\lambda_1$ and $\lambda_2$ are listed in Table 8.6 (Gutierrez *et al.*, 1989). Comparisons of frequency coefficients, which are obtained by different numerical methods, can be found in Gutierrez *et al.* (1989).

### 8.4.2   Circular arch with constant radial load

The uniform circular arch with clamped ends carrying a constant radial load is shown in Fig. 8.9.

**TABLE 8.6.** Hingeless uniform circular arch: frequency parameters for first and second mode of vibration

| $\beta_0$ (degrees) | Antisymmetric mode $\lambda_1$ | Symmetric mode $\lambda_2$ |
|---|---|---|
| 10 | 61.59 | 110.94 |
| 20 | 61.35 | 110.78 |
| 30 | 60.96 | 110.51 |
| 40 | 60.42 | 110.13 |

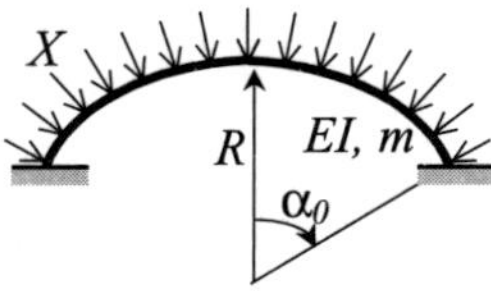


**FIGURE 8.9.**   Clamped circular arch with radial load.

The differential equation of in-plane vibration is discussed in Section 8.1.3. The fundamental frequency of vibration is obtained by the integration of the differential equation (Bezukhov *et al.*, 1969)

$$\omega = \frac{3}{4}\beta\sqrt{\frac{41\beta^4 - 20\beta^2(2+c^2) + 16(1+c^2)}{9\beta^2 + 20}}\frac{1}{R^2}\sqrt{\frac{EI}{m}} \tag{8.28}$$

where $\beta = \dfrac{180^\circ}{\alpha_0{}^\circ}$,   $c^2 = \dfrac{XR^3}{EI}$.

Vibration is unstable if $41\beta^4 - 20\beta^2(2+c^2) + 16(1+c^2) < 0$.

### 8.4.3  Non-circular arches

Notations for hingeless arches with different shapes are presented in Fig. 8.10.

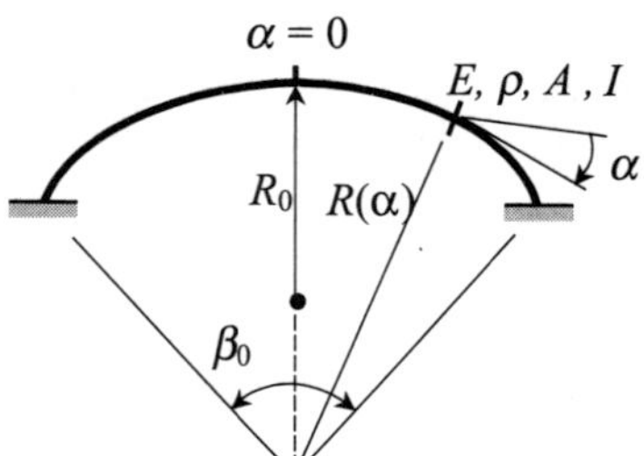


**FIGURE 8.10.**  Hingeless arches with different shapes.

The frequency of vibration of uniform symmetric arches with different shapes may be calculated by the formula

$$\omega = \frac{\lambda}{(R_0\beta_0)^2}\sqrt{\frac{EI}{\rho A}}$$

where the central angle of the arch is $\beta_0$ (in radians); the radius of curvature $R_0$ is shown at the axis of symmetry; the cross-section area, $A$, and the second moment of inertia, $I$, are constant (Fig. 8.10).

The governing functional is presented by expression (8.17). First and second frequency parameters $\lambda_1$ and $\lambda_2$ are listed in Table 8.7. These parameters correspond to fundamental (antisymmetric) and lowest symmetric modes, respectively. For their determination, a finite element method has been used (Gutierrez *et al.*, 1989. This article contains also frequency coefficients that are obtained by using polynomial approximations and the Ritz method.)

**TABLE 8.7.** Hingeless uniform arches with different shapes of neutral line: frequency parameters for first and second modes of vibration

| Arch's shape | $\beta_0$ (degrees) | $\lambda_1$ | $\lambda_2$ |
|---|---|---|---|
| Parabola | 10 | 61.12 | 110.18 |
|  | 20 | 59.49 | 107.81 |
|  | 30 | 56.82 | 103.87 |
|  | 40 | 53.19 | 98.43 |
| Catenary | 10 | 61.28 | 110.44 |
|  | 20 | 60.11 | 108.79 |
|  | 30 | 58.18 | 106.06 |
|  | 40 | 55.54 | 102.26 |
| Spiral | 10 | 61.43 | 110.69 |
|  | 20 | 60.73 | 109.77 |
|  | 30 | 59.56 | 108.27 |
|  | 40 | 57.95 | 106.16 |
| Circle | 10 | 61.59 | 110.93 |
|  | 20 | 61.35 | 110.78 |
|  | 30 | 60.96 | 110.51 |
|  | 40 | 60.42 | 110.13 |
| Cycloid | 10 | 61.75 | 111.04 |
|  | 20 | 61.98 | 111.64 |
|  | 30 | 62.37 | 112.63 |
|  | 40 | 62.95 | 114.04 |

## 8.5 CANTILEVERED UNIFORM CIRCULAR ARCH WITH A TIP MASS

The uniform circular arch with a clamped support at one end and a tip mass at the other is presented in Fig. 8.11. Parameters $A$, $R$ and $EI$ are constant.

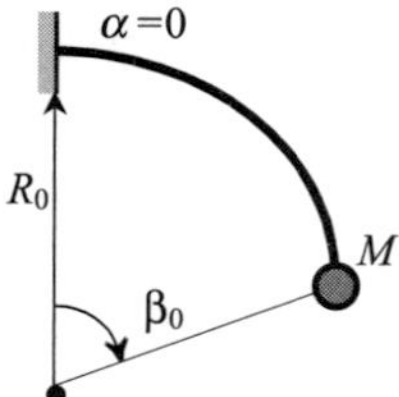

**FIGURE 8.11.**   Cantilevered uniform circular arch with a tip mass.

The fundamental frequency of the in-plane transverse vibration in the case of a constant cross-section area, may be calculated by the formula

$$\omega = \frac{\lambda_1}{(R\beta_0)^2} \sqrt{\frac{EI}{\rho A}}$$

Upper and lower bounds for the fundamental frequency coefficient, $\lambda_1$, in terms of the non-dimensional parameter $M^* = \dfrac{M}{\rho A R \beta_0}$ and angle $\beta_0$ are listed in Table 8.8.

Upper bounds are determined using the Rayleigh–Ritz method (the governing functional is presented by formula (8.17) for the case when $R = $ constant); lower bounds are obtained using the Dunkerley method (Laura *et al.*, 1987).

**TABLE 8.8.**   Cantilevered uniform circular arch with a trip mass: upper and lower bounds for fundamental frequency parameter $\lambda$

| $\beta_0$ | Bounds | $M^* = 0.0$ | 0.20 | 0.40 | 0.60 | 0.80 | 1.00 |
|---|---|---|---|---|---|---|---|
| 5 | Upper | 3.517 | — | — | — | — | 1.558 |
|   | Lower | 3.465 | — | — | — | — | 1.550 |
| 10 | Upper | 3.51 | 2.61 | 2.16 | 1.89 | 1.70 | 1.559 |
|   | Lower | 3.466 | 2.58 | 2.15 | 1.88 | 1.69 | 1.551 |
| 20 | Upper | 3.52 | — | — | — | — | 1.563 |
|   | Lower | 3.473 | — | — | — | — | 1.554 |
| 30 | Upper | 3.53 | 2.63 | 2.18 | 1.90 | 1.71 | 1.569 |
|   | Lower | 3.483 | 2.59 | 2.16 | 1.89 | 1.70 | 1.561 |
| 40 | Upper | 3.55 | — | — | — | — | 1.579 |
|   | Lower | 3.498 | — | — | — | — | 1.570 |
| 50 | Upper | 3.57 | 2.66 | 2.21 | 1.93 | 1.73 | 1.591 |
|   | Lower | 3.517 | 2.62 | 2.19 | 1.91 | 1.72 | 1.582 |
| 60 | Upper | 3.59 | — | — | — | — | 1.608 |
|   | Lower | 3.54 | — | — | — | — | 1.597 |
| 70 | Upper | 3.62 | 2.71 | 2.25 | 1.97 | 1.77 | 1.627 |
|   | Lower | 3.568 | 2.67 | 2.23 | 1.95 | 1.76 | 1.615 |
| 80 | Upper | 3.66 | — | — | — | — | 1.652 |
|   | Lower | 3.60 | — | — | — | — | 1.635 |
| 90 | Upper | 3.709 | 2.78 | 2.32 | 2.03 | 1.83 | 1.68 |
|   | Lower | 3.637 | 2.74 | 2.29 | 2.00 | 1.80 | 1.659 |
| 180 | Upper | 4.435 | 3.48 | 2.96 | 2.62 | 2.38 | 2.192 |
|   | Lower | 4.203 | 3.27 | 2.77 | 2.44 | 2.21 | 2.041 |
| 270 | Upper | 5.84 | 4.90 | 4.18 | 3.71 | 3.37 | 3.092 |
|   | Lower | 5.286 | 4.33 | 3.76 | 3.37 | 3.08 | 2.854 |

## 8.6  CANTILEVERED NON-CIRCULAR ARCHES WITH A TIP MASS

Figure 8.12 presents a non-circular arch of non-uniform cross-section, rigidly clamped at one end and carrying a concentrated mass at the other. The tangential displacement at any point of an arch is $v = V_0(\alpha) \exp(i\omega t)$.

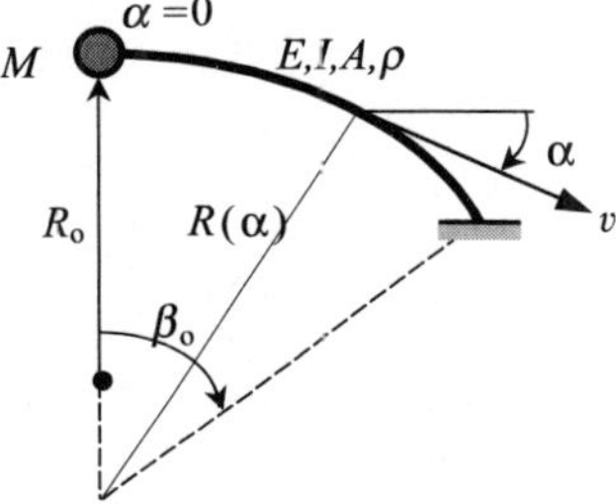


**FIGURE 8.12.** Cantilevered non-circular non-uniform arch with a tip mass.

The governing functional is presented by the formula

$$J[V_0] = \frac{E}{2} \int_0^{\beta_0} \frac{I(\alpha)}{R^3(\alpha)} \left[ V_0''' + V_0' - \frac{R'(\alpha)}{R(\alpha)}(V_0'' + V_0) \right]^2 d\alpha$$

$$- \frac{\rho\omega^2}{2} \int_0^{\beta_0} A(\alpha)R(\alpha) \cdot (V_0'^2 + V_0^2) d\alpha - \frac{M}{2}\omega^2(V_0'^2 + V_0^2)\Big|_{\alpha=0}$$

where $V_0$ is the tangential displacement amplitude, $V' = dV/d\alpha$.

### 8.6.1  Arch of uniform cross-section

A non-circular arch of uniform cross-section, rigidly clamped at one end and carrying a concentrated mass at the other, is presented in Fig. 8.12. The fundamental frequency of the in-plane transverse vibration of elastic arches is

$$\omega_1 = \frac{\lambda_1}{(R_0\beta_0)^2} \sqrt{\frac{EI}{\rho A}}$$

where $\rho$ = the mass density of the arch material

$\beta_0$ = central angle (radians)

$R_0$ = radius of an arch at $\alpha = 0$

The fundamental frequency coefficients, $\lambda_1$, in terms of the non-dimensional mass $M^* = \dfrac{M}{\rho A R_0 \beta_0}$ and angle $\beta_0$ for arches with different shapes are listed in Table 8.9. The Rayleigh optimization method has been used to obtain the results (Rossi *et al.*, 1989).

The above mentioned reference contains a comparison of the fundamental frequency coefficients $\lambda_1$, which are obtained using different numerical methods.

**TABLE 8.9.** Cantilevered uniform non-circular arch with a tip mass: fundamental frequency parameter $\lambda$

| Arch's shapes | $\beta°\backslash M^*$ | 0.0 | 0.2 | 0.4 | 0.6 | 0.8 | 1.0 |
|---|---|---|---|---|---|---|---|
| Parabola | 10 | 3.41 | 2.54 | 2.11 | 1.84 | 1.66 | 1.52 |
|  | 20 | 3.11 | 2.34 | 1.95 | 1.71 | 1.54 | 1.41 |
|  | 30 | 2.65 | 2.01 | 1.69 | 1.49 | 1.34 | 1.23 |
|  | 40 | 2.02 | 1.60 | 1.36 | 1.20 | 1.09 | 1.00 |
| Catenary | 10 | 3.44 | 2.56 | 2.13 | 1.86 | 1.67 | 1.53 |
|  | 20 | 3.24 | 2.43 | 2.02 | 1.77 | 1.59 | 1.46 |
|  | 30 | 2.90 | 2.21 | 1.85 | 1.62 | 1.46 | 1.34 |
|  | 40 | 2.46 | 1.90 | 1.61 | 1.42 | 1.28 | 1.18 |
| Spiral | 10 | 3.48 | 2.59 | 2.15 | 1.87 | 1.68 | 1.54 |
|  | 20 | 3.38 | 2.52 | 2.10 | 1.83 | 1.65 | 1.51 |
|  | 30 | 3.21 | 2.41 | 2.01 | 1.76 | 1.58 | 1.45 |
|  | 40 | 2.97 | 2.26 | 1.89 | 1.65 | 1.49 | 1.37 |
| Cycloid | 10 | 3.55 | 2.63 | 2.19 | 1.91 | 1.71 | 1.57 |
|  | 20 | 3.67 | 2.71 | 2.25 | 1.96 | 1.76 | 1.61 |
|  | 30 | 3.87 | 2.85 | 2.36 | 2.05 | 1.84 | 1.69 |
|  | 40 | 4.19 | 3.06 | 2.52 | 2.19 | 1.97 | 1.80 |

## 8.6.2  Arch with a discontinuously varying cross-section

A non-circular arch of non-uniform cross section, rigidly clamped at one end and carrying a concentrated mass at the other is presented in Fig. 8.13.

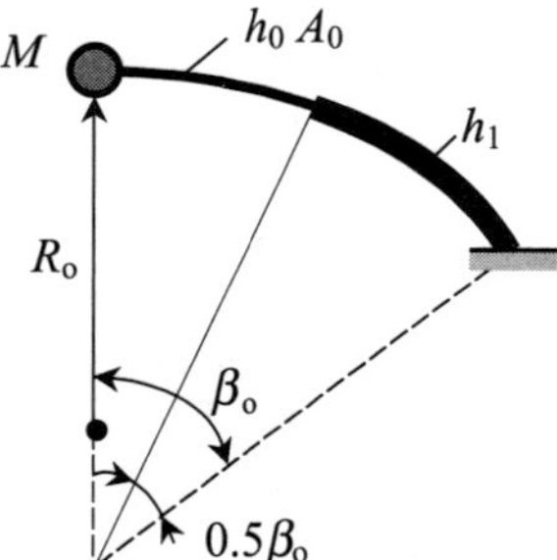


**FIGURE 8.13.**  Cantilevered non-circular arch of discontinuously varying cross-section.

Expressions for tangential and radial displacements $v(\alpha, t)$, $w(\alpha, t)$ and the corresponding governing functional are presented in Section 8.1.3.

The fundamental frequency in-plane transverse vibration is

$$\omega_1 = \frac{\lambda_1}{(R_0\beta_0)^2}\sqrt{\frac{EI_0}{\rho A_0}}$$

where $\beta_0$ (radians) is the central angle. The frequency coefficient, $\lambda_1$, for different non-dimensional mass, $M^* = \dfrac{M}{\rho A_0 R_0 \beta_0}$, geometry ratio, $h_1/h_0$, and various equations of the neutral line of the arches are presented in Tables 8.10(a) and (b). The finite element method has been used to obtain numerical results (Rossi *et al.*, 1989. This article contains results that are obtained using a polynomial approximation (a one-term solution and a three-term solution) and an optimization approach.)

**TABLE 8.10.**  (a) Cantilevered non-circular arches of discontinuously varying cross-section with a tip mass: fundamental frequency parameter $\lambda$ for $h_1/h_0 = 1.25$

| Arch's shapes | $\beta°\backslash M^*$ | 0.0 | 0.2 | 0.4 | 0.6 | 0.8 | 1.0 |
|---|---|---|---|---|---|---|---|
| Parabola | 10 | 4.63 | 3.42 | 2.83 | 2.47 | 2.21 | 2.02 |
|  | 20 | 4.23 | 3.16 | 2.63 | 2.30 | 2.06 | 1.89 |
|  | 30 | 3.58 | 2.74 | 2.30 | 2.01 | 1.82 | 1.67 |
|  | 40 | 2.74 | 2.17 | 1.84 | 1.63 | 1.48 | 1.36 |
| Catenary | 10 | 4.68 | 3.45 | 2.85 | 2.49 | 2.23 | 2.04 |
|  | 20 | 4.41 | 3.28 | 2.72 | 2.37 | 2.13 | 1.95 |
|  | 30 | 3.96 | 2.99 | 2.50 | 2.19 | 1.97 | 1.80 |
|  | 40 | 3.36 | 2.59 | 2.18 | 1.92 | 1.73 | 1.59 |
| Spiral | 10 | 4.73 | 3.48 | 2.88 | 2.50 | 2.25 | 2.06 |
|  | 20 | 4.59 | 3.40 | 2.81 | 2.45 | 2.20 | 2.01 |
|  | 30 | 4.37 | 3.26 | 2.70 | 2.36 | 2.12 | 1.94 |
|  | 40 | 4.05 | 3.06 | 2.50 | 2.23 | 2.01 | 1.84 |
| Cycloid | 10 | 4.82 | 3.54 | 2.92 | 2.54 | 2.28 | 2.09 |
|  | 20 | 4.98 | 3.64 | 3.00 | 2.61 | 2.34 | 2.14 |
|  | 30 | 5.25 | 3.81 | 3.14 | 2.73 | 2.44 | 2.23 |
|  | 40 | 5.65 | 4.07 | 3.34 | 2.89 | 2.59 | 2.37 |

**TABLE 8.10.**  (b) Cantilevered non-circular arches of discontinuously varying cross-section with a tip mass: Fundamental frequency parameter $\lambda$ for $h_1/h_0 = 10/6$

| Arch's shapes | $\beta°\backslash M^*$ | 0.0 | 0.2 | 0.4 | 0.6 | 0.8 | 1.0 |
|---|---|---|---|---|---|---|---|
| Parabola | 10 | 6.67 | 4.80 | 3.93 | 3.40 | 3.04 | 2.78 |
|  | 20 | 6.13 | 4.48 | 3.69 | 3.20 | 2.87 | 2.62 |
|  | 30 | 5.24 | 3.94 | 3.27 | 2.86 | 2.57 | 2.35 |
|  | 40 | 4.04 | 3.17 | 2.68 | 2.36 | 2.13 | 1.96 |
| Catenary | 10 | 6.73 | 4.84 | 3.96 | 3.43 | 3.06 | 2.80 |
|  | 20 | 6.37 | 4.63 | 3.80 | 3.29 | 2.95 | 2.69 |
|  | 30 | 5.77 | 4.27 | 3.53 | 3.07 | 2.75 | 2.52 |
|  | 40 | 4.93 | 3.75 | 3.13 | 2.74 | 2.47 | 2.26 |
| Spiral | 10 | 6.79 | 4.87 | 3.98 | 3.45 | 3.08 | 2.81 |
|  | 20 | 6.62 | 4.77 | 3.91 | 3.39 | 3.03 | 2.77 |
|  | 30 | 6.32 | 4.60 | 3.78 | 3.28 | 2.94 | 2.68 |
|  | 40 | 5.90 | 4.35 | 3.59 | 3.13 | 2.80 | 2.56 |
| Cycloid | 10 | 6.92 | 4.94 | 4.04 | 3.49 | 3.12 | 2.85 |
|  | 20 | 7.12 | 5.07 | 4.13 | 3.57 | 3.19 | 2.91 |
|  | 30 | 7.47 | 5.27 | 4.29 | 3.70 | 3.31 | 3.02 |
|  | 40 | 7.99 | 5.58 | 4.52 | 3.90 | 3.47 | 3.17 |

Note: if parameter $M^*$ is fixed while $h^*$ is varying, then the fundamental frequency coefficient $\lambda_1$ for the cycloid arch as a function of parameter $h^* = h_1/h_0$ is presented in Table 8.11 (for $M^* = 1.0$). The finite element method has been used (Rossi *et al.*, 1989).

**TABLE 8.11.** Cantilevered cycloid arch of discontinuously varying cross-section with a tip mass: fundamental frequency parameter $\lambda$ for $M^* = 1$

| | $\beta°\backslash h^*$ | 1.30 | 1.40 | 1.50 | 1.60 |
|---|---|---|---|---|---|
| | 10 | 2.187 | 2.381 | 2.565 | 2.740 |
| Cycloid | 20 | 2.242 | 2.438 | 2.625 | 2.801 |
| | 30 | 2.336 | 2.536 | 2.726 | 2.904 |
| | 40 | 2.474 | 2.680 | 2.873 | 3.053 |

### 8.6.3  Arch of continuously varying cross-section

An arch of continuously varying cross-section, rigidly clamped at one end and carrying a concentrated mass $M$ at the other, is presented in Fig. 8.14.

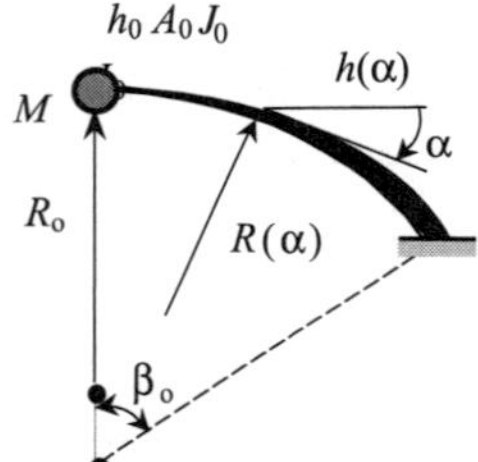


FIGURE 8.14. Cantilevered arch with continuously varying cross-section.

The height of the cross-section for any angle $\alpha$ is determined by the following formula

$$h(\alpha) = h_0\left(1 + \eta\frac{\alpha}{\beta_0}\right)$$

where $h_0$ is the height of cross-section at $\alpha = 0$; and the $\eta$ parameter, which represents the increase of the height of the cross-section at given $\alpha$, can be any number.

The fundamental frequency of vibration of an elastic arch with central angle $\beta_0$ (rad) may be calculated by the formula

$$\omega_1 = \frac{\lambda_1}{(R_0\beta_0)^2}\sqrt{\frac{EI_0}{\rho A_0}}$$

Frequency coefficients, $\lambda$, for various equations of the neutral line, opening angle, $\beta°$, parameter $\eta$ and non-dimensional mass, $M^* = \dfrac{M}{\rho A_0 R_0 \beta_0}$, are presented in Tables 8.12(a) and (b).

**TABLE 8.12.** (a) Cantilevered non-circular arches of discontinuously varying cross-section with a tip mass: Fundamental frequency parameter $\lambda$ for $\eta = 0.2$

| Arch's shapes | $\beta°\backslash M^*$ | 0.0 | 0.2 | 0.4 | 0.6 | 0.8 | 1.0 |
|---|---|---|---|---|---|---|---|
| Parabola | 10 | 4.18 | 3.13 | 2.60 | 2.27 | 2.04 | 1.87 |
| | 20 | 3.81 | 2.88 | 2.41 | 2.11 | 1.90 | 1.74 |
| | 30 | 3.23 | 2.49 | 2.10 | 1.85 | 1.67 | 1.53 |
| | 40 | 2.48 | 1.97 | 1.68 | 1.49 | 1.35 | 1.25 |
| Catenary | 10 | 4.22 | 3.15 | 2.62 | 2.29 | 2.06 | 1.88 |
| | 20 | 3.98 | 2.99 | 2.49 | 2.18 | 1.96 | 1.80 |
| | 30 | 3.58 | 2.73 | 2.29 | 2.01 | 1.81 | 1.66 |
| | 40 | 3.03 | 2.36 | 1.99 | 1.76 | 1.59 | 1.46 |
| Spiral | 10 | 4.27 | 3.18 | 2.64 | 2.31 | 2.07 | 1.90 |
| | 20 | 4.15 | 3.10 | 2.58 | 2.26 | 2.03 | 1.86 |
| | 30 | 3.94 | 2.97 | 2.48 | 2.17 | 1.95 | 1.79 |
| | 40 | 3.66 | 2.79 | 2.33 | 2.05 | 1.84 | 1.69 |
| Cycloid | 10 | 4.35 | 3.24 | 2.69 | 2.34 | 2.11 | 1.93 |
| | 20 | 4.49 | 3.33 | 2.76 | 2.41 | 2.16 | 1.98 |
| | 30 | 4.74 | 3.49 | 2.89 | 2.52 | 2.26 | 2.07 |
| | 40 | 5.11 | 3.74 | 3.08 | 2.68 | 2.40 | 2.20 |

**TABLE 8.12.** (b) Cantilevered non-circular arches of discontinuously varying cross-section with a tip mass: Fundamental frequency parameter $\lambda$ for $\eta = 0.4$

| Arch's shapes | $\beta°\backslash M^*$ | 0.0 | 0.2 | 0.4 | 0.6 | 0.8 | 1.0 |
|---|---|---|---|---|---|---|---|
| Parabola | 10 | 4.97 | 3.72 | 3.10 | 2.71 | 2.43 | 2.23 |
| | 20 | 4.54 | 3.44 | 2.88 | 2.52 | 2.27 | 2.08 |
| | 30 | 3.85 | 2.98 | 2.51 | 2.21 | 2.00 | 1.84 |
| | 40 | 2.97 | 2.37 | 2.02 | 1.80 | 1.63 | 1.50 |
| Catenary | 10 | 5.02 | 3.75 | 3.12 | 2.73 | 2.45 | 2.25 |
| | 20 | 4.73 | 3.57 | 2.98 | 2.60 | 2.34 | 2.15 |
| | 30 | 4.26 | 3.26 | 2.73 | 2.40 | 2.16 | 1.99 |
| | 40 | 3.62 | 2.83 | 2.39 | 2.11 | 1.91 | 1.76 |
| Spiral | 10 | 5.07 | 3.79 | 3.15 | 2.75 | 2.47 | 2.26 |
| | 20 | 4.93 | 3.70 | 3.08 | 2.69 | 2.42 | 2.22 |
| | 30 | 4.69 | 3.55 | 2.96 | 2.59 | 2.33 | 2.14 |
| | 40 | 4.36 | 3.33 | 2.79 | 2.45 | 2.21 | 2.03 |
| Cycloid | 10 | 5.17 | 3.85 | 3.20 | 2.79 | 2.51 | 2.30 |
| | 20 | 5.33 | 3.96 | 3.28 | 2.87 | 2.57 | 2.36 |
| | 30 | 5.62 | 4.15 | 3.43 | 2.99 | 2.68 | 2.46 |
| | 40 | 6.04 | 4.43 | 3.65 | 3.18 | 2.85 | 2.60 |

These results have been obtained by the finite element method; 12 prismatic beam elements have been used (Rossi *et al.*, 1989). This article contains results that are obtained using the Rayleigh optimization method.

*Note:*   The varying cross-section for symmetric arches may be presented in the analytical form by the following expression (Darkov, 1989)

$$I_x = \frac{I_c}{\left[1 - (1-n)\dfrac{x}{l_1}\right]\cos\,\varphi_x}$$

where $x$ = abscissa of any point on a neutral line, referred to the coordinate origin which is located at a centroid of a crown section

$\quad I_c$ = second moment of inertia of a crown section

$\quad I_x$ = second moment of inertia of a cross-section area, which is located at a distance $x$ from the coordinate origin

$\quad \varphi_x$ = angle between the tangent to the neutral line of the arch and the horizontal

$\quad l_1$ = one half of the arch span

The value of parameter $n$ is given by formula

$$n = \frac{I_c}{I_0 \cos\,\varphi_0}$$

where $I_0$, $\varphi_0$ correspond to the cross-section at the support.

## 8.7   ARCHES OF DISCONTINUOUSLY VARYING CROSS-SECTION

This section is devoted to the in-plane vibration of non-circular symmetric and non-symmetric arches of non-uniform cross-section with different boundary conditions. Figure 8.15 presents the notation of a non-circular arch of non-uniform cross-section.

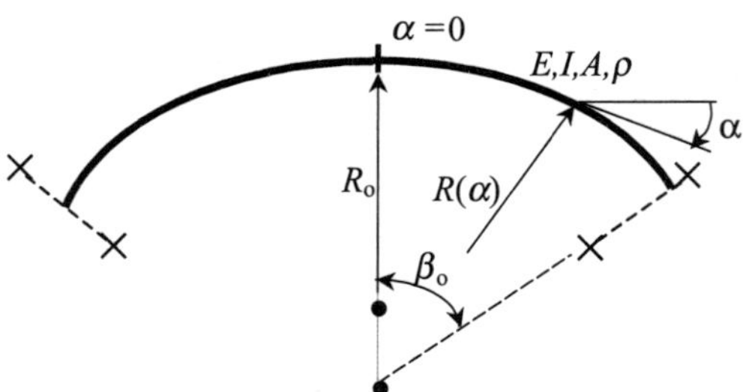


**FIGURE 8.15.**   Arch geometry (boundary conditions are not shown).

The relationships $R(\alpha)$ for different types of arch geometry are presented in Table 8.1. The mathematical model and governing functional are presented in Section 8.1.3. Numerical results for different arch shapes and boundary conditions, and types of discontinuously varying cross-sections, are presented in Tables 8.13–8.19 (Gutierrez *et al.*, 1989).

### 8.7.1  Pinned–pinned arches

Two types of symmetric pinned–pinned arches of discontinuously varying cross-section are presented in Fig. 8.16.

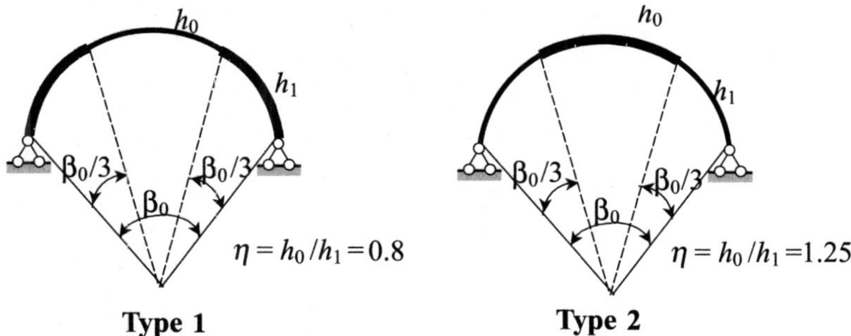


**FIGURE 8.16.**  Pinned–pinned arches of discontinuously varying cross section.

The frequency of an in-plane vibration of an elastic arch for different arch shapes is

$$\omega = \frac{\lambda}{(R_0\beta_0)^2}\sqrt{\frac{EI_0}{\rho A_0}}$$

where the parameter $\lambda$ for the first and second modes of pinned–pinned arches for both types 1 and 2 are listed in Table 8.13. The finite element method has been used.

### 8.7.2  Clamped–clamped arches

Two types of symmetric clamped–clamped arches of discontinuously varying cross-section are presented in Fig. 8.17.

The frequency of in-plane vibration of elastic arches is

$$\omega = \frac{\lambda}{(R_0\beta_0)^2}\sqrt{\frac{EI_0}{\rho A_0}}$$

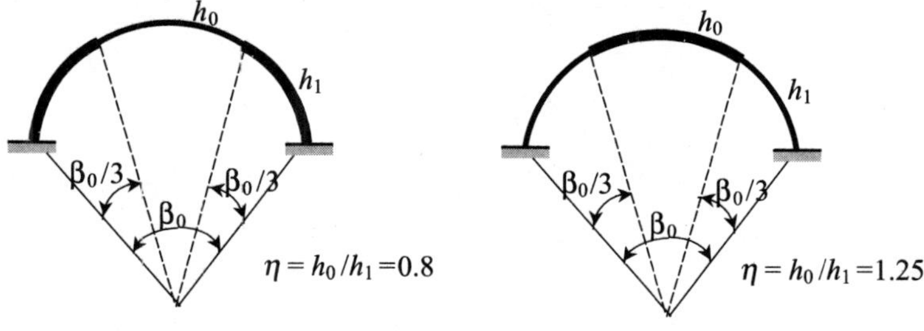


**FIGURE 8.17.**  Clamped–clamped arches of discontinuously varying cross-section.

**TABLE 8.13.** Pinned–pinned non-circular symmetric arches of discontinuously varying cross-section: frequency parameters $\lambda$ for first and second mode of vibration (Figure 8.16)

| Arch's shape | $\beta_0$ (degrees) | First mode | | Second mode | |
|---|---|---|---|---|---|
| | | Type 1 | Type 2 | Type 1 | Type 2 |
| Parabola | 10 | 45.93 | 32.17 | 100.47 | 71.25 |
| | 20 | 44.70 | 31.21 | 98.17 | 69.69 |
| | 30 | 42.70 | 29.66 | 94.37 | 67.10 |
| | 40 | 39.97 | 27.56 | 89.13 | 63.52 |
| Catenary | 10 | 46.04 | 32.26 | 100.66 | 71.39 |
| | 20 | 45.14 | 31.55 | 99.10 | 70.32 |
| | 30 | 43.66 | 30.39 | 96.45 | 68.50 |
| | 40 | 41.63 | 28.81 | 92.76 | 65.98 |
| Spiral | 10 | 46.15 | 32.34 | 100.90 | 71.54 |
| | 20 | 45.57 | 31.89 | 100.04 | 70.95 |
| | 30 | 44.63 | 31.13 | 98.54 | 69.92 |
| | 40 | 43.32 | 30.10 | 96.45 | 68.48 |
| Circle | 10 | 46.26 | 32.43 | 101.17 | 71.73 |
| | 20 | 46.01 | 32.23 | 100.98 | 71.59 |
| | 30 | 45.61 | 31.89 | 100.66 | 71.35 |
| | 40 | 45.05 | 31.43 | 100.21 | 71.01 |
| Cycloid | 10 | 46.37 | 32.51 | 101.33 | 71.82 |
| | 20 | 46.45 | 32.57 | 101.85 | 72.16 |
| | 30 | 46.60 | 32.66 | 102.71 | 72.72 |
| | 40 | 46.82 | 32.80 | 103.95 | 73.52 |

where $\beta_0$ is the central angle (in radians). The parameter $\lambda$, for the first and second modes of clamped–clamped arches for both types 1 and 2, is presented in Table 8.14. The finite element method has been used.

### 8.7.3 Non-symmetric arches

Non-symmetric arches with different shapes of discontinuously varying cross-section are presented in Fig. 8.18.

The fundamental frequency of in-plane vibration of elastic pinned and clamped arches is

$$\omega = \frac{\lambda}{(R_0\beta_0)^2}\sqrt{\frac{EI_0}{\rho A_0}}$$

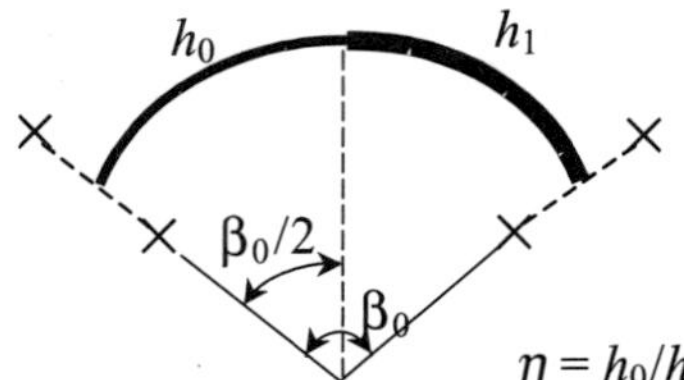


**FIGURE 8.18.** Non-symmetric arch (boundary conditions are not shown).

**TABLE 8.14.** Clamped–clamped non-circular symmetric arches of discontinuously varying cross-section: frequency parameters $\lambda$ for first and second mode of vibration (Figure 8.17)

| Arch's shape | $\beta_0$ (degrees) | First mode | | Second mode | |
|---|---|---|---|---|---|
| | | Type 1 | Type 2 | Type 1 | Type 2 |
| Parabola | 10 | 71.30 | 50.96 | 133.52 | 92.58 |
| | 20 | 69.52 | 49.17 | 130.63 | 90.63 |
| | 30 | 66.59 | 46.86 | 125.83 | 87.38 |
| | 40 | 62.59 | 43.73 | 119.18 | 82.89 |
| Catenary | 10 | 71.47 | 50.72 | 133.62 | 92.68 |
| | 20 | 70.20 | 49.70 | 131.83 | 91.43 |
| | 30 | 68.10 | 48.02 | 128.51 | 89.17 |
| | 40 | 65.21 | 45.73 | 123.87 | 86.03 |
| Spiral | 10 | 71.65 | 50.86 | 133.92 | 92.89 |
| | 20 | 70.89 | 50.24 | 133.03 | 92.24 |
| | 30 | 69.63 | 49.21 | 131.21 | 90.99 |
| | 40 | 67.88 | 47.79 | 128.65 | 89.23 |
| Circle | 10 | 71.82 | 50.99 | 134.43 | 93.19 |
| | 20 | 71.58 | 50.77 | 134.26 | 93.05 |
| | 30 | 71.17 | 50.41 | 133.95 | 92.82 |
| | 40 | 70.62 | 49.91 | 133.51 | 92.49 |
| Cycloid | 10 | 71.99 | 51.13 | 134.55 | 93.29 |
| | 20 | 72.27 | 51.31 | 135.29 | 93.76 |
| | 30 | 72.74 | 51.63 | 136.52 | 94.56 |
| | 40 | 73.42 | 52.09 | 138.26 | 95.69 |

The parameter $\lambda$ for arches with different shapes is presented in Table 8.15. Polynomial approximations and the Ritz method are used.

## 8.7.4　Arches of continuously varying cross-section

Symmetric arches of continuously varying cross-section are presented in Figs. 8.19(a) and (b).

The width of the arch is constant, the height $h_1$ for the left and right parts of the arch is

$$h_1 = h_0\left(1 - \eta\frac{\alpha}{\beta}\right) \text{ and } h_1 = h_0\left(1 + \eta\frac{\alpha}{\beta}\right)$$

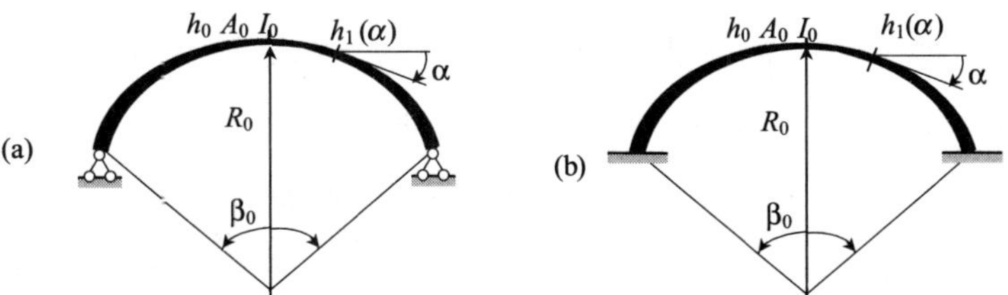


**FIGURE 8.19.** Symmetric arches of continuously varying cross-section. (a) Pinned–pinned arch; (b) clamped–clamped arch.

**TABLE 8.15.** Non-circular non-symmetrical arches of discontinuously varying cross-section with different boundary conditions: fundamental frequency parameter $\lambda$ (Figure 8.18)

| Arch's shapes | $\beta_0$ (degrees) | Pinned–pinned $\eta = 0.8$ | Clamped–clamped $\eta = 0.8$ | Pinned at left end, clamped at right end $\eta = 0.8$ | $\eta = 1.0$ |
|---|---|---|---|---|---|
| Parabola | 10 | 44.22 | 68.87 | 56.49 | 49.15 |
|  | 20 | 42.94 | 67.17 | 54.88 | 47.83 |
|  | 30 | 40.89 | 64.25 | 52.23 | 45.69 |
|  | 40 | 38.05 | 60.06 | 48.66 | 42.61 |
|  | 50 | 34.52 | 54.91 | 44.27 | 38.88 |
|  | 60 | 30.19 | 48.58 | 38.98 | 34.40 |
| Catenary | 10 | 44.36 | 69.10 | 56.63 | 49.31 |
|  | 20 | 43.35 | 67.82 | 55.46 | 48.24 |
|  | 30 | 41.81 | 65.66 | 53.55 | 46.81 |
|  | 40 | 39.79 | 62.80 | 50.99 | 44.58 |
|  | 50 | 37.09 | 58.99 | 47.66 | 41.80 |
|  | 60 | 33.94 | 54.47 | 43.81 | 38.47 |
| Spiral | 10 | 44.45 | 69.22 | 56.78 | 49.31 |
|  | 20 | 43.86 | 68.46 | 56.07 | 48.82 |
|  | 30 | 42.89 | 67.23 | 54.88 | 47.74 |
|  | 40 | 41.47 | 65.42 | 53.29 | 46.47 |
|  | 50 | 39.82 | 63.18 | 51.26 | 44.72 |
|  | 60 | 37.78 | 60.43 | 48.84 | 42.61 |
| Circle | 10 | 44.54 | 69.39 | 56.92 | 49.47 |
|  | 20 | 44.27 | 69.10 | 56.63 | 49.31 |
|  | 30 | 43.90 | 68.64 | 56.21 | 48.82 |
|  | 40 | 43.26 | 68.05 | 55.64 | 48.33 |
|  | 50 | 42.56 | 67.35 | 54.91 | 47.66 |
|  | 60 | 41.66 | 66.39 | 54.03 | 46.98 |
| Cycloid | 10 | 44.63 | 69.51 | 57.06 | 49.63 |
|  | 20 | 44.72 | 69.80 | 57.27 | 49.71 |
|  | 30 | 44.90 | 70.14 | 57.55 | 49.96 |
|  | 40 | 45.07 | 70.71 | 58.03 | 50.20 |
|  | 50 | 45.25 | 71.38 | 58.58 | 50.67 |
|  | 60 | 45.60 | 72.22 | 59.33 | 51.14 |

The current angle $0 \leq \alpha \leq \beta$, $\beta = 0.5\beta_0$ and parameter $\eta$ is any number. The fundamental and second frequency of in-plane vibration may be calculated by $\omega = \dfrac{\lambda}{(R_0\beta_0)^2}\sqrt{\dfrac{EI_0}{\rho A_0}}$. The frequency coefficients $\lambda_1$ and $\lambda_2$, as a function of parameter $\eta$, central angle $\beta_0$, various equations of the neutral line and different boundary conditions, are presented in Tables 8.16(a) and (b). The polynomial approximation and Ritz method have been applied (Gutierres *et al.*, 1989).

### 8.7.5   Pinned–pinned symmetric arches

The design diagram is presented in Fig. 8.19(a). Fundamental and second frequency coefficients, $\lambda_1$ and $\lambda_2$, as a function of the parameter $\eta$, central angle $\beta_0$ and various equations of the neutral line are presented in Table 8.16(a).

**TABLE 8.16.** (a) Pinned–pinned non-circular symmetric arches of continuously varying cross-section: Frequency parameters for first and second modes of vibration (Figure 8.19a)

| | $\beta_0$ (degrees) | Fundamental mode | | | | Second mode | | | |
|---|---|---|---|---|---|---|---|---|---|
| | | $\eta = 0.1$ | 0.2 | 0.3 | 0.4 | $\eta = 0.1$ | 0.2 | 0.3 | 0.4 |
| Parabola | 10 | 39.27 | 39.76 | 40.53 | 41.56 | 88.15 | 95.59 | 97.02 | 101.45 |
| | 20 | 38.15 | 38.63 | 39.39 | 40.40 | 86.19 | 90.54 | 94.88 | 99.21 |
| | 30 | 36.33 | 36.80 | 37.54 | 38.53 | 82.95 | 87.16 | 91.34 | 95.51 |
| | 40 | 33.87 | 34.32 | 35.03 | 35.98 | 78.48 | 82.48 | 86.46 | 90.42 |
| Catenary | 10 | 39.37 | 39.86 | 40.63 | 41.66 | 88.35 | 92.80 | 97.24 | 101.68 |
| | 20 | 38.55 | 39.03 | 39.79 | 40.81 | 86.99 | 91.37 | 95.74 | 100.12 |
| | 30 | 37.20 | 37.67 | 38.42 | 39.42 | 84.72 | 89.00 | 93.26 | 97.52 |
| | 40 | 35.36 | 35.81 | 36.55 | 37.51 | 81.57 | 85.71 | 89.82 | 93.93 |
| Spiral | 10 | 39.47 | 39.96 | 40.73 | 41.76 | 88.55 | 93.01 | 97.46 | 101.91 |
| | 20 | 38.08 | 39.43 | 40.20 | 41.22 | 87.78 | 92.21 | 96.62 | 101.03 |
| | 30 | 36.08 | 38.55 | 39.31 | 40.32 | 86.50 | 90.86 | 95.21 | 99.56 |
| | 40 | 36.88 | 37.35 | 38.10 | 39.08 | 84.70 | 88.99 | 93.25 | 97.52 |
| Circle | 10 | 39.57 | 40.06 | 40.84 | 41.86 | 88.75 | 93.22 | 97.68 | 102.14 |
| | 20 | 39.34 | 39.83 | 40.60 | 41.62 | 88.58 | 93.04 | 97.49 | 101.94 |
| | 30 | 38.97 | 39.45 | 40.22 | 41.23 | 88.29 | 92.74 | 97.18 | 101.62 |
| | 40 | 38.45 | 38.93 | 39.69 | 40.69 | 87.89 | 92.33 | 96.75 | 101.17 |
| Cycloid | 10 | 39.67 | 40.16 | 40.94 | 41.97 | 88.95 | 93.43 | 97.90 | 102.37 |
| | 20 | 39.74 | 40.23 | 31.01 | 42.04 | 89.38 | 93.88 | 98.37 | 102.87 |
| | 30 | 39.87 | 40.36 | 41.13 | 42.16 | 90.11 | 94.64 | 99.17 | 103.71 |
| | 40 | 40.06 | 40.54 | 41.32 | 42.34 | 91.14 | 95.73 | 100.31 | 104.90 |

### 8.7.6   Clamped-clamped symmetric arch

The design diagram is presented in Fig. 8.19(b). Fundamental and second frequency coefficients, $\lambda_1$ and $\lambda_2$, as a function of the parameter $\eta$, central angle $\beta_0$ and various equations of the neutral line are presented in Table 8.16(b).

### 8.7.7   Arch structures of continuously varying cross-section

Different types of continuously varying cross-sections of elastic arch structures are presented in Fig. 8.20. Cases 1 and 2 present arches of symmetric varying cross-section, cases 3 and 4 present arches of non-symmetric varying cross-section.

**TABLE 8.16.** (b) Clamped–clamped non-circular symmetric arches of continuously varying cross-section: Frequency parameters for first and second modes of vibration (Figure 8.19b)

| | $\beta_0$ (degrees) | Fundamental mode | | | | Second mode | | | |
|---|---|---|---|---|---|---|---|---|---|
| | | $\eta = 0.1$ | 0.2 | 0.3 | 0.4 | $\eta = 0.1$ | 0.2 | 0.3 | 0.4 |
| Parabola | 10 | 61.49 | 62.57 | 64.27 | 66.48 | 116.79 | 123.36 | 129.90 | 136.44 |
| | 20 | 59.86 | 60.91 | 62.58 | 64.76 | 114.28 | 120.71 | 127.11 | 133.52 |
| | 30 | 57.18 | 58.20 | 59.82 | 61.92 | 110.13 | 116.34 | 122.52 | 128.70 |
| | 40 | 53.53 | 54.51 | 56.05 | 58.06 | 104.39 | 110.31 | 116.19 | 122.06 |
| Catenary | 10 | 61.65 | 62.73 | 64.43 | 66.65 | 117.06 | 123.63 | 130.19 | 136.75 |
| | 20 | 60.47 | 61.54 | 63.22 | 65.41 | 115.32 | 121.80 | 128.27 | 134.73 |
| | 30 | 58.54 | 59.59 | 61.23 | 63.37 | 112.44 | 118.77 | 125.08 | 131.39 |
| | 40 | 55.89 | 56.90 | 58.49 | 60.57 | 108.43 | 114.56 | 120.66 | 126.75 |
| Spiral | 10 | 61.81 | 62.89 | 64.60 | 66.82 | 117.32 | 123.91 | 130.48 | 137.05 |
| | 20 | 61.10 | 62.17 | 63.87 | 66.07 | 116.36 | 122.90 | 129.43 | 135.95 |
| | 30 | 59.93 | 60.99 | 62.66 | 64.84 | 114.77 | 121.23 | 127.67 | 134.11 |
| | 40 | 58.31 | 59.35 | 60.99 | 63.13 | 112.55 | 118.90 | 125.22 | 131.54 |
| Circle | 10 | 61.96 | 63.05 | 64.76 | 66.98 | 117.58 | 124.18 | 130.77 | 137.36 |
| | 20 | 61.72 | 62.81 | 64.51 | 66.74 | 117.41 | 124.01 | 130.59 | 137.18 |
| | 30 | 61.33 | 62.41 | 64.11 | 66.33 | 117.13 | 123.72 | 130.29 | 136.87 |
| | 40 | 60.79 | 61.87 | 63.56 | 65.77 | 116.74 | 123.22 | 129.88 | 136.44 |
| Cycloid | 10 | 62.12 | 63.21 | 64.92 | 67.15 | 117.84 | 124.46 | 131.06 | 137.67 |
| | 20 | 62.36 | 63.45 | 65.17 | 67.40 | 118.47 | 125.12 | 130.76 | 138.41 |
| | 30 | 62.76 | 63.85 | 65.59 | 67.84 | 119.52 | 126.24 | 132.94 | 139.66 |
| | 40 | 63.34 | 64.45 | 66.19 | 68.46 | 121.02 | 127.83 | 134.63 | 141.45 |

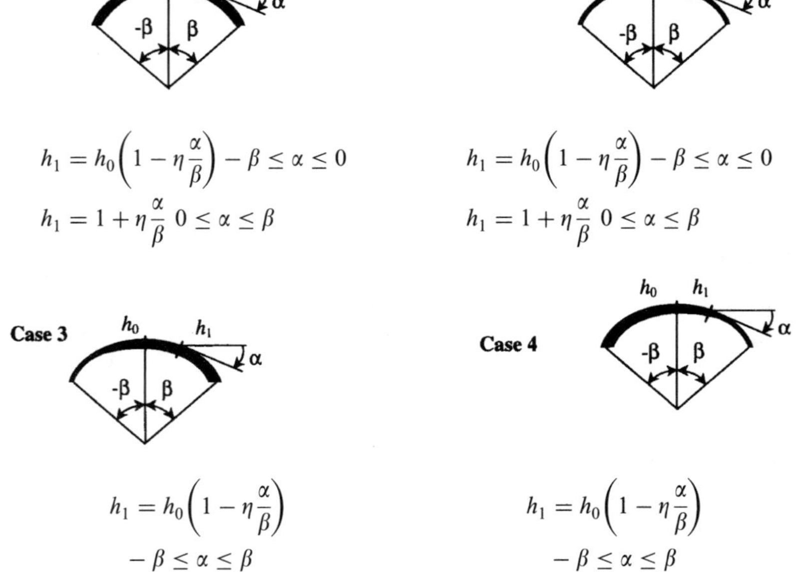



**Case 1**

$$h_1 = h_0\left(1 - \eta\frac{\alpha}{\beta}\right) \quad -\beta \le \alpha \le 0$$

$$h_1 = 1 + \eta\frac{\alpha}{\beta} \quad 0 \le \alpha \le \beta$$

**Case 2**

$$h_1 = h_0\left(1 - \eta\frac{\alpha}{\beta}\right) \quad -\beta \le \alpha \le 0$$

$$h_1 = 1 + \eta\frac{\alpha}{\beta} \quad 0 \le \alpha \le \beta$$

**Case 3**

$$h_1 = h_0\left(1 - \eta\frac{\alpha}{\beta}\right)$$

$$-\beta \le \alpha \le \beta$$

**Case 4**

$$h_1 = h_0\left(1 - \eta\frac{\alpha}{\beta}\right)$$

$$-\beta \le \alpha \le \beta$$

**FIGURE 8.20.** Different types of non-uniform cross-section of arch structures. Boundary conditions are not shown. Parameter $\eta$ is any positive number.

The frequency of the in-plane vibration is

$$\omega = \frac{\lambda}{(R_0\beta_0)^2}\sqrt{\frac{EI}{\rho A_0}}$$

where $\beta_0 = 2\beta$ is the angle of opening. Fundamental frequency parameters $\lambda$ are predicted in Tables 8.17–8.19 (Gutierres et al., 1989). Polynomial approximations and the Ritz method are used.

**TABLE 8.17.** Pinned–clamped non-circular arches of symmetric continuously varying cross-section: Fundamental frequency parameter (Figure 8.20)

| Arch's shape | $\beta_0$ (degrees) | Pinned–clamped, Case 1 | | | | Pinned–clamped, Case 2 | | | |
|---|---|---|---|---|---|---|---|---|---|
| | | $\eta = 0.1$ | 0.2 | 0.3 | 0.4 | $\eta = 0.1$ | 0.2 | 0.3 | 0.4 |
| Parabola | 10 | 51.76 | 54.47 | 57.13 | 59.80 | 46.21 | 43.45 | 40.39 | 37.09 |
| | 20 | 50.43 | 53.06 | 55.71 | 58.31 | 44.98 | 42.16 | 39.09 | 35.88 |
| | 30 | 48.16 | 50.75 | 53.36 | 55.85 | 42.89 | 40.10 | 37.09 | 33.96 |
| | 40 | 45.07 | 47.58 | 49.96 | 52.46 | 40.00 | 37.20 | 34.29 | 31.11 |
| | 50 | 41.18 | 43.54 | 45.78 | 48.08 | 36.22 | 33.58 | 30.85 | 27.71 |
| | 60 | 36.44 | 38.67 | 40.89 | 42.98 | 31.87 | 29.39 | 26.68 | 23.83 |
| Catenary | 10 | 51.84 | 54.55 | 57.27 | 59.93 | 46.39 | 43.45 | 40.39 | 37.20 |
| | 20 | 50.91 | 53.59 | 56.21 | 58.78 | 45.51 | 42.52 | 39.59 | 36.22 |
| | 30 | 49.31 | 51.92 | 54.55 | 57.06 | 44.00 | 41.18 | 38.05 | 34.87 |
| | 40 | 47.07 | 49.63 | 52.07 | 54.62 | 41.76 | 39.09 | 36.11 | 32.98 |
| | 50 | 44.18 | 46.64 | 49.07 | 51.45 | 39.09 | 36.44 | 33.58 | 30.33 |
| | 60 | 40.79 | 43.08 | 45.34 | 47.58 | 35.88 | 33.34 | 30.33 | 27.42 |
| Spiral | 10 | 52.00 | 54.69 | 57.41 | 60.06 | 46.56 | 43.63 | 40.59 | 37.31 |
| | 20 | 51.45 | 54.03 | 56.78 | 59.33 | 45.86 | 43.08 | 40.00 | 36.77 |
| | 30 | 50.43 | 53.06 | 55.64 | 58.31 | 44.98 | 42.14 | 39.09 | 35.77 |
| | 40 | 49.07 | 51.61 | 54.18 | 56.71 | 43.63 | 40.79 | 37.84 | 34.64 |
| | 50 | 47.24 | 49.80 | 52.30 | 54.84 | 41.95 | 39.19 | 36.33 | 33.10 |
| | 60 | 45.16 | 47.66 | 50.04 | 52.46 | 40.00 | 37.20 | 34.40 | 31.24 |
| Circle | 10 | 52.15 | 54.84 | 57.48 | 60.20 | 46.56 | 43.72 | 40.69 | 37.41 |
| | 20 | 51.92 | 54.55 | 57.27 | 60.00 | 46.39 | 43.51 | 40.39 | 37.20 |
| | 30 | 51.53 | 54.25 | 56.85 | 59.53 | 46.04 | 43.08 | 40.20 | 36.87 |
| | 40 | 50.91 | 53.66 | 56.28 | 58.99 | 45.51 | 42.70 | 39.59 | 36.33 |
| | 50 | 50.27 | 52.99 | 55.57 | 58.24 | 44.81 | 41.95 | 38.98 | 35.77 |
| | 60 | 49.47 | 52.07 | 54.77 | 57.41 | 44.09 | 41.37 | 38.36 | 35.10 |
| Cycloid | 10 | 52.23 | 54.99 | 57.61 | 60.33 | 46.73 | 43.81 | 40.79 | 37.52 |
| | 20 | 52.38 | 55.13 | 57.82 | 60.53 | 46.81 | 44.00 | 40.98 | 37.63 |
| | 30 | 52.61 | 55.35 | 58.03 | 60.72 | 47.07 | 44.18 | 41.18 | 37.84 |
| | 40 | 52.91 | 55.71 | 58.37 | 61.12 | 47.41 | 44.45 | 41.37 | 38.15 |
| | 50 | 53.36 | 56.14 | 58.85 | 61.64 | 47.74 | 44.81 | 41.76 | 38.57 |
| | 60 | 53.81 | 56.63 | 59.46 | 62.22 | 48.24 | 45.34 | 42.23 | 38.98 |

**TABLE 8.18.** Pinned–pinned and pinned–clamped non-circular arches of non-symmetric continuously varying cross-section: Fundamental frequency parameter (Figure 8.20)

| Arch's shape | $\beta_0$ (degrees) | Pinned–pinned, Case 3 | | | | Pinned–clamped, Case 3 | | | |
|---|---|---|---|---|---|---|---|---|---|
| | | $\eta = 0.1$ | 0.2 | 0.3 | 0.4 | $\eta = 0.1$ | 0.2 | 0.3 | 0.4 |
| Parabola | 10 | 38.88 | 38.57 | 38.15 | 37.41 | 49.47 | 49.71 | 49.63 | 48.99 |
| | 20 | 37.73 | 37.52 | 37.09 | 36.33 | 48.24 | 48.33 | 48.16 | 47.66 |
| | 30 | 36.00 | 35.77 | 35.21 | 34.52 | 45.95 | 46.13 | 45.86 | 45.34 |
| | 40 | 33.46 | 33.22 | 32.74 | 32.12 | 42.89 | 42.98 | 42.80 | 42.33 |
| | 50 | 30.33 | 33.06 | 29.66 | 29.12 | 39.09 | 39.19 | 38.98 | 38.57 |
| | 60 | 26.68 | 26.38 | 25.92 | 25.61 | 34.52 | 34.64 | 34.64 | 34.40 |
| Catenary | 10 | 38.98 | 38.78 | 38.36 | 37.52 | 49.63 | 49.88 | 49.80 | 49.15 |
| | 20 | 38.15 | 37.94 | 37.52 | 36.77 | 48.66 | 48.82 | 48.74 | 48.16 |
| | 30 | 36.87 | 36.66 | 36.11 | 35.44 | 47.07 | 47.15 | 47.07 | 46.56 |
| | 40 | 35.10 | 34.64 | 34.29 | 33.58 | 44.90 | 44.90 | 44.81 | 44.27 |
| | 50 | 32.74 | 32.37 | 31.87 | 31.36 | 41.52 | 42.14 | 41.95 | 41.47 |
| | 60 | 29.79 | 29.66 | 29.25 | 28.70 | 38.67 | 38.78 | 38.67 | 38.26 |
| Spiral | 10 | 39.09 | 38.78 | 38.36 | 37.63 | 49.80 | 50.04 | 49.96 | 49.31 |
| | 20 | 38.57 | 38.36 | 37.94 | 37.09 | 49.15 | 49.39 | 49.31 | 48.66 |
| | 30 | 37.73 | 37.52 | 36.98 | 36.33 | 48.16 | 48.33 | 48.24 | 47.66 |
| | 40 | 36.55 | 36.22 | 35.77 | 35.21 | 46.73 | 46.98 | 46.81 | 46.30 |
| | 50 | 35.10 | 34.87 | 34.29 | 33.58 | 45.07 | 45.25 | 44.98 | 44.45 |
| | 60 | 33.22 | 32.98 | 32.49 | 31.87 | 42.89 | 43.08 | 42.89 | 42.42 |
| Circle | 10 | 39.19 | 38.98 | 38.57 | 37.73 | 49.88 | 50.12 | 49.96 | 49.47 |
| | 20 | 38.88 | 38.78 | 38.15 | 37.52 | 49.63 | 49.88 | 49.80 | 49.15 |
| | 30 | 38.67 | 38.36 | 37.94 | 37.20 | 49.31 | 49.55 | 49.47 | 48.81 |
| | 40 | 38.05 | 37.84 | 37.31 | 36.66 | 48.82 | 49.07 | 48.99 | 48.33 |
| | 50 | 37.41 | 37.20 | 36.66 | 36.00 | 48.16 | 48.41 | 48.33 | 47.83 |
| | 60 | 36.66 | 36.44 | 36.00 | 35.32 | 47.32 | 47.58 | 47.49 | 46.98 |
| Cycloid | 10 | 39.29 | 38.98 | 38.57 | 37.84 | 50.04 | 50.27 | 50.12 | 49.63 |
| | 20 | 39.29 | 39.09 | 38.57 | 37.94 | 50.20 | 50.43 | 50.35 | 49.80 |
| | 30 | 39.49 | 39.29 | 38.78 | 38.05 | 50.35 | 50.67 | 50.75 | 50.12 |
| | 40 | 39.59 | 39.39 | 38.98 | 38.26 | 50.75 | 51.14 | 51.22 | 50.59 |
| | 50 | 39.79 | 39.59 | 39.19 | 38.57 | 51.22 | 51.76 | 51.84 | 51.30 |
| | 60 | 40.10 | 39.90 | 39.49 | 38.98 | 51.76 | 52.46 | 52.68 | 52.15 |

**TABLE 8.19.** Clamped–clamped and pinned–clamped non-circular arches of non-symmetric continuously varying cross-section: Fundamental frequency parameter (Figure 8.20)

| Arch's shape | $\beta_0$ (degrees) | Clamped–clamped, Case 3 | | | | Pinned–clamped, Case 4 | | | |
|---|---|---|---|---|---|---|---|---|---|
| | | $\eta = 0.1$ | 0.2 | 0.3 | 0.4 | $\eta = 0.1$ | 0.2 | 0.3 | 0.4 |
| Parabola | 10 | 60.53 | 60.46 | 59.73 | 58.37 | 48.49 | 47.74 | 46.46 | 45.34 |
| | 20 | 59.05 | 58.85 | 58.10 | 56.78 | 47.15 | 46.47 | 45.43 | 44.09 |
| | 30 | 56.49 | 56.21 | 55.42 | 54.11 | 45.07 | 44.36 | 43.26 | 41.85 |
| | 40 | 52.91 | 52.65 | 51.76 | 50.43 | 42.14 | 41.28 | 40.29 | 38.88 |
| | 50 | 48.41 | 40.08 | 47.15 | 46.14 | 38.26 | 37.52 | 36.55 | 35.32 |
| | 60 | 42.98 | 42.70 | 41.85 | 40.86 | 33.70 | 32.98 | 32.00 | 30.05 |
| Catenary | 10 | 60.59 | 60.59 | 59.93 | 58.65 | 48.58 | 47.91 | 46.81 | 45.43 |
| | 20 | 59.59 | 59.46 | 58.78 | 57.41 | 47.66 | 46.90 | 45.95 | 44.45 |
| | 30 | 57.75 | 57.55 | 56.85 | 55.49 | 46.04 | 45.34 | 44.36 | 42.89 |
| | 40 | 55.20 | 54.99 | 54.18 | 52.83 | 44.00 | 43.17 | 42.14 | 40.79 |
| | 50 | 52.00 | 51.69 | 50.75 | 49.55 | 41.18 | 40.39 | 39.39 | 38.05 |
| | 60 | 48.08 | 47.66 | 46.81 | 45.78 | 37.84 | 37.09 | 36.11 | 34.87 |
| Spiral | 10 | 60.79 | 60.79 | 60.13 | 58.78 | 48.74 | 48.00 | 46.98 | 45.51 |
| | 20 | 60.13 | 60.06 | 59.33 | 58.10 | 48.16 | 47.41 | 46.30 | 44.98 |
| | 30 | 58.99 | 58.92 | 58.24 | 56.85 | 47.15 | 46.39 | 45.43 | 43.90 |
| | 40 | 57.55 | 57.41 | 56.56 | 55.20 | 45.86 | 45.07 | 44.00 | 42.61 |
| | 50 | 55.57 | 55.35 | 54.55 | 53.21 | 44.09 | 43.26 | 42.33 | 40.98 |
| | 60 | 53.21 | 52.92 | 52.07 | 50.83 | 42.04 | 41.18 | 40.20 | 38.88 |
| Circle | 10 | 60.92 | 60.92 | 60.26 | 58.92 | 48.82 | 48.08 | 47.07 | 45.69 |
| | 20 | 60.72 | 60.66 | 60.00 | 58.65 | 48.58 | 47.83 | 46.81 | 45.43 |
| | 30 | 60.33 | 60.26 | 59.59 | 58.31 | 48.24 | 47.49 | 46.47 | 45.07 |
| | 40 | 59.80 | 59.73 | 59.05 | 57.82 | 47.74 | 46.90 | 45.95 | 44.54 |
| | 50 | 59.12 | 59.05 | 58.37 | 57.13 | 47.07 | 46.21 | 45.25 | 43.81 |
| | 60 | 58.31 | 58.24 | 57.68 | 56.35 | 46.21 | 45.51 | 44.54 | 43.08 |
| Cycloid | 10 | 61.05 | 61.05 | 60.39 | 59.05 | 48.90 | 48.24 | 47.15 | 45.78 |
| | 20 | 61.35 | 61.25 | 60.66 | 59.33 | 49.07 | 48.33 | 47.32 | 45.95 |
| | 30 | 61.57 | 61.57 | 61.05 | 59.80 | 49.31 | 48.49 | 47.49 | 46.13 |
| | 40 | 62.03 | 62.16 | 61.64 | 60.46 | 49.55 | 48.82 | 47.91 | 46.47 |
| | 50 | 62.54 | 62.86 | 62.48 | 61.31 | 49.96 | 49.23 | 48.24 | 46.81 |
| | 60 | 63.30 | 63.62 | 63.56 | 62.41 | 50.35 | 49.71 | 48.74 | 47.49 |

## REFERENCES

Blevins, R.D. (1979) *Formulas for Natural Frequency and Mode Shape* (New York: Van Nostrand Reinhold).

Bolotin, V.V. (1978) *Vibration of Linear Systems*, vol. 1, In *Handbook: Vibration in Tecnnik*, vol 1–6 (Moscow: Mashinostroenie) (In Russian).

Borg, S.F. and Gennaro, J.J. (1959) *Advanced Structural Analysis* (New Jersey: Van Nostrand).

De Rosa, M.A. (1991) The influence of the support flexibilities on the vibration frequencies of arches. *Journal of Sound and Vibration*, **146**(1), 162–169.

Federhofer, K. (1934) Sitzungsberichte d. Akad. d. Wiss.

Filipich, C.P., Laura, P.A.A. and Cortinez, V.H. (1987) In-plane vibrations of an arch of variable cross section elastially restrained against rotation at one end and carrying a concentrated mass, at the other. *Applied Acoustics*, **21**, 241–246.

Gutierrez, R.H., Laura, P.A.A., Rossi, R.E., Bertero, R. and Villaggi, A. (1989) In-plane vibrations of non-circular arcs of non-uniform cross-section. *Journal of Sound and Vibration*, **129**(2), 181–200.

Laura, P.A.A., Filipich, C.P. and Cortinez, V.H. (1987) In-plane vibrations of an elastically cantilevered circular arc with a tip mass. *Journal of Sound and Vibration*, **115**(3), 437–446.

Laura, P.A.A., Verniere De Irassar, P.L., Carnicer, R. and Bertero, R. (1988a) A note on vibrations of a circumferential arch with thickness varying in a discontinuous fashion. *Journal of Sound and Vibration*, **120**(1), 95–105.

Laura, P.A.A., Bambill, E., Filipich, C.P. and Rossi, R.E. (1988b) A note on free flexural vibrations of a non-uniform elliptical ring in its plane. *Journal of Sound and Vibration*, **126**(2), 249–254.

Lee, B.K. and Wilson, J.F. (1989) Free vibrations of arches with variable curvature. *Journal of Sound and Vibration*, **136**(1), 75–89.

Maurizi, M.J., Rossi, R.E. and Belles, P.M. (1991) Lowest natural frequency of clamped circular arcs of linearly tapered width. *Journal of Sound and Vibration*, **144**(2), 357–361.

Romanelli, E. and Laura, P.A. (1972) Fundamental frequencies of non-circular, elastic hinged arcs. *Journal of Sound and Vibration*, **24**(1), 17–22.

Rossi, R.E., Laura, P.A.A. and Verniere De Irassar, P.L. (1989) In-plane vibrations of cantilevered non-circular arcs of non-uniform cross-section with a tip mass. *Journal of Sound and Vibration*, **129**(2), 201–213.


## FURTHER READING

Abramovitz, M. and Stegun, I.A. (1970) *Handbook of Mathematical Functions* (New York: Dover).

Bezukhov, N.I., Luzhin, O.V. and Kolkunov, N.V. (1969) *Stability and Structural Dynamics* (Moscow, Stroizdat).

Chang, T.C. and Volterra, E. (1969) Upper and lower bounds or frequencies of elastic arcs. *The Journal of the Acoustical Society of America*, **46**(5) (Part 2), 1165–1174.

Collatz, L. (1963) *Eigenwertaufgaben mit technischen Anwendungen* (Leipzig: Geest and Portig).

Darkov, A. (1984) *Structural Mechanics* English translation (Moscow: Mir Publishers).

Den Hartog, J.P. (1928) The lowest natural frequencies of circular arcs. *Philosophical Magazine Series* **75**, 400–408.

Ewins, D.J. (1985) *Modal Testing: Theory and Practice* (New York: Wiley).

Filipich, C.P. and Laura, P.A.A. (1988) First and second natural frequencies of hinged and clamped circular arcs: a discussion of a classical paper. *Journal of Sound and Vibration*, **125**, 393–396.

Filipich, C.P. and Rosales, M.B. (1990) In-plane vibration of symmetrically supported circumferential rings. *Journal of Sound and Vibration*, **136**(2), 305–314.

Flugge, W. (1962) *Handbook of Engineering Mechanics* (New York: McGraw-Hill).

Irie, T., Yamada, G. and Tanaka, K. (1982) Free out-of-plane vibration of arcs. *ASME Journal of Applied Mechanics*, **49**, 439–441.

Laura, P.A.A. and Maurizi, M.J. (1987) Recent research on vibrations of arch-type structures. *The Shock and Vibration Digest*, **19**(1), 6–9.

Laura, P.A.A. and Verniere De Irassar, P.L. (1988) A note on in-plane vibrations of arch-type structures of non-uniform cross-section: the case of linearly varying thickness. *Journal of Sound and Vibration*, **124**, 1–12.

Morgaevsky, A.B. (1940) Vibrations of parabolic arches. *Scientific Transactions*, Vol. IV, Metallurgical Institute, Dnepropetrovsk, (in Russian).

Pfeiffer, F. (1934) *Vibration of Elastic Systems* (Moscow–Leningrad, ONT1) p. 154. Translated from (1928) *Mechanik Der Elastischen Korper: Handbuch Der Physik*, Band IV (Berlin).

Rabinovich, I.M. (1954) Eigenvalues and eigenfunctions of parabolic and other arches. *Investigation Theory of Structures*, Gosstroizdat, Moscow, Vol. V. (In Russian).

Rzhanitsun A.R. (1982) Structural mechanics (Moscow: Vushaya Shkola).

Sakiyama, T. (1985) Free vibrations of arches with variable cross section and non-symmetrical axis. *Journal of Sound and Vibration*, **102**, 448–452.

Suzuki, K., Aida, H. and Takahashi, S. (1978) Vibrations of curved bars perpendicular to their planes. *Bulletin of the Japanese Society of Mechanical Engineering*, **21**, 1685–1695.

Volterra, E. and Morell, J.D. (1960) A note on the lowest natural frequencies of elastic arcs. *American Society of Mechanical Engineering Journal of Applied Mechanics* 27, 4744–4746.

Volterra, E. and Morell, J.D. (1961) Lowest natural frequencies of elastic hinged arcs. *Journal of the Acoustical Society of America*, 33(12), 1787–1790.

Volterra, E. and Morell, J.D. (1961) Lowest natural frequencies of elastic arcs outside the plane of initial curvature. *ASME Journal of Applied Mechanics*, **28**, 624–627.

Wang, T.M. (1972) Lowest natural frequencies of clamped parabolic arcs. *Proceedings of the American Society of Civil Engineering* **98**(ST1), 407–411.

Wang, T.M. and Moore, J.A. (1973) Lowest natural extensional frequency of clamped elliptic arcs. *Journal of Sound and Vibration*, **30**, 1–7.

Wang, T.M. (1975) Effect of variable curvature on fundamental frequency of clamped parabolic arcs. *Journal of Sound and Vibration*, **41**, 247–251.

Wasserman, Y. (1978) Spatial symmetrical vibrations and stability of circular arches with flexibly supported ends. *Journal of Sound and Vibration*, **59**, 181–194.

Weaver, W., Timoshenko, S.P. and Young, D.H. (1990) *Vibration Problems in Engineering*, 5th edn (New York: Wiley).

Wolf, J.A. (1971) Natural frequencies of circular arches. *Proceedings of the American Society of Civil Engineering* **97**(ST9), 2337–2349.

Young, W.C. (1989) *Roark's Formulas for Stress and Strain*, 6th edn (New York: McGraw-Hill).

# EIGENFUNCTIONS AND THEIR DERIVATIVES FOR ONE-SPAN BEAMS WITH DIFFERENT BOUNDARY CONDITIONS

**A.1**  Clamped–free beam.
First and second eigenfunctions

| $\alpha = \dfrac{x}{l}$ | $X_1$ | $X_1'$ | $X_1''$ | $X_1'''$ | $X_2$ | $X_2'$ | $X_2''$ | $X_2'''$ |
|---|---|---|---|---|---|---|---|---|
| 0 | 0 | 0 | 7.0318 | −9.6790 | 0 | 0 | 44.074 | −210.71 |
| 0.025 | 0.0022 | 0.1728 | 6.7899 | −9.6790 | 0.0132 | 1.0364 | 38.804 | −210.65 |
| 0.05 | 0.0086 | 0.3395 | 6.5479 | −9.6775 | 0.0507 | 1.9403 | 33.544 | −210.30 |
| 0.075 | 0.0191 | 0.5001 | 6.3062 | −9.6738 | 0.1092 | 2.7136 | 28.295 | −209.34 |
| 0.1 | 0.0335 | 0.6547 | 6.0643 | −9.6658 | 0.1853 | 3.3553 | 23.084 | −207.58 |
| 0.125 | 0.0518 | 0.8034 | 5.8227 | −9.6525 | 0.2759 | 3.8681 | 17.925 | −204.79 |
| 0.15 | 0.0738 | 0.9460 | 5.5815 | −9.6335 | 0.3776 | 4.2526 | 12.855 | −200.83 |
| 0.175 | 0.0991 | 1.0823 | 5.3415 | −9.6063 | 0.4874 | 4.5117 | 7.896 | −195.59 |
| 0.2 | 0.1278 | 1.2129 | 5.1016 | −9.5718 | 0.6021 | 4.6486 | 3.087 | −188.98 |
| 0.225 | 0.1596 | 1.3376 | 4.8627 | −9.5270 | 0.7188 | 4.6676 | −1.542 | −180.95 |
| 0.25 | 0.1946 | 1.4562 | 4.6251 | −9.4727 | 0.8346 | 4.5733 | −5.948 | −171.53 |
| 0.275 | 0.2324 | 1.5689 | 4.3889 | −9.4066 | 0.9466 | 4.3720 | −10.106 | −160.71 |
| 0.3 | 0.2729 | 1.6756 | 4.1551 | −9.3288 | 1.0524 | 4.0704 | −13.976 | −148.56 |
| 0.325 | 0.3161 | 1.7766 | 3.9228 | −9.2378 | 1.1494 | 3.6761 | −17.205 | −135.20 |
| 0.35 | 0.3617 | 1.8718 | 3.6931 | −9.1328 | 1.2355 | 3.1970 | −20.726 | −120.71 |
| 0.375 | 0.4097 | 1.9613 | 3.4662 | −9.0139 | 1.3086 | 2.6431 | −23.551 | −105.25 |
| 0.4 | 0.4598 | 2.0452 | 3.2423 | −8.8796 | 1.3671 | 2.0226 | −25.981 | −89.00 |
| 0.425 | 0.5118 | 2.1234 | 3.0225 | −8.7298 | 1.4093 | 1.3475 | −27.996 | −72.13 |
| 0.45 | 0.5659 | 2.1963 | 2.8063 | −8.5632 | 1.4340 | 0.6266 | −29.585 | −54.85 |
| 0.475 | 0.6217 | 2.2638 | 2.5943 | −8.3796 | 1.4403 | −0.1279 | −30.737 | −37.39 |
| 0.5 | 0.6791 | 2.3261 | 2.3872 | −8.1785 | 1.4274 | −0.9065 | −31.455 | −19.95 |
| 0.525 | 0.7379 | 2.3831 | 2.1858 | −7.9597 | 1.3949 | −1.6977 | −31.737 | −2.81 |
| 0.55 | 0.7982 | 2.4353 | 1.9898 | −7.7226 | 1.3425 | −2.4899 | −31.599 | 13.82 |
| 0.575 | 0.8596 | 2.4828 | 1.7995 | −7.4657 | 1.2704 | −3.2744 | −31.052 | 29.7 |
| 0.6 | 0.9223 | 2.5255 | 1.6162 | −7.1904 | 1.1790 | −4.0395 | −30.121 | 44.58 |
| 0.625 | 0.9860 | 2.5637 | 1.4401 | −6.8953 | 1.0686 | −4.7770 | −28.835 | 58.26 |
| 0.65 | 1.0504 | 2.5974 | 1.2720 | −6.5819 | 0.9404 | −5.4782 | −27.222 | 70.46 |
| 0.675 | 1.1158 | 2.6272 | 1.1116 | −6.2478 | 0.7952 | −6.1360 | −25.324 | 81.01 |
| 0.7 | 1.1818 | 2.6531 | 0.9596 | −5.8925 | 0.6344 | −6.7437 | −23.193 | 89.73 |
| 0.725 | 1.2482 | 2.6754 | 0.8170 | −5.5160 | 0.4585 | −7.2936 | −20.856 | 96.35 |
| 0.75 | 1.3156 | 2.6940 | 0.6840 | −5.1203 | 0.2704 | −7.7861 | −18.379 | 100.73 |
| 0.775 | 1.3830 | 2.7096 | 0.5612 | −4.7040 | 0.0695 | −8.2108 | −15.844 | 102.88 |
| 0.8 | 1.4509 | 2.7223 | 0.4490 | −4.2658 | −0.1400 | −8.5771 | −13.260 | 102.41 |
| 0.825 | 1.5192 | 2.7321 | 0.3480 | −3.8063 | −0.3582 | −8.8768 | −10.733 | 99.40 |
| 0.85 | 1.5877 | 2.7396 | 0.2591 | −3.3269 | −0.5830 | −9.1162 | −8.305 | 93.65 |
| 0.875 | 1.6561 | 2.7452 | 0.1821 | −2.8255 | −0.8137 | −9.2930 | −6.077 | 85.24 |
| 0.9 | 1.7248 | 2.7489 | 0.1179 | −2.3028 | −1.0475 | −9.4204 | −4.075 | 73.91 |
| 0.925 | 1.7936 | 2.7512 | 0.0670 | −1.7589 | −1.2842 | −9.4998 | −2.403 | 59.79 |
| 0.95 | 1.8623 | 2.7525 | 0.0299 | −1.1936 | −1.5224 | −9.5426 | −1.115 | 42.76 |
| 0.975 | 1.9313 | 2.7528 | 0.0076 | −0.6078 | −1.7612 | −9.5589 | −0.290 | 22.40 |
| 1 | 2 | 2.7528 | 0 | 0 | −2 | −9.5610 | 0 | 0 |

**A.1**   (*cont'd*) Clamped–free beam.
Third and fourth eigenfunctions

| $\alpha = \dfrac{x}{l}$ | $X_3$ | $X_3'$ | $X_3''$ | $X_3'''$ | $X_4$ | $X_4'$ | $X_4''$ | $X_4'''$ |
|---|---|---|---|---|---|---|---|---|
| 0 | 0 | 0 | 123.403 | −986.57 | 0 | 0 | 241.82 | −2659.1 |
| 0.025 | 0.0360 | 2.783 | 99.196 | −967.41 | 0.0686 | 5.215 | 175.39 | −2650.5 |
| 0.05 | 0.1341 | 4.962 | 75.086 | −959.75 | 0.2470 | 8.775 | 109.68 | −2595.6 |
| 0.075 | 0.2791 | 6.539 | 51.306 | −940.41 | 0.4939 | 10.718 | 46.28 | −2461.7 |
| 0.1 | 0.4561 | 7.531 | 28.195 | −905.60 | 0.7700 | 11.127 | −12.59 | −2231.0 |
| 0.125 | 0.6509 | 7.958 | 6.175 | −853.01 | 1.0387 | 10.146 | −64.43 | −1899.7 |
| 0.15 | 0.8496 | 7.853 | −14.291 | −781.61 | 1.2675 | 7.985 | −106.81 | −1476.5 |
| 0.175 | 1.0395 | 7.260 | −32.745 | −691.6 | 1.4303 | 4.902 | −137.65 | −981.3 |
| 0.2 | 1.2090 | 6.236 | −48.734 | −584.34 | 1.5076 | 1.209 | −155.49 | −441.7 |
| 0.225 | 1.3484 | 4.847 | −61.842 | −462.38 | 1.4884 | −2.757 | −159.65 | 109.0 |
| 0.25 | 1.4490 | 3.171 | −71.754 | −328.96 | 1.3703 | −6.659 | −150.26 | 634.3 |
| 0.275 | 1.5051 | 1.289 | −78.229 | −187.98 | 1.1587 | −10.167 | −128.43 | 1099.1 |
| 0.3 | 1.5125 | −0.711 | −81.128 | −43.98 | 0.8675 | −12.993 | −96.08 | 1471.6 |
| 0.325 | 1.4693 | −2.738 | −80.444 | 98.36 | 0.5167 | −14.905 | −55.94 | 1727.1 |
| 0.35 | 1.3761 | −4.703 | −76.268 | 234.09 | 0.1315 | −15.749 | −10.87 | 1844.7 |
| 0.375 | 1.2353 | −6.524 | −68.831 | 358.75 | −0.2612 | −15.438 | 35.20 | 1822.4 |
| 0.4 | 1.0518 | −8.121 | −58.458 | 467.91 | −0.6316 | −14.001 | 78.94 | 1657.4 |
| 0.425 | 0.8320 | −9.428 | −45.606 | 557.92 | −0.9522 | −11.543 | 117.05 | 1365.2 |
| 0.45 | 0.5832 | −10.382 | −30.769 | 625.4 | −1.2013 | −8.225 | 146.37 | 969.6 |
| 0.475 | 0.3159 | −10.953 | −14.532 | 668.22 | −1.3584 | −4.317 | 164.92 | 498.0 |
| 0.5 | 0.0390 | −11.103 | 2.428 | 685.24 | −1.4133 | −0.096 | 171.10 | −12.5 |
| 0.525 | −0.2361 | −10.828 | 19.497 | 675.83 | −1.3664 | 4.165 | 163.91 | −518.0 |
| 0.55 | −0.4988 | −10.134 | 36.017 | 640.63 | −1.2127 | 8.05 | 144.91 | −992.3 |
| 0.575 | −0.7391 | −9.040 | 51.339 | 581.49 | −0.9690 | 11.319 | 114.91 | −1393.5 |
| 0.6 | −0.9477 | −7.582 | 64.911 | 500.93 | −0.6540 | 13.722 | 76.10 | −1691.9 |
| 0.625 | −1.1155 | −5.814 | 76.250 | 402.28 | −0.2917 | 15.074 | 13.35 | −1865.8 |
| 0.65 | −1.2355 | −3.797 | 84.953 | 289.80 | 0.0900 | 15.267 | −16.05 | −1902.8 |
| 0.675 | −1.3038 | −1.591 | 90.662 | 168.78 | 0.4618 | 14.279 | −62.75 | −1801.2 |
| 0.7 | −1.3144 | 0.712 | 93.347 | 43.51 | 0.7946 | 12.171 | −105.02 | −1570.0 |
| 0.725 | −1.2681 | 3.052 | 92.840 | −79.5 | 1.0622 | 9.088 | −140.20 | −1228.3 |
| 0.75 | −1.1656 | 5.355 | 89.230 | −194.41 | 1.2426 | 5.241 | −165.75 | −804.2 |
| 0.775 | −1.0044 | 7.514 | 83.041 | −298.12 | 1.3200 | 0.894 | −180.02 | −340.6 |
| 0.8 | −0.7941 | 9.487 | 74.472 | −383.98 | 1.2856 | −3.66 | −182.31 | 147.1 |
| 0.825 | −0.532 | 11.222 | 64.027 | −447.31 | 1.1378 | −8.124 | −172.94 | 593.3 |
| 0.85 | −0.2327 | 12.678 | 52.332 | −483.97 | 0.8825 | −12.221 | −153.25 | 965.7 |
| 0.875 | 0.0994 | 13.834 | 40.082 | −490.55 | 0.5318 | −15.72 | −125.58 | 1226.8 |
| 0.9 | 0.4565 | 14.684 | 28.074 | −464.26 | 0.1028 | −18.46 | −93.11 | 1344.9 |
| 0.925 | 0.8313 | 15.247 | 17.158 | −403.08 | −0.3842 | −20.368 | −59.75 | 1294.9 |
| 0.95 | 1.2168 | 15.559 | 8.221 | −305.66 | −0.9088 | −21.476 | −29.92 | 1059.4 |
| 0.975 | 1.6076 | 15.682 | 2.183 | −171.38 | −1.4525 | −21.933 | −8.40 | 628.4 |
| 1 | 2.0000 | 15.700 | 0 | 0 | −2.0000 | −22.013 | 0 | 0 |

**A.2**   Pinned–pinned beam.
First and second eigenfunctions

| $\alpha = \dfrac{x}{l}$ | $X_1$ | $X_1'$ | $X_1''$ | $X_1'''$ | $X_2$ | $X_2'$ | $X_2''$ | $X_2''''$ |
|---|---|---|---|---|---|---|---|---|
| 0 | 0 | 4.4429 | 0 | −43.849 | 0 | 8.8857 | 0 | −350.79 |
| 0.025 | 0.1110 | 4.4291 | −1.095 | −43.714 | 0.2212 | 8.7762 | −8.734 | −346.48 |
| 0.05 | 0.2213 | 4.3881 | −2.184 | −43.310 | 0.4370 | 8.4508 | −17.253 | −333.62 |
| 0.075 | 0.3301 | 4.3201 | −3.258 | −42.638 | 0.6420 | 7.9171 | −25.347 | −312.55 |
| 0.1 | 0.4370 | 4.2253 | −4.313 | −41.703 | 0.8313 | 7.1887 | −32.817 | −283.80 |
| 0.125 | 0.5412 | 4.1047 | −5.341 | −40.511 | 1 | 6.2831 | −39.479 | −248.05 |
| 0.15 | 0.6420 | 3.9568 | −6.337 | −39.070 | 1.1441 | 5.2228 | −45.168 | −206.19 |
| 0.175 | 0.7389 | 3.7882 | −7.293 | −37.387 | 1.2601 | 4.034 | −49.746 | −159.25 |
| 0.2 | 0.8313 | 3.5943 | −8.204 | −35.475 | 1.3450 | 2.7458 | −53.098 | −108.40 |
| 0.225 | 0.9185 | 3.3784 | −9.065 | −33.343 | 1.3968 | 1.3000 | −55.142 | −54.87 |
| 0.25 | 1 | 3.1416 | −9.87 | −31.006 | 1.4142 | 0 | −55.830 | 0 |
| 0.275 | 1.0754 | 2.8854 | −10.613 | −28.478 | 1.3968 | −1.3900 | −55.142 | 54.57 |
| 0.3 | 0.1441 | 2.6115 | −11.292 | −25.774 | 1.3450 | −2.7458 | −53.098 | 108.40 |
| 0.325 | 1.2058 | 2.3214 | −11.901 | −22.910 | 1.2601 | −4.0340 | −49.746 | 159.25 |
| 0.35 | 1.2601 | 2.0171 | −12.436 | −19.908 | 1.1441 | −5.2228 | −45.168 | 206.19 |
| 0.375 | 1.3066 | 1.7002 | −12.895 | −16.779 | 1 | −6.2831 | −39.479 | 248.05 |
| 0.4 | 1.3450 | 1.3729 | −13.275 | −13.549 | 0.8313 | −7.1887 | −32.817 | 283.80 |
| 0.425 | 1.3751 | 1.0372 | −13.572 | −10.236 | 0.6420 | −7.9171 | −25.347 | 312.55 |
| 0.45 | 1.3968 | 0.6949 | −13.786 | −6.859 | 0.4370 | −8.4508 | −17.253 | 333.62 |
| 0.475 | 1.4098 | 0.3486 | −13.915 | −3.441 | 0.2212 | −8.7762 | −8.734 | 346.48 |
| 0.5 | 1.4142 | 0 | −13.958 | 0 | 0 | −8.8857 | 0 | 350.79 |
| 0.525 | 1.4098 | −0.3486 | −13.915 | 3.441 | −0.2212 | −8.7762 | 8.734 | 346.48 |
| 0.55 | 1.3968 | −0.6949 | −13.786 | 6.859 | −0.437 | −8.4508 | 17.253 | 333.62 |
| 0.575 | 1.3751 | −1.0372 | −13.572 | 10.236 | −0.642 | −7.9171 | 25.347 | 312.55 |
| 0.6 | 1.3450 | −1.3729 | −13.275 | 13.549 | −0.8313 | −7.1887 | 32.817 | 283.80 |
| 0.625 | 1.3066 | −1.7002 | −12.895 | 16.779 | −1 | −6.2831 | 39.479 | 248.05 |
| 0.65 | 1.2601 | −2.0171 | −12.436 | 19.908 | −1.1441 | −5.2228 | 45.168 | 206.19 |
| 0.675 | 1.2058 | −2.3214 | −11.91 | 22.910 | −1.2601 | −4.0340 | 49.746 | 159.25 |
| 0.7 | 1.1441 | −2.6115 | −11.292 | 25.774 | −1.3450 | −2.7458 | 53.098 | 108.40 |
| 0.725 | 1.0754 | −2.8854 | −10.613 | 28.478 | −1.3968 | −1.3900 | 55.142 | 54.87 |
| 0.75 | 1 | −3.1416 | −9.87 | 31.006 | −1.4142 | 0 | 55.830 | 0 |
| 0.775 | 0.9185 | −3.3784 | −9.065 | 33.343 | −1.3968 | 1.3900 | 55.142 | −54.87 |
| 0.8 | 0.8313 | −3.5943 | −8.204 | 35.475 | −1.3450 | 2.7458 | 53.098 | −108.40 |
| 0.825 | 0.7839 | −3.7882 | −7.293 | 37.387 | −1.2601 | 4.0340 | 49.746 | −159.25 |
| 0.85 | 0.642 | −3.9586 | −6.337 | 39.070 | −1.1441 | 5.2228 | 45.168 | −206.19 |
| 0.875 | 0.5412 | −4.1047 | −5.341 | 40.511 | −1 | 6.2831 | 39.479 | −248.05 |
| 0.9 | 0.437 | −4.2253 | −4.313 | 41.703 | −0.8313 | 7.1887 | 32.817 | −283.80 |
| 0.925 | 0.3301 | −4.3201 | −3.258 | 42.638 | −0.6420 | 7.9171 | 25.347 | −312.55 |
| 0.95 | 0.2213 | −4.3881 | −2.184 | 43.310 | −0.4370 | 8.4508 | 17.253 | −333.62 |
| 0.975 | 0.111 | −4.4291 | −1.095 | 43.714 | −0.2212 | 8.7762 | 8.734 | −346.48 |
| 1 | 0 | −4.4429 | 0 | 43.849 | 0 | 8.8857 | 0 | −350.79 |

**A.2** (*cont'd*) Pinned–pinned beam.
Third and fourth eigenfunctions

| $\alpha = \dfrac{x}{l}$ | $X_3$ | $X_3'$ | $X_3''$ | $X_3'''$ | $X_4$ | $X_4'$ | $X_4''$ | $X_4'''$ |
|---|---|---|---|---|---|---|---|---|
| 0 | 0 | 13.329 | 0 | −1183.9 | 0 | 17.771 | 0 | −2806.3 |
| 0.025 | 0.3301 | 12.960 | −29.33 | −1151.2 | 0.4370 | 16.901 | −69.01 | −2669.0 |
| 0.05 | 0.6420 | 11.876 | −57.03 | −1054.9 | 0.8313 | 14.377 | −131.27 | −2270.4 |
| 0.075 | 0.9185 | 10.134 | −81.58 | −900.3 | 1.1441 | 10.445 | −180.68 | −1649.5 |
| 0.1 | 1.1441 | 7.834 | −101.63 | −695.9 | 1.3450 | 5.491 | −212.40 | −867.2 |
| 0.125 | 1.3066 | 5.101 | −116.06 | −453.1 | 1.4142 | 0.000 | −223.32 | 0.0 |
| 0.15 | 1.3968 | 2.085 | −124.07 | −185.2 | 1.3450 | −5.491 | −212.40 | 867.2 |
| 0.175 | 1.4098 | −1.046 | −125.23 | 92.9 | 1.1441 | −10.445 | −180.68 | 1649.5 |
| 0.2 | 1.3450 | −4.119 | −119.47 | 365.9 | 0.8313 | −14.377 | −131.27 | 2270.4 |
| 0.225 | 1.2058 | −6.964 | −107.11 | 618.6 | 0.4370 | −16.901 | −69.01 | 2669.0 |
| 0.25 | 1.0000 | −9.425 | −88.82 | 837.2 | 0.0000 | −17.771 | 0.000 | 2806.3 |
| 0.275 | 0.7386 | −11.364 | −65.64 | 1009.5 | −0.4370 | −16.901 | 69.01 | 2669.0 |
| 0.3 | 0.4370 | −12.676 | −38.82 | 1126.0 | −0.8313 | −14.377 | 131.27 | 2270.4 |
| 0.325 | 0.1110 | −13.287 | −9.86 | 1180.3 | −1.1441 | −10.445 | 180.68 | 1649.5 |
| 0.35 | −0.2212 | −13.165 | 19.65 | 1169.3 | −1.3450 | −5.491 | 212.40 | 867.2 |
| 0.375 | −0.5412 | −12.314 | 48.07 | 1093.8 | −1.4142 | 0 | 223.32 | 0.0 |
| 0.4 | −0.8313 | −10.783 | 73.84 | 957.8 | −1.3450 | 5.491 | 212.40 | −867.2 |
| 0.425 | −1.0754 | −8.656 | 95.52 | 768.9 | −1.1441 | 10.445 | 180.68 | −1649.5 |
| 0.45 | −1.2601 | −6.051 | 111.93 | 537.5 | −0.8313 | 14.377 | 131.27 | −2270.4 |
| 0.475 | −1.3751 | −3.111 | 122.15 | 276.4 | −0.4370 | 16.901 | 69.01 | −2669.0 |
| 0.5 | −1.4142 | 0.000 | 125.62 | 0.0 | 0.0000 | 17.771 | 0.00 | −2806.0 |
| 0.525 | −1.3751 | 3.111 | 122.15 | −276.4 | 0.4370 | 16.901 | −69.01 | −2669.0 |
| 0.55 | −1.2601 | 6.051 | 111.93 | −537.5 | 0.8313 | 14.377 | −131.27 | −2270.4 |
| 0.575 | −1.0754 | 8.656 | 95.52 | −768.9 | 1.1441 | 10.445 | −180.68 | −1649.5 |
| 0.6 | −0.8313 | 10.783 | 73.84 | −957.8 | 1.3450 | 5.491 | −212.40 | −867.2 |
| 0.625 | −0.5412 | 12.314 | 48.07 | −1093.8 | 1.4142 | 0.000 | −223.32 | 0.0 |
| 0.65 | −0.2212 | 13.165 | 19.65 | −1169.3 | 1.3450 | −5.491 | −212.40 | 867.2 |
| 0.675 | 0.1110 | 13.287 | −9.86 | −1180.3 | 1.1441 | −10.445 | −180.68 | 1649.5 |
| 0.7 | 0.4370 | 12.676 | −38.82 | −1126.0 | 0.8313 | −14.377 | −131.27 | 2270.4 |
| 0.725 | 0.7389 | 11.364 | −65.64 | −1009.5 | 0.4370 | −16.901 | −69.01 | 2669.0 |
| 0.75 | 1.000 | 9.425 | −88.82 | −837.2 | 0.0000 | −17.771 | 0.00 | 2806.3 |
| 0.775 | 1.2058 | 6.964 | −107.11 | −618.6 | −0.4370 | −16.901 | 69.01 | 2669.0 |
| 0.8 | 1.3450 | 4.119 | −119.47 | −365.9 | −0.8313 | −14.377 | 131.27 | 2270.4 |
| 0.825 | 1.4098 | 1.046 | −125.23 | −92.9 | −1.1441 | −10.445 | 180.68 | 1649.5 |
| 0.85 | 1.3968 | −2.085 | −124.07 | 185.2 | −1.3450 | −5.491 | 212.40 | 867.2 |
| 0.875 | 1.3066 | −5.101 | −116.06 | 453.1 | −1.4142 | 0.000 | 223.32 | 0.0 |
| 0.9 | 1.1441 | −7.834 | −101.63 | 695.9 | −1.3450 | 5.491 | 212.40 | −867.2 |
| 0.925 | 0.9185 | −10.134 | −81.58 | 900.3 | −1.1441 | 10.445 | 180.68 | −1649.5 |
| 0.95 | 0.642 | −11.876 | −57.03 | 1054.9 | −0.8313 | 14.377 | 131.27 | −2270.4 |
| 0.975 | 0.3301 | −12.96 | −29.33 | 1151.2 | −0.4370 | 16.901 | 69.01 | −2669.0 |
| 1 | 0.0000 | −13.329 | 0.00 | 1183.9 | 0.000 | 17.771 | 0.00 | −2806.3 |

**A.3**   Pinned–clamped beam.
First and second eigenfunctions

| $\alpha = \dfrac{x}{l}$ | $X_1$ | $X_1'$ | $X_1''$ | $X_1'''$ | $X_2$ | $X_2'$ | $X_2''$ | $X_2'''$ |
|---|---|---|---|---|---|---|---|---|
| 0 | 0 | 5.7102 | 0 | −83.26 | 0 | 9.9844 | 0 | −500.08 |
| 0.025 | 0.1425 | 5.6841 | −2.078 | −82.83 | 0.2483 | 9.8285 | −12.438 | −492.31 |
| 0.05 | 0.2838 | 5.6063 | −4.135 | −81.57 | 0.4889 | 9.3657 | −24.487 | −469.25 |
| 0.075 | 0.4224 | 5.4777 | −6.15 | −79.47 | 0.7141 | 8.6105 | −35.776 | −431.61 |
| 0.1 | 0.5572 | 5.2993 | −8.102 | −76.56 | 0.9172 | 7.5861 | −45.965 | −380.56 |
| 0.125 | 0.6870 | 5.0732 | −9.971 | −72.86 | 1.0915 | 6.3246 | −54.707 | −317.71 |
| 0.15 | 0.8105 | 4.8017 | −11.738 | −68.41 | 1.2317 | 4.8650 | −61.760 | −245.02 |
| 0.175 | 0.9267 | 4.4874 | −13.384 | −63.24 | 1.3335 | 3.2527 | −66.894 | −164.77 |
| 0.2 | 1.0346 | 4.1335 | −14.895 | −57.41 | 1.3935 | 1.5375 | −69.955 | −79.44 |
| 0.225 | 1.1331 | 3.7439 | −16.251 | −50.96 | 1.4099 | −0.2270 | −70.847 | 8.27 |
| 0.25 | 1.2215 | 3.3224 | −17.438 | −43.96 | 1.3822 | −1.9865 | −69.545 | 95.63 |
| 0.275 | 1.2990 | 2.8735 | −18.444 | −36.49 | 1.3111 | −3.6863 | −66.091 | 179.90 |
| 0.3 | 1.3650 | 2.4019 | −19.258 | −28.54 | 1.1988 | −5.2740 | −60.597 | 258.43 |
| 0.325 | 1.4189 | 1.9122 | −19.869 | −20.26 | 1.0487 | −6.7006 | −53.239 | 328.76 |
| 0.35 | 1.4605 | 1.4101 | −20.269 | −11.7 | 0.8655 | −7.9222 | −44.247 | 388.65 |
| 0.375 | 1.4894 | 0.9006 | −20.452 | −2.93 | 0.6547 | −8.9017 | −33.910 | 436.21 |
| 0.4 | 1.5055 | 0.3893 | −20.414 | 5.98 | 0.4227 | −9.6062 | −22.552 | 469.92 |
| 0.425 | 1.5089 | −0.1183 | −20.153 | 14.94 | 0.1766 | −10.0238 | −10.537 | 488.68 |
| 0.45 | 1.4997 | −0.6164 | −19.667 | 23.89 | −0.0760 | −10.1337 | 1.752 | 491.83 |
| 0.475 | 1.4782 | −1.0997 | −18.959 | 32.74 | −0.3275 | −9.9373 | 13.923 | 479.22 |
| 0.5 | 1.4449 | −1.5626 | −18.031 | 41.43 | −0.5703 | −9.4422 | 25.584 | 451.13 |
| 0.525 | 1.4003 | −1.9995 | −16.890 | 49.89 | −0.7973 | −8.6660 | 36.358 | 408.36 |
| 0.55 | 1.3451 | −2.4053 | −15.540 | 58.05 | −1.0015 | −7.6340 | 45.889 | 352.09 |
| 0.575 | 1.2803 | −2.7748 | −13.989 | 65.86 | −1.1771 | −6.3840 | 53.863 | 283.94 |
| 0.6 | 1.2067 | −3.1032 | −12.250 | 73.26 | −1.3192 | −4.957 | 60.005 | 205.58 |
| 0.625 | 1.1255 | −3.3858 | −10.331 | 80.19 | −1.4239 | −3.401 | 64.096 | 120.05 |
| 0.65 | 1.0379 | −3.6184 | −8.244 | 86.62 | −1.4786 | −1.771 | 65.963 | 28.95 |
| 0.675 | 0.9451 | −3.7967 | −6.004 | 92.51 | −1.5123 | −0.122 | 65.516 | −64.91 |
| 0.7 | 0.8485 | −3.9174 | −3.623 | 97.85 | −1.4951 | 1.486 | 62.714 | −158.97 |
| 0.725 | 0.7497 | −3.9769 | −1.116 | 102.60 | −1.4388 | 2.994 | 57.586 | −250.72 |
| 0.75 | 0.6502 | −3.9722 | 1.501 | 106.76 | −1.3467 | 4.346 | 50.218 | −337.82 |
| 0.775 | 0.5516 | −3.9009 | 4.216 | 110.33 | −1.2233 | 5.488 | 40.752 | −418.18 |
| 0.8 | 0.4557 | −3.7605 | 7.013 | 113.32 | −1.0744 | 6.368 | 29.384 | −490.00 |
| 0.825 | 0.3642 | −3.5498 | 9.878 | 115.75 | −0.9074 | 6.943 | 16.334 | −551.91 |
| 0.85 | 0.2789 | −3.2664 | 12.797 | 117.66 | −0.7302 | 7.173 | 1.875 | −603.04 |
| 0.875 | 0.2015 | −2.9095 | 15.756 | 119.08 | −0.5519 | 7.027 | −13.725 | −643.04 |
| 0.9 | 0.1340 | −2.4783 | 18.747 | 120.08 | −0.3822 | 6.480 | −30.186 | −672.11 |
| 0.925 | 0.0782 | −1.9720 | 21.758 | 120.70 | −0.2314 | 5.513 | −47.246 | −691.13 |
| 0.95 | 0.0360 | −1.3904 | 24.788 | 121.03 | −0.1102 | 4.114 | −64.671 | −701.61 |
| 0.975 | 0.0093 | −0.7331 | 27.807 | 121.16 | −0.0294 | 2.278 | −82.247 | −705.73 |
| 1 | 0 | 0 | 30.836 | 121.18 | 0 | 0 | −99.929 | −706.35 |

**A.3**   (*cont'd*) Pinned–clamped beam.
Third and fourth eigenfunctions

| $\alpha = \dfrac{x}{l}$ | $X_3$ | $X_3'$ | $X_3''$ | $X_3'''$ | $X_4$ | $X_4'$ | $X_4''$ | $X_4'''$ |
|---|---|---|---|---|---|---|---|---|
| 0 | 0 | 14.440 | 0 | −1505.2 | 0 | 18.882 | 0 | −3366.1 |
| 0.025 | 0.3571 | 13.972 | −37.22 | −1456.4 | 0.4633 | 17.840 | −82.60 | −3180.4 |
| 0.05 | 0.6910 | 12.599 | −72.03 | −1313.2 | 0.8755 | 14.828 | −156.09 | −2643.4 |
| 0.075 | 0.9802 | 10.398 | −102.17 | −1085.0 | 1.1911 | 10.181 | −212.33 | −1814.8 |
| 0.1 | 1.2059 | 7.546 | −125.69 | −786.4 | 1.3751 | 4.408 | −245.14 | −785.9 |
| 0.125 | 1.3534 | 4.193 | −141.07 | −436.8 | 1.4047 | −1.851 | −250.89 | 329.9 |
| 0.15 | 1.4133 | 0.569 | −147.29 | −58.9 | 1.2843 | −7.905 | −228.96 | 1409.3 |
| 0.175 | 1.3816 | −3.092 | −143.98 | 322.8 | 1.0914 | −13.087 | −181.74 | 2333.0 |
| 0.2 | 1.2603 | −6.552 | −131.33 | 683.7 | 0.642 | −16.825 | −114.47 | 2999.1 |
| 0.225 | 1.0576 | −9.588 | −110.18 | 1000.3 | 0.1937 | −18.704 | −34.55 | 3334.3 |
| 0.25 | 0.7862 | −12.001 | −81.86 | 1252.1 | −0.2759 | −18.520 | 49.17 | 3301.3 |
| 0.275 | 0.4639 | −13.636 | −48.24 | 1422.8 | −0.7151 | −16.292 | 127.46 | 2903.9 |
| 0.3 | 0.1117 | −14.387 | −11.48 | 1501.5 | −1.0754 | −12.264 | 191.69 | 2185.9 |
| 0.325 | −0.2476 | −14.204 | 26.02 | 1482.9 | −1.3171 | −6.884 | 234.74 | 1226.5 |
| 0.35 | −0.5908 | −13.099 | 61.86 | 1368.4 | −1.4133 | −0.774 | 251.88 | 131.8 |
| 0.375 | −0.8955 | −11.144 | 93.70 | 1165.4 | −1.3535 | 5.479 | 241.21 | −977.8 |
| 0.4 | −1.1419 | −8.465 | 119.5 | 887.1 | 1.1444 | 11.094 | 203.90 | −1979.3 |
| 0.425 | −1.3141 | −5.234 | 137.58 | 551.6 | −0.8091 | 15.484 | 144.06 | −2762.6 |
| 0.45 | −1.4008 | −1.660 | 146.78 | 180.8 | −0.3845 | 18.164 | 68.26 | −3241.2 |
| 0.475 | −1.3962 | 2.026 | 146.53 | −201.2 | 0.0823 | 18.837 | −15.00 | −3362.4 |
| 0.5 | −1.3005 | 5.587 | 136.84 | −569.5 | 0.5399 | 17.429 | −96.70 | −3112.9 |
| 0.525 | −1.1197 | 8.795 | 118.36 | −900.2 | 0.9376 | 14.091 | −167.78 | −2520.4 |
| 0.55 | −0.8654 | 11.433 | 92.32 | −1171.4 | 1.2314 | 9.194 | −220.40 | −1650.7 |
| 0.575 | −0.5537 | 13.362 | 60.44 | −1365.2 | 1.3887 | 3.274 | −248.80 | −599.9 |
| 0.6 | 0.2044 | 14.433 | 24.82 | −1468.8 | 1.3920 | −3.018 | −249.86 | 515.2 |
| 0.625 | 0.1599 | 14.592 | −12.14 | −1474.9 | 1.2405 | −8.991 | −223.53 | 1570.9 |
| 0.65 | 0.5175 | 13.833 | −48.10 | −1382.4 | 0.9506 | −13.991 | −172.80 | 2449.5 |
| 0.675 | 0.8448 | 12.216 | −80.52 | −1196.4 | 0.5537 | −17.474 | −103.36 | 3053.0 |
| 0.7 | 1.1221 | 9.855 | −107.23 | −927.8 | 0.0927 | −19.068 | −23.02 | 3312.3 |
| 0.725 | 1.3328 | 6.917 | −126.36 | −592.7 | −0.3825 | −18.609 | 59.13 | 3196.5 |
| 0.75 | 1.4677 | 3.612 | −136.48 | −210.8 | −0.8212 | −16.174 | 133.74 | 2714.3 |
| 0.775 | 1.5123 | 0.176 | −136.69 | 195.6 | −1.1772 | −12.059 | 192.18 | 1913.7 |
| 0.8 | 1.4747 | −3.137 | −126.68 | 603.3 | −1.4142 | −6.760 | 227.43 | 875.4 |
| 0.825 | 1.3586 | −6.075 | −106.70 | 989.8 | −1.1506 | −0.919 | 234.83 | −296.1 |
| 0.85 | 1.1762 | −8.396 | −77.53 | 1335.6 | −1.4617 | 4.733 | 212.47 | −1486.3 |
| 0.875 | 0.9457 | −9.885 | −40.40 | 1624.6 | −1.2816 | 9.462 | 161.29 | −2583.9 |
| 0.9 | 0.6904 | −10.326 | 3.14 | 1847.1 | −1.0020 | 12.586 | 84.82 | −3496.3 |
| 0.925 | 0.4372 | −9.689 | 51.37 | 2000.0 | −1.6704 | 13.538 | −11.46 | −4162.3 |
| 0.95 | 0.2163 | −7.769 | 102.58 | 2087.6 | −1.3467 | 11.903 | −121.07 | −4563.6 |
| 0.975 | 0.0596 | −4.547 | 155.31 | 2123.2 | −0.0990 | 7.427 | −237.70 | −4733.3 |
| 1 | 0 | 0 | 208.49 | 2128.7 | 0 | 0 | −356.53 | −4760.3 |

**A.4**  Pinned–free beam.
First and second eigenfunctions

| $\alpha = \dfrac{x}{l}$ | $X_1$ | $X_1'$ | $X_1''$ | $X_1'''$ | $X_2$ | $X_2'$ | $X_2''$ | $X_2'''$ |
|---|---|---|---|---|---|---|---|---|
| 0 | 0 | 5.4003 | 0 | −88.040 | 0 | 10.008 | 0 | −498.87 |
| 0.025 | 0.1348 | 5.3729 | −2.197 | −87.638 | 0.2489 | 9.853 | −12.407 | −491.08 |
| 0.05 | 0.2682 | 5.2907 | −4.376 | −86.440 | 0.4901 | 9.391 | −24.426 | −467.96 |
| 0.075 | 0.3989 | 5.1544 | −6.514 | −84.455 | 0.7160 | 8.638 | −35.683 | −430.23 |
| 0.1 | 0.5255 | 4.9656 | −8.592 | −81.707 | 0.9198 | 7.616 | −45.826 | −379.03 |
| 0.125 | 0.6467 | 4.7255 | −10.592 | −78.221 | 1.0949 | 6.359 | −54.536 | −316.00 |
| 0.15 | 0.7613 | 4.4367 | −12.497 | −74.033 | 1.2361 | 4.904 | −61.543 | −243.07 |
| 0.175 | 0.8682 | 4.1016 | −14.288 | −69.186 | 1.3388 | 3.298 | −66.626 | −162.52 |
| 0.2 | 0.9661 | 3.7234 | −15.952 | −63.733 | 1.4001 | 1.590 | −69.625 | −76.82 |
| 0.225 | 1.0540 | 3.3053 | −17.471 | −57.724 | 1.4179 | −0.165 | −70.447 | 11.34 |
| 0.25 | 1.1310 | 2.8510 | −18.833 | −51.226 | 1.3919 | −1.914 | −69.061 | 99.25 |
| 0.275 | 1.1963 | 2.3651 | −20.029 | −44.304 | 1.3227 | 3.601 | −65.509 | 184.19 |
| 0.3 | 1.2490 | 1.8512 | −21.046 | −37.032 | 1.2188 | −5.172 | −59.898 | 263.51 |
| 0.325 | 1.2887 | 1.3142 | −21.878 | −29.484 | 1.0655 | −6.580 | −52.400 | 334.80 |
| 0.35 | 1.3146 | 0.7589 | −22.517 | −21.742 | 0.8856 | −7.778 | −43.243 | 395.83 |
| 0.375 | 1.3265 | 0.1900 | −22.963 | −13.887 | 0.6786 | −8.730 | −32.709 | 444.77 |
| 0.4 | 1.3240 | −0.3876 | −23.212 | −6.003 | 0.4514 | −9.405 | −21.118 | 480.12 |
| 0.425 | 1.3071 | −0.9690 | −23.264 | 1.822 | 0.2109 | −9.780 | −8.826 | 500.84 |
| 0.45 | 1.2756 | −1.5493 | −23.122 | 9.504 | −0.0351 | −9.844 | −3.797 | 506.33 |
| 0.475 | 1.2297 | −2.1236 | −22.791 | 16.956 | −0.2787 | −9.591 | 16.364 | 496.51 |
| 0.5 | 1.1695 | −2.6872 | −22.277 | 24.092 | −0.5120 | −9.029 | 28.498 | 471.78 |
| 0.525 | 1.0954 | −3.2360 | −21.589 | 30.829 | −0.7276 | −8.173 | 39.834 | 432.97 |
| 0.55 | 1.0079 | −3.7654 | −20.739 | 37.085 | −0.9184 | −7.047 | 50.040 | 381.45 |
| 0.575 | 0.9073 | −4.2718 | −19.740 | 42.783 | −1.0780 | −5.683 | 58.815 | 318.97 |
| 0.6 | 0.7945 | −4.7512 | −18.606 | 47.846 | −1.2009 | −4.120 | 65.914 | 247.65 |
| 0.625 | 0.6700 | −5.2010 | −17.354 | 52.211 | −1.2827 | −2.403 | 71.144 | 169.93 |
| 0.65 | 0.5347 | −5.6182 | −16.002 | 55.789 | −1.3202 | −0.579 | 74.380 | 88.46 |
| 0.675 | 0.3894 | −6.0005 | −14.571 | 58.539 | −1.3112 | 1.299 | 75.561 | 6.10 |
| 0.7 | 0.2350 | −6.3462 | −13.082 | 60.398 | −1.2552 | 3.182 | 74.701 | −74.23 |
| 0.725 | 0.0724 | −6.6542 | −11.559 | 61.316 | −1.1525 | 5.018 | 71.891 | −149.61 |
| 0.75 | −0.0974 | −6.9242 | −10.025 | 61.245 | −1.0051 | 6.761 | 67.286 | −217.16 |
| 0.775 | −0.2735 | −7.1556 | −8.506 | 60.146 | −0.8156 | 8.369 | 61.120 | −274.19 |
| 0.8 | −0.4549 | −7.3497 | −7.027 | 57.984 | −0.5880 | 9.807 | 53.686 | −318.18 |
| 0.825 | −0.6407 | −7.5076 | −5.615 | 54.730 | −0.3269 | 11.046 | 45.339 | −346.89 |
| 0.85 | −0.8300 | −7.6313 | −4.299 | 50.326 | −0.0375 | 12.069 | 36.485 | −358.39 |
| 0.875 | −1.0219 | −7.7235 | −3.107 | 44.860 | 0.2747 | 12.870 | 27.577 | −351.09 |
| 0.9 | −1.2159 | −7.7879 | −2.066 | 38.212 | 0.6041 | 13.452 | 19.096 | −323.75 |
| 0.925 | −1.4112 | −7.8283 | −1.205 | 30.406 | 0.9456 | 13.832 | 11.564 | −275.43 |
| 0.95 | −1.6072 | −7.8500 | −0.556 | 21.437 | 1.2943 | 14.042 | 5.505 | −205.57 |
| 0.975 | −1.8035 | −7.8581 | −0.143 | 11.303 | 1.6466 | 14.124 | 1.468 | −113.80 |
| 1 | −2.0000 | −7.8593 | −0.000 | 0.000 | 2.0000 | 14.137 | 0.000 | 0.00 |

**A.4**  (*cont'd*) Pinned–free beam.
Third and fourth eigenfunctions

| $\alpha = \dfrac{x}{l}$ | $X_3$ | $X_3'$ | $X_3''$ | $X_3'''$ | $X_4$ | $X_4'$ | $X_4''$ | $X_4'''$ |
|---|---|---|---|---|---|---|---|---|
| 0 | 0 | 14.438 | 0 | −1505.3 | 0 | 18.882 | 0 | −3366.1 |
| 0.025 | 0.3571 | 13.971 | −37.23 | −1456.6 | 0.4633 | 17.840 | −82.60 | −3180.3 |
| 0.05 | 0.6910 | 12.597 | −72.04 | −1313.4 | 0.8755 | 14.828 | −156.08 | −2643.4 |
| 0.075 | 0.9801 | 10.408 | −102.19 | −1085.2 | 1.1911 | 10.180 | −212.33 | −1814.8 |
| 0.1 | 1.2057 | 7.543 | −125.71 | −786.6 | 1.3751 | 4.408 | −245.14 | −785.7 |
| 0.125 | 1.3532 | 4.190 | −141.09 | −437.1 | 1.4074 | −1.851 | −250.89 | 329.9 |
| 0.15 | 1.4129 | 0.565 | −147.34 | −59.3 | 1.2843 | −7.905 | −228.95 | 1409.3 |
| 0.175 | 1.3811 | −3.097 | −144.02 | 322.4 | 1.0195 | −13.087 | −181.73 | 2333.0 |
| 0.2 | 1.2598 | −6.558 | −131.39 | 683.1 | 0.6421 | −16.824 | −114.45 | 2999.2 |
| 0.225 | 1.0569 | −9.595 | −110.26 | 999.5 | 0.1938 | −18.704 | −34.54 | 3334.4 |
| 0.25 | 0.7852 | −12.011 | −81.95 | 1251.1 | −0.2759 | −18.519 | 49.19 | 3301.4 |
| 0.275 | 0.4627 | −13.648 | −48.36 | 1421.5 | −0.7150 | −16.289 | 127.49 | 2904.2 |
| 0.3 | 0.1102 | −14.399 | −11.65 | 1499.8 | −1.0753 | −12.262 | 191.72 | 2186.4 |
| 0.325 | −0.2496 | −14.225 | 25.81 | 1480.7 | −1.3168 | −6.880 | 234.79 | 1227.1 |
| 0.35 | −0.5934 | −13.126 | 61.59 | 1365.6 | −1.4129 | −0.739 | 251.94 | 132.5 |
| 0.375 | −0.8989 | −11.179 | 93.35 | 1161.8 | −1.3531 | 5.484 | 241.30 | −976.5 |
| 0.4 | −1.1463 | −8.510 | 119.04 | 882.4 | −1.1438 | 11.103 | 204.02 | −1977.8 |
| 0.425 | −1.3197 | −5.292 | 136.99 | 545.6 | −0.8082 | 15.497 | 144.24 | −2760.5 |
| 0.45 | −1.4080 | −1.734 | 146.03 | 173.0 | −0.3832 | 18.182 | 68.55 | −3288.2 |
| 0.475 | −1.4056 | 1.930 | 145.55 | −211.3 | 0.0842 | 18.861 | −14.68 | −3358.2 |
| 0.5 | −1.3126 | 5.464 | 135.57 | −582.4 | 0.5425 | 17.462 | −96.25 | −3106.9 |
| 0.525 | −1.1354 | 8.635 | 116.73 | −916.8 | 0.9412 | 14.138 | −167.15 | −2512.0 |
| 0.55 | −0.8856 | 11.236 | 90.22 | −1192.9 | 1.2363 | 9.259 | −219.52 | −1638.9 |
| 0.575 | −0.5798 | 13.096 | 57.72 | −1393.0 | 1.3956 | 3.365 | −247.57 | −583.6 |
| 0.6 | −0.2381 | 14.090 | 21.31 | −1504.7 | 1.4016 | −2.890 | −248.15 | 538.0 |
| 0.625 | 0.1164 | 14.148 | −16.67 | −1521.1 | 1.2539 | −8.812 | −221.15 | 1602.7 |
| 0.65 | 0.4614 | 13.260 | −53.95 | −1442.1 | 0.9693 | −13.741 | −169.47 | 2494.1 |
| 0.675 | 0.7724 | 11.476 | −88.07 | −1273.5 | 0.5798 | −17.125 | −98.71 | 3115.1 |
| 0.7 | 1.0286 | 8.900 | −116.98 | −1027.3 | 0.1292 | −18.581 | −16.53 | 3399.0 |
| 0.725 | 1.2120 | 6.251 | −138.94 | −721.1 | −0.3316 | −17.931 | 68.19 | 3317.6 |
| 0.75 | 1.3091 | 2.022 | −152.71 | −376.6 | −0.7502 | −15.226 | 146.40 | 2883.3 |
| 0.775 | 1.3112 | −1.876 | −157.65 | −18.4 | −1.078 | −10.734 | 209.85 | 2149.6 |
| 0.8 | 1.2152 | −5.787 | −153.74 | 327.0 | −1.2758 | −4.911 | 252.11 | 1204.9 |
| 0.825 | 1.0236 | −9.495 | −141.63 | 633.3 | −1.3173 | 1.661 | 269.29 | 163.9 |
| 0.85 | 0.7438 | −12.811 | −122.61 | 875.2 | −1.1918 | 8.337 | 260.59 | −843.9 |
| 0.875 | 0.3876 | −15.584 | −98.59 | 1030.5 | −0.9047 | 14.495 | 228.47 | −1686.9 |
| 0.9 | −0.0301 | −17.718 | −71.97 | 1080.3 | −0.4758 | 19.612 | 178.63 | −2243.9 |
| 0.925 | −0.4927 | −19.184 | −45.58 | 1010.1 | 0.0643 | 23.348 | 119.52 | −2413.5 |
| 0.95 | −0.9840 | −20.025 | −22.55 | 809.9 | 0.6791 | 25.600 | 61.81 | −2121.9 |
| 0.975 | −1.4898 | −20.367 | −6.21 | 474.0 | 1.3333 | 26.551 | 17.65 | −1324.0 |
| 1 | −2.000 | −20.420 | 0.00 | 0.0 | 2.000 | 26.703 | 0.00 | 0.0 |

**A.5**  Clamped–clamped beam.
First and second eigenfunctions

| $\alpha = \dfrac{x}{l}$ | $X_1$ | $X_1'$ | $X_1''$ | $X_1'''$ | $X_2$ | $X_2'$ | $X_2''$ | $X_2'''$ |
|---|---|---|---|---|---|---|---|---|
| 0 | 0 | 0 | 44.745 | −207.96 | 0 | 0 | 123.34 | −969.35 |
| 0.025 | 0.0134 | 1.0516 | 39.559 | −207.89 | 0.0360 | 2.780 | 99.12 | −968.19 |
| 0.05 | 0.0514 | 1.9738 | 34.376 | −207.51 | 0.1340 | 4.956 | 74.99 | −960.55 |
| 0.075 | 0.1114 | 2.7730 | 29.167 | −206.50 | 0.2789 | 6.533 | 51.18 | −941.22 |
| 0.1 | 0.1891 | 3.4366 | 24.036 | −204.66 | 0.4557 | 7.522 | 28.05 | −906.50 |
| 0.125 | 0.2818 | 3.9728 | 18.964 | −201.72 | 0.6502 | 7.944 | 6.01 | −853.89 |
| 0.15 | 0.3863 | 4.3835 | 13.981 | −197.56 | 0.8485 | 7.835 | −14.48 | −782.77 |
| 0.175 | 0.5005 | 4.6733 | 9.078 | −191.98 | 1.0379 | 7.236 | −32.97 | −692.95 |
| 0.2 | 0.6194 | 4.8407 | 4.373 | −185.00 | 1.2067 | 6.207 | −49.00 | −586.02 |
| 0.225 | 0.7410 | 4.8931 | −0.140 | −176.51 | 1.3452 | 4.811 | −62.15 | −464.49 |
| 0.25 | 0.8626 | 4.8755 | −4.420 | −166.50 | 1.4448 | 3.126 | −72.12 | −331.49 |
| 0.275 | 0.9824 | 4.6732 | −8.468 | −154.86 | 1.5007 | 1.233 | −78.66 | −191.10 |
| 0.3 | 1.0960 | 4.4146 | −12.170 | −141.89 | 1.5055 | −0.778 | −81.65 | −47.85 |
| 0.325 | 1.2019 | 4.0687 | −15.547 | −127.47 | 1.4605 | −2.819 | −81.08 | 93.56 |
| 0.35 | 1.2983 | 3.6429 | −18.523 | −111.91 | 1.3651 | −4.802 | −77.04 | 228.25 |
| 0.375 | 1.3838 | 3.1434 | −21.130 | −95.00 | 1.2216 | −6.644 | −69.75 | 351.62 |
| 0.4 | 1.4555 | 2.5882 | −23.279 | −77.26 | 1.0346 | −8.266 | −59.58 | 459.21 |
| 0.425 | 1.5126 | 1.9854 | −24.976 | −58.71 | 0.8107 | −9.603 | −46.96 | 547.16 |
| 0.45 | 1.5542 | 1.3459 | −26.204 | −39.56 | 0.5572 | −10.597 | −32.43 | 612.49 |
| 0.475 | 1.5796 | 0.6764 | −26.951 | −19.80 | 0.2839 | −11.213 | −16.55 | 652.53 |
| 0.5 | 1.5882 | 0.000 | −27.198 | 0.00 | 0.000 | −11.422 | 0.00 | 665.93 |
| 0.525 | 1.5796 | −0.6764 | −26.951 | 19.80 | −0.2839 | −11.213 | 16.55 | 652.53 |
| 0.55 | 1.5542 | −1.3459 | −26.204 | 39.56 | −0.5572 | −10.597 | 32.43 | 612.49 |
| 0.575 | 1.5126 | −1.9854 | −24.976 | 58.71 | −0.8107 | −9.603 | 46.96 | 547.16 |
| 0.6 | 1.4555 | −2.5882 | −23.279 | 77.26 | −1.0346 | −8.266 | 59.58 | 459.21 |
| 0.625 | 1.3838 | −3.1434 | −21.130 | 95.00 | −1.2216 | −6.644 | 69.75 | 351.62 |
| 0.65 | 1.2983 | −3.6429 | −18.523 | 111.91 | −1.3651 | −4.802 | 77.04 | 228.25 |
| 0.675 | 1.2019 | −4.0687 | −15.547 | 127.47 | −1.4605 | −2.819 | 81.08 | 93.56 |
| 0.7 | 1.0960 | −4.4146 | −12.170 | 141.89 | −1.5055 | −0.778 | 81.65 | −47.85 |
| 0.725 | 0.9824 | −4.6732 | −8.468 | 154.86 | −1.5007 | 1.233 | 78.66 | −191.10 |
| 0.75 | 0.8626 | −4.8355 | −4.420 | 166.50 | −1.4448 | 3.126 | 72.12 | −331.49 |
| 0.775 | 0.7410 | −4.8931 | −0.140 | 176.51 | −1.3452 | 4.811 | 62.15 | −464.49 |
| 0.8 | 0.6194 | −4.8407 | 4.373 | 185.00 | −1.2067 | 6.207 | 49.00 | −586.02 |
| 0.825 | 0.5005 | −4.6733 | 9.078 | 191.98 | −1.0379 | 7.236 | 32.97 | −692.95 |
| 0.85 | 0.3863 | −4.3835 | 13.981 | 197.56 | −0.8485 | 7.835 | 14.48 | −782.77 |
| 0.875 | 0.2818 | −3.9728 | 18.964 | 201.72 | −0.6502 | 7.944 | −6.01 | −853.89 |
| 0.9 | 0.1891 | −3.4366 | 24.036 | 204.66 | −0.4557 | 7.522 | −28.05 | −906.50 |
| 0.925 | 0.1114 | −2.7730 | 29.167 | 206.50 | −0.2789 | 6.533 | −51.18 | −941.22 |
| 0.95 | 0.0514 | −1.9738 | 34.376 | 207.51 | −0.1340 | 4.956 | −74.99 | −960.55 |
| 0.975 | 0.0134 | −1.0516 | 39.559 | 207.89 | −0.0360 | 2.780 | −99.12 | −968.19 |
| 1.000 | 0.000 | 0.000 | 44.745 | 207.96 | 0.000 | 0.000 | −123.34 | −969.35 |

**A.5**  (*cont'd*) Clamped–clamped beam.
Third and fourth eigenfunctions

| $\alpha = \dfrac{x}{l}$ | $X_3$ | $X_3'$ | $X_3''$ | $X_3'''$ | $X_4$ | $X_4'$ | $X_4''$ | $X_4'''$ |
|---|---|---|---|---|---|---|---|---|
| 0 | 0 | 0 | 241.81 | −2658.8 | 0 | 0 | 399.72 | −5650.9 |
| 0.025 | 0.0686 | 5.215 | 175.40 | −2650.2 | 0.1102 | 8.228 | 258.69 | −5613.0 |
| 0.05 | 0.2468 | 8.776 | 109.69 | −2595.1 | 0.3822 | 12.959 | 120.74 | −5376.9 |
| 0.075 | 0.4939 | 10.718 | 46.30 | −2461.4 | 0.7302 | 14.346 | −7.50 | −4824.4 |
| 0.1 | 0.7701 | 11.128 | −12.57 | −2230.8 | 1.0745 | 12.736 | −117.52 | −3919.9 |
| 0.125 | 1.0388 | 10.148 | −64.39 | −1899.5 | 1.3468 | 8.692 | −200.86 | −2702.7 |
| 0.15 | 1.2676 | 7.988 | −106.78 | −1476.4 | 1.4951 | 2.971 | −250.87 | −1271.7 |
| 0.175 | 1.4305 | 4.906 | −137.60 | −981.2 | 1.4886 | −3.540 | −263.86 | 231.7 |
| 0.2 | 1.5079 | 1.214 | −155.44 | −441.5 | 1.3192 | −9.913 | −240.01 | 1646.9 |
| 0.225 | 1.4888 | −2.751 | −159.60 | 109.3 | 1.0016 | −15.270 | −183.55 | 2816.6 |
| 0.25 | 1.3708 | −6.652 | −150.20 | 634.7 | 0.5702 | −18.881 | −102.37 | 3609.6 |
| 0.275 | 1.1596 | −10.158 | −128.36 | 1099.8 | 0.0762 | −20.271 | −6.96 | 3934.1 |
| 0.3 | 0.8686 | −12.981 | −95.99 | 1472.6 | −0.4221 | −19.227 | 90.33 | 3757.7 |
| 0.325 | 0.5181 | −14.891 | −55.72 | 1727.5 | −0.8656 | −15.843 | 176.97 | 3109.5 |
| 0.35 | 0.1332 | −15.730 | −10.73 | 1846.8 | −1.1989 | −10.547 | 242.37 | 2067.7 |
| 0.375 | −0.2589 | −15.416 | 35.42 | 1824.1 | −1.3822 | −3.973 | 278.18 | 765.0 |
| 0.4 | −0.6287 | −13.973 | 79.24 | 1661.0 | −1.3932 | 3.070 | 279.89 | −636.6 |
| 0.425 | −0.9482 | −11.508 | 117.47 | 1369.9 | −1.2342 | 9.747 | 246.55 | −1953.2 |
| 0.45 | −1.1965 | −8.176 | 146.92 | 967.1 | −0.9189 | 15.197 | 183.48 | −3039.6 |
| 0.475 | −1.3521 | −4.252 | 165.66 | 506.6 | −0.4901 | 18.749 | 97.71 | −3750.5 |
| 0.5 | −1.4052 | 0 | 172.09 | 0 | 0 | 19.981 | 0 | −3998.2 |
| 0.525 | −1.3521 | 4.252 | 165.66 | −506.6 | 0.4901 | 18.749 | −97.71 | −3750.5 |
| 0.55 | −1.1965 | 8.176 | 146.92 | −976.1 | 0.9189 | 15.197 | −183.48 | −3039.6 |
| 0.575 | −0.9483 | 11.508 | 117.47 | −1369.9 | 1.2342 | 9.747 | −246.55 | −1953.2 |
| 0.6 | −0.6287 | 13.973 | 79.24 | −1661 | 1.3932 | 3.070 | −279.89 | −636.6 |
| 0.625 | −0.2589 | 15.416 | 35.42 | −1824.1 | 1.3822 | −3.973 | −278.18 | 765.0 |
| 0.65 | 0.1332 | 15.730 | −10.73 | −1846.8 | 1.1989 | −10.547 | −242.37 | 2067.7 |
| 0.675 | 0.5181 | 14.891 | −55.72 | −1727.5 | 0.8656 | −15.843 | −176.97 | 3109.5 |
| 0.7 | 0.8686 | 12.981 | −95.99 | −1472.6 | 0.4221 | −19.227 | −90.33 | 3757.7 |
| 0.725 | 1.1596 | 10.158 | −128.36 | −1099.8 | −0.0762 | −20.271 | 6.96 | 3934.1 |
| 0.75 | 1.3708 | 6.652 | −150.20 | −634.7 | −0.5702 | −18.881 | 102.37 | 3609.6 |
| 0.775 | 1.4888 | 2.751 | −159.60 | −109.3 | −1.0016 | −15.270 | 183.55 | 2816.6 |
| 0.8 | 1.5079 | −1.214 | −155.44 | 441.5 | −1.3192 | −9.913 | 240.01 | 1646.9 |
| 0.825 | 1.4305 | −4.906 | −137.60 | 981.2 | −1.4886 | −3.540 | 263.86 | 231.7 |
| 0.85 | 1.2676 | −7.988 | −106.78 | 1476.1 | −1.4951 | 2.970 | 250.87 | −1271.7 |
| 0.875 | 1.0388 | −10.148 | −64.39 | 1899.5 | −1.3468 | 8.962 | 200.86 | −2702.7 |
| 0.9 | 0.7701 | −11.128 | −12.57 | 2230.8 | −1.0745 | 12.736 | 117.52 | −3919.9 |
| 0.925 | 0.4939 | −10.718 | 46.30 | 2461.4 | −0.7302 | 14.346 | 7.50 | −4824.4 |
| 0.95 | 0.2468 | −8.776 | 109.69 | 2595.1 | −0.3822 | 12.959 | −120.74 | −5376.9 |
| 0.975 | 0.0686 | −5.215 | 175.40 | 2650.2 | −0.1102 | 8.228 | −258.69 | −5613.0 |
| 1 | 0 | 0 | 241.81 | 2658.8 | 0 | 0 | −399.72 | −5650.9 |

**A.6**   Free–free beam.
First and second eigenfunctions

| $\alpha = \dfrac{x}{l}$ | $X_1$ | $X_1'$ | $X_1''$ | $X_1'''$ | $X_2$ | $X_2'$ | $X_2''$ | $X_2'''$ |
|---|---|---|---|---|---|---|---|---|
| 0 | −2.0000 | 9.2945 | 0.000 | 0.00 | −2.0000 | 15.720 | 0.000 | 0 |
| 0.025 | −1.7681 | 9.2921 | −0.300 | −23.53 | −1.6072 | 15.700 | −2.201 | −171.46 |
| 0.05 | −1.5364 | 9.2750 | −1.150 | −44.16 | −1.2160 | 15.575 | −8.261 | −305.63 |
| 0.075 | −1.3035 | 9.2302 | −2.492 | −62.04 | −0.8300 | 15.262 | −17.198 | −402.88 |
| 0.1 | −1.0744 | 9.1471 | −4.231 | −76.89 | −0.4549 | 14.700 | −28.104 | −463.85 |
| 0.125 | −0.8476 | 9.0163 | −6.305 | −88.88 | −0.0975 | 13.846 | −40.097 | −489.92 |
| 0.15 | −0.6249 | 8.8303 | −8.642 | −98.07 | 0.2349 | 12.693 | −52.326 | −483.16 |
| 0.175 | −0.4058 | 8.5811 | −11.197 | −104.56 | 0.5347 | 11.237 | −64.007 | −446.28 |
| 0.2 | −0.1954 | 8.2688 | −13.857 | −108.30 | 0.7945 | 9.503 | −74.416 | −382.78 |
| 0.225 | 0.0062 | 7.8892 | −16.579 | −109.47 | 1.0078 | 7.532 | −82.956 | −296.67 |
| 0.25 | 0.1976 | 7.4417 | −19.299 | −108.19 | 1.1695 | 5.375 | −89.101 | −192.79 |
| 0.275 | 0.3785 | 6.9220 | −21.979 | −104.55 | 1.2756 | 3.098 | −92.487 | −76.04 |
| 0.3 | 0.5440 | 6.3420 | −24.521 | −98.77 | 1.3240 | 0.776 | −92.846 | 47.99 |
| 0.325 | 0.6949 | 5.6976 | −26.891 | −91.03 | 1.3147 | −1.517 | −90.070 | 173.86 |
| 0.35 | 0.8280 | 5.0020 | −29.047 | −81.50 | 1.2492 | −3.701 | −84.183 | 296.15 |
| 0.375 | 0.9445 | 4.2462 | −30.959 | −70.33 | 1.1311 | −5.702 | −75.333 | 409.76 |
| 0.4 | 1.0405 | 3.4536 | −32.562 | −57.91 | 0.9662 | −7.446 | −63.806 | 509.78 |
| 0.425 | 1.1164 | 2.6246 | −33.841 | −44.42 | 0.7615 | −8.872 | −49.995 | 592.20 |
| 0.45 | 1.1713 | 1.7680 | −34.772 | −30.11 | 0.5259 | −9.932 | −34.365 | 653.49 |
| 0.475 | 1.2047 | 0.8850 | −35.341 | −15.13 | 0.2683 | −10.581 | −17.507 | 691.48 |
| 0.5 | 1.2157 | 0.0000 | −35.531 | 0.00 | 0.0000 | −10.799 | 0.000 | 704.41 |
| 0.525 | 1.2047 | −0.8850 | −35.341 | 15.13 | −0.2683 | −10.581 | 17.507 | 691.48 |
| 0.55 | 1.1713 | −1.7680 | −34.772 | 30.11 | −0.5259 | −9.932 | 34.365 | 653.49 |
| 0.575 | 1.1164 | −2.6246 | −33.841 | 44.42 | −0.7615 | −8.872 | 49.995 | 592.20 |
| 0.6 | 1.0405 | −3.4536 | −32.562 | 57.91 | −0.9662 | −7.446 | 63.806 | 509.78 |
| 0.625 | 0.9445 | −4.2462 | −30.959 | 70.33 | −1.1311 | −5.702 | 75.333 | 409.76 |
| 0.65 | 0.8280 | −5.0020 | −29.047 | 81.50 | −1.2492 | −3.701 | 84.183 | 296.15 |
| 0.675 | 0.6949 | −5.6976 | −26.891 | 91.03 | −1.3147 | −1.517 | 90.070 | 173.86 |
| 0.7 | 0.5440 | −6.3420 | −24.521 | 98.77 | −1.3240 | 0.776 | 92.846 | 47.99 |
| 0.725 | 0.3785 | −6.9220 | −21.979 | 104.55 | −1.2756 | 3.098 | 92.487 | −76.04 |
| 0.75 | 0.1976 | −7.4417 | −19.299 | 108.19 | −1.1695 | 5.375 | 89.101 | −192.79 |
| 0.775 | 0.0062 | −7.8892 | −16.579 | 109.47 | −1.0078 | 7.532 | 82.956 | −296.67 |
| 0.8 | −0.1954 | −8.2688 | −13.857 | 108.30 | −0.7945 | 9.503 | 74.416 | −382.78 |
| 0.825 | −0.4058 | −8.5811 | −11.197 | 104.56 | −0.5347 | 11.237 | 64.007 | −446.28 |
| 0.85 | −0.6249 | −8.8303 | −8.642 | 98.07 | −0.2349 | 12.693 | 52.326 | −483.16 |
| 0.875 | −0.8476 | −9.0163 | −6.305 | 88.88 | 0.0975 | 13.846 | 40.097 | −489.92 |
| 0.9 | −1.0744 | −9.1471 | −4.231 | 76.89 | 0.4549 | 14.700 | 28.104 | −463.85 |
| 0.925 | −1.3035 | −9.2302 | −2.492 | 62.04 | 0.8300 | 15.262 | 17.198 | −402.88 |
| 0.95 | −1.5364 | −9.2750 | −1.150 | 44.16 | 1.2160 | 15.575 | 8.261 | −305.63 |
| 0.975 | −1.7681 | −9.2921 | −0.300 | 23.53 | 1.6072 | 15.700 | 2.201 | −171.46 |
| 1 | −2.0000 | −9.2945 | 0.000 | 0.00 | 2.0000 | 15.720 | 0.000 | 0.00 |

**A.6**  (*cont'd*) Free–free beam.
Third and fourth eigenfunctions

| $\alpha = \dfrac{x}{l}$ | $X_3$ | $X_3'$ | $X_3''$ | $X_3'''$ | $X_4$ | $X_4'$ | $X_4''$ | $X_4'''$ |
|---|---|---|---|---|---|---|---|---|
| 0 | −2.0000 | 21.991 | 0.00 | 0.0 | −2.0000 | 28.274 | 0.000 | 0.000 |
| 0.025 | −1.4507 | 21.920 | −8.30 | −630.5 | −1.2944 | 28.085 | −22.02 | −1664.5 |
| 0.05 | −0.9073 | 21.465 | −29.84 | −1061.0 | −0.6041 | 26.904 | −76.39 | −2590.0 |
| 0.075 | −0.3829 | 20.358 | −59.72 | −1295.9 | −0.0375 | 24.139 | −145.94 | −2867.2 |
| 0.1 | 0.1039 | 18.451 | −93.10 | −1345.4 | + 0.5880 | 19.613 | −214.75 | −2545.4 |
| 0.125 | 0.5326 | 15.711 | −125.59 | −1227.0 | 1.0050 | 13.524 | −269.16 | −1737.1 |
| 0.15 | 0.8832 | 12.211 | −153.26 | −965.7 | 1.2552 | 6.363 | −298.80 | −593.9 |
| 0.175 | 1.1381 | 8.115 | −172.95 | −593.1 | 1.3202 | −1.159 | −297.51 | 707.5 |
| 0.2 | 1.2857 | 3.651 | −182.31 | −146.8 | 1.2009 | −8.240 | −263.65 | 1981.2 |
| 0.225 | 1.3200 | −0.904 | −180.01 | 332.6 | 0.9184 | −14.093 | −200.17 | 3051.8 |
| 0.25 | 1.2423 | −5.250 | −165.74 | 804.2 | 0.5122 | −18.061 | −113.95 | 3773.6 |
| 0.275 | 1.0617 | −9.096 | −140.20 | 1228.1 | 0.0348 | −19.684 | −15.23 | 4051.3 |
| 0.3 | 0.7939 | −12.180 | −105.02 | 1569.5 | −0.4520 | −18.802 | 84.35 | 3842.7 |
| 0.325 | 0.4609 | −14.288 | −62.64 | 1800.3 | −0.8855 | −15.559 | 173.00 | 3166.4 |
| 0.35 | 0.0887 | −15.275 | −16.11 | 1901.8 | −1.2127 | −10.346 | 239.62 | 2170.8 |
| 0.375 | −0.2929 | −15.087 | 31.30 | 1863.9 | −1.3919 | −3.828 | 276.24 | 794.1 |
| 0.4 | −0.6554 | −13.739 | 76.01 | 1689.3 | −1.4005 | 3.185 | 278.44 | −613.6 |
| 0.425 | −0.9716 | −11.330 | 114.65 | 1391.3 | −1.2336 | 9.773 | 246.66 | −1951.6 |
| 0.45 | −1.2153 | −8.073 | 144.66 | 988.5 | −0.9181 | 15.209 | 183.65 | −3037.2 |
| 0.475 | −1.3702 | −4.190 | 163.48 | 514.1 | −0.4889 | 18.766 | 97.95 | −3747.1 |
| 0.5 | −1.4233 | 0.000 | 169.89 | 0.0 | 0.000 | 20.005 | 0.00 | −3993.4 |
| 0.525 | −1.3702 | 4.190 | 163.48 | −514.1 | 0.4889 | 18.766 | −97.95 | −3747.1 |
| 0.55 | −1.2153 | 8.073 | 144.66 | −988.5 | 0.9181 | 15.209 | −183.65 | −3037.2 |
| 0.575 | −0.9716 | 11.330 | 114.65 | −1391.3 | 1.2336 | 9.773 | −246.66 | −1951.6 |
| 0.6 | −0.6554 | 13.739 | 76.01 | −1689.3 | 1.4005 | 3.185 | −278.44 | −613.6 |
| 0.625 | −0.2929 | 15.087 | 31.30 | −1863.9 | 1.3919 | −3.828 | −276.24 | 794.1 |
| 0.65 | 0.0887 | 15.275 | −16.11 | −1901.8 | 1.2127 | −10.346 | −239.62 | 2107.8 |
| 0.675 | 0.4609 | 14.288 | −62.64 | −1800.3 | 0.8855 | −15.559 | −173.00 | 3166.4 |
| 0.7 | 0.7939 | 12.180 | −105.02 | −1569.5 | 0.4520 | −18.802 | −84.35 | 3842.7 |
| 0.725 | 1.0617 | 9.096 | −140.20 | −1228.1 | −0.0348 | −19.684 | 15.23 | 4051.3 |
| 0.75 | 1.2423 | 5.250 | −165.74 | −804.2 | −0.5122 | −18.061 | 113.95 | 3773.6 |
| 0.775 | 1.320 | 0.904 | −180.01 | −332.6 | −0.9184 | −14.093 | 200.17 | 3051.8 |
| 0.8 | 1.2857 | −3.651 | −182.31 | 146.8 | −1.2009 | −8.240 | 263.65 | 1981.2 |
| 0.825 | 1.1381 | −8.115 | −172.95 | 593.1 | −1.3202 | −1.159 | 297.51 | 707.5 |
| 0.85 | 0.8832 | −12.211 | −153.26 | 965.7 | −1.2552 | 6.363 | 298.80 | −593.9 |
| 0.875 | 0.5326 | −15.711 | −125.59 | 1227.0 | −1.0050 | 13.524 | 269.16 | −1737.1 |
| 0.9 | 0.1039 | −18.451 | −93.10 | 1345.4 | −0.5880 | 19.613 | 214.75 | −2545.4 |
| 0.925 | −0.3829 | −20.358 | −59.72 | 1295.9 | −0.0375 | 24.139 | 145.94 | −2867.2 |
| 0.95 | −0.9073 | −21.465 | −29.84 | 1061.0 | 0.6041 | 26.904 | 76.39 | −2590.0 |
| 0.975 | −1.4507 | −21.920 | −8.30 | 630.5 | 1.2944 | 28.085 | 22.02 | −1644.5 |
| 1 | −2.0000 | −21.991 | 0.00 | 0.0 | 2.0000 | 28.274 | 0.00 | 0 |

# EIGENFUNCTIONS AND THEIR DERIVATIVES FOR MULTISPAN BEAMS WITH EQUAL LENGTH AND DIFFERENT BOUNDARY CONDITIONS

**B.1** Two–span beam.
First and second eigenfunctions

**First Span**

| $\alpha = \dfrac{x}{l}$ | $X_1$ | $X_1'$ | $X_1''$ | $X_1'''$ | $X_2$ | $X_2'$ | $X_2''$ | $X_2'''$ | |
|---|---|---|---|---|---|---|---|---|---|
| 0 | 0 | 3.142 | 0 | −31.02 | 0 | 4.038 | 0 | −58.89 | 1 |
| 0.05 | 0.1565 | 3.103 | −1.545 | −30.64 | 0.2007 | 3.965 | −2.925 | −57.70 | 0.95 |
| 0.1 | 0.3091 | 2.988 | −3.051 | −29.50 | 0.3941 | 3.748 | −5.731 | −54.15 | 0.90 |
| 0.15 | 0.4540 | 2.799 | −4.482 | −27.64 | 0.5732 | 3.396 | −8.303 | −48.38 | 0.85 |
| 0.2 | 0.5879 | 2.542 | −5.803 | −25.09 | 0.7316 | 2.923 | −10.54 | −40.60 | 0.80 |
| 0.25 | 0.7072 | 2.222 | −6.981 | −21.93 | 0.8638 | 2.349 | −12.33 | −31.09 | 0.75 |
| 0.3 | 0.8091 | 1.846 | −7.989 | −18.23 | 0.9652 | 1.698 | −13.62 | −20.18 | 0.70 |
| 0.35 | 0.8911 | 1.426 | −8.797 | −14.08 | 1.0330 | 0.9967 | −14.34 | −8.268 | 0.65 |
| 0.4 | 0.9511 | 0.9704 | −9.388 | −9.58 | 1.0650 | 0.2748 | −14.44 | 4.237 | 0.60 |
| 0.45 | 0.9877 | 0.4909 | −9.751 | −4.847 | 1.0600 | −0.4366 | −13.91 | 16.91 | 0.55 |
| 0.5 | 1 | 0 | −9.872 | 0 | 1.0220 | −1.106 | −12.75 | 29.32 | 0.50 |
| 0.55 | 0.9877 | −0.4909 | −9.751 | 4.847 | 0.9511 | −1.702 | −10.99 | 41.08 | 0.45 |
| 0.6 | 0.9511 | −0.9704 | −9.388 | 9.580 | 0.8532 | −2.195 | −8.66 | 51.38 | 0.40 |
| 0.65 | 0.8911 | −1.426 | −8.797 | 14.08 | 0.7337 | −2.559 | −5.827 | 61.28 | 0.35 |
| 0.7 | 0.8091 | −1.846 | −7.989 | 18.23 | 0.5998 | −2.770 | −2.558 | 69.22 | 0.30 |
| 0.75 | 0.7072 | −2.222 | −6.981 | 21.93 | 0.4596 | −2.809 | 1.068 | 75.52 | 0.25 |
| 0.8 | 0.5879 | −2.542 | −5.803 | 25.09 | 0.3220 | −2.659 | 4.967 | 80.16 | 0.20 |
| 0.85 | 0.4540 | −2.799 | −4.482 | 27.64 | 0.1970 | −2.309 | 9.058 | 83.23 | 0.15 |
| 0.9 | 0.3091 | −2.988 | −3.051 | 29.50 | 0.0946 | −1.751 | 13.27 | 84.93 | 0.10 |
| 0.95 | 0.1565 | −3.103 | −1.545 | 30.64 | 0.0254 | −0.9815 | 17.53 | 85.61 | 0.05 |
| 1 | 0 | −3.142 | 0 | 31.02 | 0 | 0 | 21.82 | 85.71 | 0 |
| | $-X_1$ | $X_1'$ | $-X_1''$ | $X_1'''$ | $X_2$ | $-X_2'$ | $X_2''$ | $-X_2'''$ | $\alpha$ |

**Second Span**

Third and fourth eigenfunctions    **First Span**

| $\alpha$ | $X_3$ | $X_3'$ | $X_3''$ | $X_3'''$ | $X_4$ | $X_4'$ | $X_4''$ | $X_4'''$ | |
|---|---|---|---|---|---|---|---|---|---|
| 0 | 0 | 6.283 | 0 | −2480 | 0 | 7.060 | 0 | −353.6 | 1 |
| 0.05 | 0.309 | 5.976 | −12.20 | −235.9 | 0.3456 | 6.623 | −17.32 | −331.8 | 0.95 |
| 0.1 | 0.5878 | 5.083 | −23.20 | −200.7 | 0.6485 | 5.364 | −32.50 | −269.1 | 0.90 |
| 0.15 | 0.8090 | 3.693 | −31.94 | −145.8 | 0.8709 | 3.440 | −43.67 | −173.3 | 0.85 |
| 0.2 | 0.9511 | 1.942 | −37.54 | −76.65 | 0.9853 | 1.087 | −49.47 | −56.17 | 0.80 |
| 0.25 | 1 | 0 | −39.48 | 0 | 0.9773 | −1.405 | −49.18 | 67.64 | 0.75 |
| 0.3 | 0.9511 | −1.942 | −37.54 | 76.65 | 0.8476 | −3.73 | −42.85 | 182.8 | 0.70 |
| 0.35 | 0.8090 | −3.693 | −31.94 | 145.8 | 0.6119 | −5.602 | −31.29 | 274.8 | 0.65 |
| 0.4 | 0.5878 | −5.083 | −23.20 | 200.7 | 0.2988 | −6.795 | −15.94 | 332.3 | 0.60 |
| 0.45 | 0.3090 | −5.976 | −12.20 | 235.9 | −0.0537 | −7.166 | 1.242 | 347.8 | 0.55 |
| 0.5 | 0 | −6.283 | 0 | 248.0 | −0.4033 | −6.676 | 18.09 | 319.0 | 0.50 |
| 0.55 | −0.3090 | −5.976 | 12.20 | 235.9 | −0.7082 | −5.398 | 32.45 | 249.0 | 0.45 |
| 0.6 | −0.5878 | −5.083 | 23.20 | 200.7 | −0.9328 | −3.504 | 42.43 | 145.5 | 0.40 |
| 0.65 | −0.8090 | −3.693 | 31.94 | 145.8 | −1.0530 | −1.251 | 46.65 | 20.44 | 0.35 |
| 0.7 | −0.9511 | −1.942 | 37.54 | 76.65 | −1.0570 | 1.051 | 44.35 | −112.5 | 0.30 |
| 0.75 | −1 | 0 | 39.48 | 0 | −0.9522 | 3.074 | 35.51 | −238.9 | 0.25 |
| 0.8 | −0.9511 | 1.942 | 37.54 | −76.65 | −0.7597 | 4.503 | 20.77 | −346.5 | 0.20 |
| 0.85 | −0.8090 | 3.693 | 31.94 | −145.8 | −0.5162 | 5.072 | 1.319 | −426.5 | 0.15 |
| 0.9 | −0.5878 | 5.083 | 23.20 | −200.7 | −0.2702 | 4.581 | −21.35 | −475.3 | 0.10 |
| 0.95 | −0.309 | 5.976 | 12.20 | −235.9 | −0.0778 | 2.908 | −45.74 | −496.2 | 0.05 |
| 1 | 0 | 6.283 | 0 | −248.0 | 0 | 0 | −70.67 | −499.6 | 0 |
| | $-X_3$ | $X_3'$ | $-X_3''$ | $X_3'''$ | $X_4$ | $-X_4'$ | $X_4''$ | $-X_4'''$ | $\alpha$ |

**Second Span**

## B.2  Three–span beam.

First and second eigenfunctions

**First Span**

| $\alpha$ | $X_1$ | $X_1'$ | $X_1''$ | $X_1'''$ | $X_2$ | $X_2'$ | $X_2''$ | $X_2'''$ | |
|---|---|---|---|---|---|---|---|---|---|
| 0 | 0 | 2.565 | 0 | −25.33 | 0 | 3.509 | 0 | −42.38 | 1 |
| 0.05 | 0.1277 | 2.534 | −1.261 | −25.01 | 0.1746 | 3.457 | −2.107 | −41.68 | 0.95 |
| 0.1 | 0.2523 | 2.440 | −2.491 | −24.09 | 0.3439 | 3.300 | −4.145 | −39.60 | 0.90 |
| 0.15 | 0.3707 | 2.286 | −3.660 | −22.57 | 0.5092 | 3.044 | −6.046 | −36.21 | 0.85 |
| 0.2 | 0.4800 | 2.075 | −4.738 | −20.49 | 0.6469 | 2.699 | −7.746 | −31.60 | 0.80 |
| 0.25 | 0.5774 | 1.814 | −5.700 | −17.91 | 0.7715 | 2.274 | −9.188 | −25.92 | 0.75 |
| 0.3 | 0.6606 | 1.508 | −6.522 | −14.88 | 0.8732 | 1.785 | −10.32 | −19.33 | 0.70 |
| 0.35 | 0.7276 | 1.164 | −7.183 | −11.49 | 0.9492 | 1.248 | −11.11 | −12.02 | 0.65 |
| 0.4 | 0.7766 | 0.7924 | −7.666 | −7.822 | 0.9975 | 0.6804 | 11.52 | −4.223 | 0.60 |
| 0.45 | 0.8065 | 0.4009 | −7.962 | −3.957 | 1.017 | 0.1027 | −11.53 | −3.849 | 0.55 |
| 0.5 | 0.8165 | 0 | −8.061 | 0 | 1.008 | −0.4654 | −11.13 | 11.96 | 0.50 |
| 0.55 | 0.8065 | −0.4009 | −7.962 | 3.957 | 0.971 | −1.004 | −10.33 | 19.89 | 0.45 |
| 0.6 | 0.7766 | −0.7924 | −7.666 | 7.822 | 0.9083 | −1.492 | −9.149 | 27.42 | 0.40 |
| 0.65 | 0.7276 | −1.164 | −7.183 | 11.49 | 0.8299 | −1.913 | −7.601 | 34.35 | 0.35 |
| 0.7 | 0.6606 | −1.508 | −6.522 | 14.88 | 0.7185 | −2.247 | −5.726 | 40.53 | 0.30 |
| 0.75 | 0.5774 | −1.814 | −5.700 | 17.91 | 0.6000 | −2.480 | −3.564 | 45.81 | 0.25 |
| 0.8 | 0.4800 | −2.075 | −4.738 | 20.49 | 0.4724 | −2.599 | −1.162 | 50.10 | 0.20 |
| 0.85 | 0.3707 | −2.286 | −3.660 | 22.57 | 0.3420 | −2.593 | 1.429 | 53.35 | 0.15 |
| 0.9 | 0.2523 | −2.44 | −2.491 | 24.09 | 0.2153 | −2.454 | 4.156 | 55.57 | 0.10 |
| 0.95 | 0.1277 | −2.534 | −1.261 | 25.01 | 0.0989 | −2.176 | 6.97 | 56.82 | 0.05 |
| 1 | 0 | −2.565 | 0 | 25.33 | 0 | −1.757 | 9.824 | 57.2 | 0 |
| | $X_1$ | $-X_1'$ | $X_1''$ | $-X_1'''$ | $-X_2$ | $X_2'$ | $-X_2''$ | $X_2'''$ | $\alpha$ |

**Third Span**

Third and fourth eigenfunctions    **First Span**

| $\alpha$ | $X_3$ | $X_3'$ | $X_3''$ | $X_3'''$ | $X_4$ | $X_4'$ | $X_4''$ | $X_4'''$ | |
|---|---|---|---|---|---|---|---|---|---|
| 0 | 0 | 2.834 | 0 | −49.78 | 0 | 5.130 | 0 | −202.5 | 1 |
| 0.05 | 0.1407 | 2.722 | −2.469 | −48.58 | 0.2523 | 4.879 | −9.96 | −192.6 | 0.95 |
| 0.1 | 0.2752 | 2.589 | −4.819 | −45.02 | 0.4799 | 4.151 | −18.94 | −163.8 | 0.90 |
| 0.15 | 0.3977 | 2.294 | −6.935 | −39.27 | 0.6605 | 3.015 | −26.08 | −119.0 | 0.85 |
| 0.2 | 0.5029 | 1.901 | −8.713 | −31.56 | 0.7765 | 1.585 | −30.65 | −62.59 | 0.80 |
| 0.25 | 0.5865 | 1.430 | −10.06 | −22.24 | 0.8165 | 0 | −32.23 | 0 | 0.75 |
| 0.3 | 0.6450 | 0.9031 | −10.92 | −11.71 | 0.7765 | −1.585 | −30.65 | 62.59 | 0.70 |
| 0.35 | 0.6763 | 0.3473 | −11.22 | −0.4086 | 0.6605 | −3.015 | −26.08 | 119.0 | 0.65 |
| 0.4 | 0.6797 | −0.2095 | −10.95 | 11.19 | 0.4799 | −4.151 | −18.94 | 163.8 | 0.60 |
| 0.45 | 0.6559 | −0.7384 | −10.11 | 22.61 | 0.2523 | −4.879 | −9.96 | 192.6 | 0.55 |
| 0.5 | 0.6068 | −1.211 | −8.702 | 33.41 | 0 | −5.130 | 0 | 202.5 | 0.50 |
| 0.55 | 0.5362 | −1.600 | −6.783 | 43.17 | −0.2523 | −4.879 | 9.96 | 192.6 | 0.45 |
| 0.6 | 0.4486 | −1.881 | −4.407 | 51.59 | −0.4799 | −4.151 | 18.94 | 163.8 | 0.40 |
| 0.65 | 0.3502 | −2.034 | −1.651 | 58.41 | −0.6605 | −3.015 | 26.08 | 119.0 | 0.35 |
| 0.7 | 0.2476 | −2.042 | 1.405 | 63.50 | −0.7765 | −1.585 | 30.65 | 62.59 | 0.30 |
| 0.75 | 0.1486 | −1.890 | 4.671 | 66.87 | −0.8165 | 0 | 32.23 | 0 | 0.25 |
| 0.8 | 0.0614 | −1.572 | 8.065 | 68.64 | −0.7765 | 1.585 | 30.65 | −62.59 | 0.20 |
| 0.85 | −0.0057 | −1.081 | 11.51 | 69.08 | −0.6605 | 3.015 | 26.08 | −119.0 | 0.15 |
| 0.9 | −0.0441 | −0.4212 | 14.96 | 68.61 | −0.4799 | 4.151 | 18.94 | −163.8 | 0.10 |
| 0.95 | −0.045 | 0.4122 | 18.37 | 67.79 | −0.2523 | 4.879 | 9.960 | −192.6 | 0.05 |
| 1 | 0 | 1.415 | 21.74 | 67.33 | 0 | 5.13 | 0 | −202.5 | 0 |
| | $X_3$ | $-X_3'$ | $X_3''$ | $-X_3'''$ | $-X_4$ | $X_4'$ | $-X_4''$ | $X_4'''$ | $\alpha$ |

**Third Span**

**B.2**   (*cont'd*) Three–span beam.
First and second eigenfunctions

**Second Span**

| $\alpha$ | $X_1$ | $X_1'$ | $X_1''$ | $X_1'''$ | $X_2$ | $X_2'$ | $X_2''$ | $X_2'''$ | |
|---|---|---|---|---|---|---|---|---|---|
| 0 | 0 | −2.565 | 0 | 25.33 | 0 | −1.7570 | 9.824 | −14.87 | 1 |
| 0.05 | −0.1277 | −2.534 | 1.261 | 25.01 | −0.0758 | −1.2830 | 9.097 | −15.17 | 0.95 |
| 0.1 | −0.2523 | −2.44 | 2.491 | 24.09 | −0.1285 | −0.8475 | 8.319 | −16.01 | 0.90 |
| 0.15 | −0.3707 | −2.286 | 3.660 | 22.57 | −0.1613 | −0.4520 | 7.490 | −17.18 | 0.85 |
| 0.2 | −0.4800 | −2.075 | 4.738 | 20.49 | −0.1748 | −0.0995 | 6.598 | −18.54 | 0.80 |
| 0.25 | −0.5774 | −1.814 | 5.700 | 17.91 | −0.1720 | 0.2066 | 5.636 | −19.93 | 0.75 |
| 0.3 | −0.6606 | −1.508 | 6.522 | 14.88 | −0.1550 | 0.4630 | 4.606 | −21.25 | 0.70 |
| 0.35 | −0.7276 | −1.164 | 7.183 | 11.49 | −0.1265 | 0.6662 | 3.514 | −22.38 | 0.65 |
| 0.4 | −0.7766 | −0.7924 | 7.666 | 7.822 | −0.0893 | 0.8136 | 2.372 | −23.25 | 0.60 |
| 0.45 | −0.8065 | −0.4009 | 7.962 | 3.957 | −0.0461 | 0.9028 | 1.195 | −23.79 | 0.55 |
| 0.5 | −0.8165 | 0 | 8.061 | 0 | 0 | 0.9327 | 0 | −23.98 | 0.50 |
| | $X_1$ | $-X_1'$ | $X_1''$ | $-X_1'''$ | $-X_2$ | $X_2'$ | $-X_2''$ | $X_2'''$ | $\alpha$ |

Third and fourth eigenfunctions     **Second Span**

| $\alpha$ | $X_3$ | $X_3'$ | $X_3''$ | $X_3'''$ | $X_4$ | $X_4'$ | $X_4''$ | $X_4'''$ | |
|---|---|---|---|---|---|---|---|---|---|
| 0 | 0 | 1.415 | 21.74 | −116.90 | 0 | 5.13 | 0 | −202.5 | 1 |
| 0.05 | 0.0954 | 2.353 | 15.85 | −116.00 | 0.2523 | 4.879 | −9.96 | −192.6 | 0.95 |
| 0.1 | 0.2305 | 3.001 | 10.11 | −113.30 | 0.4799 | 4.151 | −18.94 | −163.8 | 0.90 |
| 0.15 | 0.3908 | 3.367 | 4.563 | −108.00 | 0.6605 | 3.015 | −26.08 | −119.0 | 0.85 |
| 0.2 | 0.5626 | 3.463 | −0.6461 | −99.88 | 0.7765 | 1.585 | −30.65 | −62.59 | 0.80 |
| 0.25 | 0.7330 | 3.310 | −5.376 | −88.83 | 0.8165 | 0 | −32.23 | 0 | 0.75 |
| 0.3 | 0.8900 | 2.936 | −9.482 | −74.98 | 0.7765 | −1.585 | −30.65 | 62.59 | 0.70 |
| 0.35 | 1.0230 | 2.375 | −12.83 | −58.62 | 0.6605 | −3.015 | −26.08 | 119.00 | 0.65 |
| 0.4 | 1.1250 | 1.667 | −15.31 | −40.26 | 0.4799 | −4.151 | −18.94 | 163.80 | 0.60 |
| 0.45 | 1.1880 | 0.8595 | −16.83 | −20.49 | 0.2523 | −4.879 | −9.96 | 192.60 | 0.55 |
| 0.5 | 1.2100 | 0 | −17.35 | 0 | 0 | −5.13 | 0 | 202.50 | 0.50 |
| | $X_3$ | $-X_3'$ | $X_3''$ | $-X_3'''$ | $-X_4$ | $X_4'$ | $-X_4''$ | $X_4'''$ | $\alpha$ |

First and second eigenfunctions

**First Span**

| α | $X_1$ | $X_1'$ | $X_1''$ | $X_1'''$ | $X_2$ | $X_2'$ | $X_2''$ | $X_2'''$ | |
|---|---|---|---|---|---|---|---|---|---|
| 0 | 0 | 2.222 | 0 | −21.93 | 0 | 3.577 | 0 | −34.38 | 1 |
| 0.05 | 0.1106 | 2.194 | −1.092 | −21.66 | 0.1537 | 3.045 | −1.711 | −33.87 | 0.95 |
| 0.1 | 0.2185 | 2.113 | −2.157 | −20.86 | 0.3031 | 2.918 | −3.370 | −32.35 | 0.90 |
| 0.15 | 0.3211 | 1.980 | −3.170 | −19.54 | 0.4442 | 2.710 | −4.930 | −29.87 | 0.85 |
| 0.2 | 0.4157 | 1.797 | −4.104 | −17.74 | 0.5729 | 2.427 | −6.343 | −26.50 | 0.80 |
| 0.25 | 0.5001 | 1.571 | −4.937 | −15.51 | 0.6858 | 2.079 | −7.567 | −22.32 | 0.75 |
| 0.3 | 0.5721 | 1.306 | −5.648 | −12.89 | 0.7798 | 1.675 | −8.563 | −17.45 | 0.70 |
| 0.35 | 0.6301 | 1.008 | −6.220 | −9.955 | 0.8525 | 1.227 | −9.302 | −12.03 | 0.65 |
| 0.4 | 0.6725 | 0.6862 | −6.639 | −6.774 | 0.902 | 0.7492 | −9.759 | −6.20 | 0.60 |
| 0.45 | 0.6984 | 0.3472 | −6.895 | −3.427 | 0.9272 | 0.256 | −9.917 | −0.125 | 0.55 |
| 0.5 | 0.7071 | 0 | −6.981 | 0 | 0.9276 | −0.2375 | −9.770 | 6.035 | 0.50 |
| 0.55 | 0.6984 | −0.3472 | −6.895 | 3.427 | 0.9037 | −0.7159 | −9.316 | 12.12 | 0.45 |
| 0.6 | 0.6725 | −0.6862 | −6.639 | 6.774 | 0.8566 | −1.164 | −8.562 | 17.96 | 0.40 |
| 0.65 | 0.6301 | −1.008 | −6.220 | 9.955 | 0.7881 | −1.567 | −7.526 | 23.42 | 0.35 |
| 0.7 | 0.5721 | −1.306 | −5.648 | 12.89 | 0.7008 | −1.912 | −6.229 | 28.36 | 0.30 |
| 0.75 | 0.5001 | −1.571 | −4.937 | 15.51 | 0.598 | −2.186 | −4.700 | 32.67 | 0.25 |
| 0.8 | 0.4157 | −1.797 | −4.104 | 17.74 | 0.4835 | −2.379 | −2.973 | 36.26 | 0.20 |
| 0.85 | 0.3211 | −1.980 | −3.170 | 19.54 | 0.3616 | −2.481 | −1.086 | 39.07 | 0.15 |
| 0.9 | 0.2185 | −2.113 | −2.157 | 20.86 | 0.237 | −2.486 | 0.920 | 41.05 | 0.10 |
| 0.95 | 0.1106 | −2.194 | −1.092 | 21.66 | 0.1148 | −2.388 | 3.005 | 42.21 | 0.05 |
| 1 | 0 | −2.222 | 0 | 21.93 | 0 | −2.185 | 5.128 | 42.59 | 0 |
| | $-X_1$ | $X_1'$ | $-X_1''$ | $X_1'''$ | $X_2$ | $-X_2'$ | $X_2''$ | $-X_2'''$ | α |

**Fourth Span**

Third and fourth eigenfunctions  **First Span**

| α | $X_3$ | $X_3'$ | $X_3''$ | $X_3'''$ | $X_4$ | $X_4'$ | $X_4''$ | $X_4'''$ | |
|---|---|---|---|---|---|---|---|---|---|
| 0 | 0 | 2.8550 | 0 | −41.64 | 0 | 2.1 | 0 | −39.98 | 1 |
| 0.05 | 0.1419 | 2.8030 | −2.068 | −40.80 | 0.1042 | 2.05 | −1.982 | −38.45 | 0.95 |
| 0.1 | 0.2786 | 2.6500 | −4.052 | −38.29 | 0.2034 | 1.904 | −3.861 | −35.89 | 0.90 |
| 0.15 | 0.4053 | 2.4010 | −5.871 | −34.21 | 0.2931 | 1.668 | −5.539 | −30.95 | 0.85 |
| 0.2 | 0.5173 | 2.0670 | −7.450 | −28.71 | 0.3689 | 1.355 | −6.928 | −24.36 | 0.80 |
| 0.25 | 0.6108 | 1.6610 | −8.722 | −21.98 | 0.4275 | 0.9809 | −7.953 | −16.44 | 0.75 |
| 0.3 | 0.6825 | 1.2010 | −9.631 | −14.27 | 0.4663 | 0.5663 | −8.556 | −7.548 | 0.70 |
| 0.35 | 0.7303 | 0.7048 | −10.140 | −5.846 | 0.4838 | 0.133 | −8.699 | 1.904 | 0.65 |
| 0.4 | 0.7528 | 0.1943 | −10.210 | −2.996 | 0.4797 | −0.2956 | −8.364 | 11.49 | 0.60 |
| 0.45 | 0.7498 | −0.3087 | −9.835 | 11.96 | 0.4547 | −0.6955 | −7.555 | 20.78 | 0.55 |
| 0.5 | 0.7224 | −0.7818 | −9.017 | 20.73 | 0.411 | −1.044 | −6.297 | 29.39 | 0.50 |
| 0.55 | 0.6725 | −1.2030 | −7.770 | 29.05 | 0.3516 | −1.318 | −4.633 | 36.97 | 0.45 |
| 0.6 | 0.6033 | −1.5520 | −6.124 | 36.65 | 0.2807 | −1.501 | −2.622 | 43.25 | 0.40 |
| 0.65 | 0.5188 | −1.8100 | −4.120 | 43.33 | 0.2033 | −1.576 | −0.3329 | 48.05 | 0.35 |
| 0.7 | 0.4241 | −1.9590 | −1.809 | 48.94 | 0.1251 | −1.531 | 2.158 | 51.3 | 0.30 |
| 0.75 | 0.325 | −1.9860 | 0.7549 | 53.40 | 0.0523 | −1.358 | 4.772 | 53.05 | 0.25 |
| 0.8 | 0.2277 | −1.8800 | 3.512 | 56.68 | −0.0085 | −1.053 | 7.440 | 53.46 | 0.20 |
| 0.85 | 0.1393 | −1.6330 | 6.404 | 58.85 | −0.0507 | −0.6142 | 10.100 | 52.83 | 0.15 |
| 0.9 | 0.0669 | −1.2390 | 9.381 | 60.06 | −0.0677 | −0.0436 | 12.710 | 51.61 | 0.10 |
| 0.95 | 0.0179 | −0.6941 | 12.400 | 60.53 | −0.053 | 0.656 | 15.260 | 50.36 | 0.05 |
| 1 | 0 | 0 | 15.43 | 60.61 | 0 | 1.482 | 17.760 | 49.77 | 0 |
| | $-X_3$ | $X_3'$ | $-X_3''$ | $X_3'''$ | $X_4$ | $-X_4'$ | $X_4''$ | $X_4'''$ | α |

**Fourth Span**

**B.3** (*cont'd*) Four–span beam.
First and second eigenfunctions

**Second Span**

| $\alpha$ | $X_1$ | $X_1'$ | $X_1''$ | $X_1'''$ | $X_2$ | $X_2'$ | $X_2''$ | $X_2'''$ | |
|---|---|---|---|---|---|---|---|---|---|
| 0 | 0.00 | −2.222 | 0.000 | 21.93 | 0.000 | −2.185 | 5.128 | 6.026 | 1 |
| 0.05 | −0.1106 | −2.194 | 1.092 | 21.66 | −0.1027 | −1.920 | 5.431 | 5.678 | 0.95 |
| 0.1 | −0.2185 | −2.113 | 2.157 | 20.86 | −0.1918 | −1.642 | 5.693 | 4.695 | 0.90 |
| 0.15 | −0.3211 | −1.980 | 3.170 | 19.54 | −0.2667 | −1.352 | 5.892 | 3.168 | 0.85 |
| 0.2 | −0.4157 | −1.797 | 4.104 | 17.74 | −0.3268 | −1.054 | 6.002 | 1.193 | 0.80 |
| 0.25 | −0.5001 | −1.571 | 4.937 | 15.51 | −0.3721 | −0.7535 | 6.005 | −1.131 | 0.75 |
| 0.3 | −0.5721 | −1.306 | 5.648 | 12.89 | −0.4023 | −0.4557 | 5.885 | −3.705 | 0.70 |
| 0.35 | −0.6301 | −1.008 | 6.220 | 9.955 | −0.4178 | −0.1672 | 5.662 | −6.430 | 0.65 |
| 0.4 | −0.6725 | −0.6862 | 6.639 | 6.774 | −0.4192 | 0.1051 | 5.241 | −9.212 | 0.60 |
| 0.45 | −0.6984 | −0.3472 | 6.895 | 3.427 | −0.4076 | 0.3545 | 4.712 | −11.96 | 0.55 |
| 0.5 | −0.7071 | 0.0000 | 6.981 | 0.000 | −0.3843 | 0.5741 | 4.047 | −14.59 | 0.50 |
| 0.55 | −0.6984 | 0.3472 | 6.895 | −3.427 | −0.350 | 0.7546 | 3.266 | −17.06 | 0.45 |
| 0.6 | −0.6725 | 0.6862 | 6.639 | −6.774 | −0.3093 | 0.8978 | 2.349 | −19.22 | 0.40 |
| 0.65 | −0.6301 | 1.008 | 6.220 | −9.955 | −0.2619 | 0.9904 | 1.339 | −21.11 | 0.35 |
| 0.7 | −0.5721 | 1.306 | 5.648 | −12.89 | −0.2111 | 1.030 | 0.2428 | −22.68 | 0.30 |
| 0.75 | −0.5001 | 1.571 | 4.937 | −15.51 | −0.1598 | 1.013 | −0.9234 | −23.91 | 0.25 |
| 0.8 | 0.4157 | 1.797 | 4.104 | −17.74 | −0.1107 | 0.937 | −2.142 | −24.81 | 0.20 |
| 0.85 | −0.3211 | 1.980 | 3.170 | −19.54 | −0.0671 | 0.7987 | −3.399 | −25.39 | 0.15 |
| 0.9 | −0.2185 | 2.113 | 2.157 | −20.86 | −0.0319 | 0.5968 | −4.677 | −25.71 | 0.10 |
| 0.95 | −0.1106 | 2.194 | 1.092 | −21.66 | 0.0085 | 0.3306 | −5.967 | −25.84 | 0.05 |
| 1 | 0.000 | 2.222 | 0.000 | −21.93 | 0.000 | 0.000 | −7.260 | −25.86 | 0 |
| | $-X_1$ | $X_1'$ | $-X_1''$ | $X_1'''$ | $X_2$ | $-X_2'$ | $X_2''$ | $-X_2'''$ | $\alpha$ |

**Third Span**

Third and fourth eigenfunctions      **Second Span**

| $\alpha$ | $X_3$ | $X_3'$ | $X_3''$ | $X_3'''$ | $X_4$ | $X_4'$ | $X_4''$ | $X_4'''$ | |
|---|---|---|---|---|---|---|---|---|---|
| 0 | 0.0000 | 0.000 | 15.430 | −60.61 | 0.000 | 1.482 | 17.76 | −105.8 | 1 |
| 0.05 | 0.0179 | 0.6941 | 12.400 | −60.53 | 0.0940 | 2.233 | 12.39 | −104.9 | 0.95 |
| 0.1 | 0.0669 | 1.239 | 9.381 | −60.06 | 0.2189 | 2.722 | 7.21 | −101.8 | 0.90 |
| 0.15 | 0.1393 | 1.633 | 6.404 | −58.85 | 0.3619 | 2.957 | 2.25 | −96.08 | 0.85 |
| 0.2 | 0.2277 | 1.880 | 3.512 | −56.68 | 0.5107 | 2.953 | −2.35 | −87.44 | 0.80 |
| 0.25 | 0.3250 | 1.986 | 0.7549 | −53.40 | 0.6536 | 2.731 | −6.445 | −75.88 | 0.75 |
| 0.3 | 0.4241 | 1.959 | −1.809 | −48.94 | 0.7806 | 2.319 | −9.893 | −61.64 | 0.70 |
| 0.35 | 0.5188 | 1.810 | −4.120 | −43.33 | 0.8830 | 1.755 | −12.57 | −45.10 | 0.65 |
| 0.4 | 0.6033 | 1.552 | −6.124 | −36.65 | 0.9541 | 1.077 | −14.37 | −26.84 | 0.60 |
| 0.45 | 0.6725 | 1.203 | −7.770 | −29.05 | 0.9895 | 0.3328 | −15.24 | −7.516 | 0.55 |
| 0.5 | 0.7224 | 0.7818 | −9.017 | −20.73 | 0.9871 | −0.4303 | −15.12 | 12.13 | 0.50 |
| 0.55 | 0.7498 | 0.3087 | −9.835 | −11.96 | 0.9471 | −1.163 | −14.03 | 31.36 | 0.45 |
| 0.6 | 0.7528 | −0.1943 | −10.210 | −2.996 | 0.8721 | −1.818 | −12.00 | 49.44 | 0.40 |
| 0.65 | 0.7303 | −0.7048 | −10.140 | 5.845 | 0.7673 | −2.349 | −9.116 | 65.73 | 0.35 |
| 0.7 | 0.6825 | −1.201 | −9.631 | 14.27 | 0.6399 | −2.717 | −5.47 | 79.71 | 0.30 |
| 0.75 | 0.6108 | −1.661 | −8.722 | 21.98 | 0.4989 | −2.886 | −1.19 | 91.01 | 0.25 |
| 0.8 | 0.5173 | −2.067 | −7.450 | 28.71 | 0.3550 | −2.827 | 3.582 | 99.47 | 0.20 |
| 0.85 | 0.4053 | −2.401 | −5.871 | 34.21 | 0.2205 | −2.522 | 8.714 | 105.1 | 0.15 |
| 0.9 | 0.2786 | −2.650 | −4.052 | 38.29 | 0.1074 | −1.953 | 14.06 | 108.3 | 0.10 |
| 0.95 | 0.1419 | −2.803 | −2.068 | 40.80 | 0.0294 | −1.113 | 19.51 | 109.6 | 0.05 |
| 1 | 0.000 | −2.855 | 0.000 | 41.64 | 0.000 | 0.000 | 25.00 | 109.9 | 0 |
| | $-X_3$ | $X_3'$ | $-X_3''$ | $X_3'''$ | $X_4$ | $-X_4'$ | $X_4''$ | $-X_4'''$ | $\alpha$ |

**Third Span**

First and second eigenfunctions

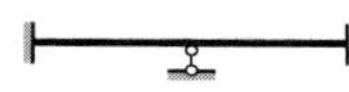

**First Span**

| $\alpha = \dfrac{x}{l}$ | $X_1$ | $X_1'$ | $X_1''$ | $X_1'''$ | $X_2$ | $X_2'$ | $X_2''$ | $X_2'''$ | |
|---|---|---|---|---|---|---|---|---|---|
| 0 | 0.000 | 0.000 | 21.69 | −85.22 | 0.000 | 0.000 | 31.64 | −147.0 | 1 |
| 0.05 | 0.0253 | 0.9778 | 17.43 | −85.12 | 0.0364 | 1.398 | 24.29 | 146.7 | 0.95 |
| 0.1 | 0.0942 | 1.743 | 13.18 | −84.45 | 0.1337 | 2.43 | 16.99 | −144.7 | 0.90 |
| 0.15 | 0.1961 | 2.297 | 8.999 | −82.75 | 0.2734 | 3.101 | 9.872 | −139.7 | 0.85 |
| 0.2 | 0.3205 | 2.644 | 4.931 | −79.70 | 0.4379 | 3.423 | 3.092 | −130.8 | 0.80 |
| 0.25 | 0.4572 | 2.793 | 1.055 | −75.08 | 0.6103 | 3.419 | −3.138 | −117.7 | 0.75 |
| 0.3 | 0.5967 | 2.755 | −2.549 | −68.81 | 0.7749 | 3.122 | −8.606 | −100.3 | 0.70 |
| 0.35 | 0.7298 | 2.544 | −5.799 | −60.92 | 0.9183 | 2.575 | −13.11 | −79.08 | 0.65 |
| 0.4 | 0.8486 | 2.182 | −8.616 | −51.52 | 1.029 | 1.830 | −16.46 | −54.63 | 0.60 |
| 0.45 | 0.9459 | 1.691 | −10.93 | −40.82 | 1.099 | 0.9499 | −18.53 | −27.91 | 0.55 |
| 0.5 | 1.016 | 1.099 | −12.68 | −29.13 | 1.123 | 0.000 | −19.23 | 0.000 | 0.50 |
| 0.55 | 1.055 | 0.4332 | −13.83 | −16.79 | 1.099 | −0.9499 | −18.53 | 27.91 | 0.45 |
| 0.6 | 1.059 | −0.247 | −14.36 | −4.193 | 1.029 | −1.830 | −16.46 | 54.63 | 0.40 |
| 0.65 | 1.027 | −0.9919 | −14.25 | 8.241 | 0.9183 | −2.575 | −13.11 | 79.08 | 0.35 |
| 0.7 | 0.9598 | −1.689 | −13.54 | 20.09 | 0.7749 | −3.122 | −8.606 | 100.3 | 0.30 |
| 0.75 | 0.8589 | −2.337 | −12.26 | 30.93 | 0.6103 | −3.419 | −3.138 | 117.7 | 0.25 |
| 0.8 | 0.7274 | −2.907 | −10.47 | 40.39 | 0.4379 | −3.423 | 3.092 | 130.8 | 0.20 |
| 0.85 | 0.6380 | −3.376 | −8.251 | 48.12 | 0.2734 | −3.101 | 9.872 | 139.7 | 0.15 |
| 0.9 | 0.3918 | −3.726 | −5.693 | 53.86 | 0.1337 | −2.430 | 16.99 | 144.7 | 0.10 |
| 0.95 | 0.1995 | −3.942 | −2.902 | 57.38 | 0.0364 | −1.398 | 24.29 | 146.7 | 0.05 |
| 1 | 0.000 | −4.015 | 0.000 | 58.52 | 0.000 | 0.000 | 31.64 | 147.0 | 0 |
| | $-X_1$ | $X_1'$ | $-X_1''$ | $X_1'''$ | $X_2$ | $-X_2'$ | $X_2''$ | $-X_2'''$ | $\alpha$ |

**Second Span**

Third and fourth eigenfunctions　　　**First Span**

| $\alpha$ | $X_3$ | $X_3'$ | $X_3''$ | $X_3'''$ | $X_4$ | $X_4'$ | $X_4''$ | $X_4'''$ | |
|---|---|---|---|---|---|---|---|---|---|
| 0 | 0.0000 | 0.000 | 70.66 | −499.5 | 0.000 | 0.000 | 87.22 | −685.5 | 1 |
| 0.05 | 0.0779 | 2.909 | 45.73 | −496.2 | 0.0947 | 3.505 | 53.03 | −679.3 | 0.95 |
| 0.1 | 0.2702 | 4.582 | 21.34 | −475.3 | 0.3222 | 5.319 | 19.84 | −641.1 | 0.90 |
| 0.15 | 0.5163 | 5.072 | −1.327 | −426.4 | 0.600 | 5.540 | −10.25 | −535.5 | 0.85 |
| 0.2 | 0.7597 | 4.503 | −20.78 | −346.5 | 0.8533 | 4.389 | −34.65 | −414.4 | 0.80 |
| 0.25 | 0.9522 | 3.073 | −35.51 | −238.9 | 1.022 | 2.210 | −51.00 | −234.4 | 0.75 |
| 0.3 | 1.057 | 1.050 | −44.35 | −112.4 | 1.064 | −0.5507 | −57.74 | −33.80 | 0.70 |
| 0.35 | 1.053 | −1.252 | −46.64 | 20.49 | 0.9652 | −3.397 | −54.47 | 161.5 | 0.65 |
| 0.4 | 0.9328 | −3.505 | −42.43 | 145.6 | 0.7315 | −5.846 | −42.13 | 324.8 | 0.60 |
| 0.45 | 0.7081 | −5.399 | −32.45 | 249.0 | 0.3940 | −7.495 | −22.91 | 433.1 | 0.55 |
| 0.5 | 0.4032 | −6.677 | −18.09 | 319.0 | 0.000 | −8.076 | 0.000 | 471.0 | 0.50 |
| 0.55 | 0.0536 | −7.166 | −1.235 | 347.8 | −0.3940 | −7.495 | 22.91 | 433.1 | 0.45 |
| 0.6 | −0.2989 | −6.795 | 15.95 | 332.3 | −0.7315 | −5.846 | 42.13 | 324.8 | 0.40 |
| 0.65 | −0.6120 | −5.602 | 31.29 | 274.8 | −0.9652 | −3.397 | 54.47 | 161.5 | 0.35 |
| 0.7 | −0.8477 | −3.729 | 42.85 | 182.7 | −1.064 | −0.5507 | 57.74 | −33.8 | 0.30 |
| 0.75 | −0.9774 | −1.404 | 49.18 | 67.58 | −1.022 | 2.210 | 51.00 | −234.4 | 0.25 |
| 0.8 | −0.9853 | 1.088 | 49.46 | −56.23 | −0.8533 | 4.389 | 34.65 | −414.4 | 0.20 |
| 0.85 | −0.8709 | 3.440 | 43.67 | −173.3 | −0.600 | 5.540 | 10.25 | −553.5 | 0.15 |
| 0.9 | −0.6484 | 5.364 | 32.49 | −269.2 | −0.3222 | 5.319 | −19.84 | −641.1 | 0.10 |
| 0.95 | −0.3456 | 6.622 | 17.31 | −331.9 | −0.0947 | 3.505 | −53.03 | −679.3 | 0.05 |
| 1 | 0.000 | 7.059 | −0.0136 | −353.5 | 0.000 | 0.000 | −87.22 | −685.5 | 0 |
| | $-X_3$ | $X_3'$ | $-X_3''$ | $X_3'''$ | $X_4$ | $-X_4'$ | $X_4''$ | $-X_4'''$ | $\alpha$ |

**Second Span**

**B.5**  Three–span beam.
First and second eigenfunctions

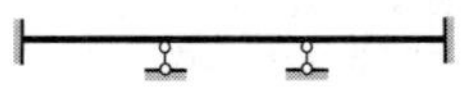

### First Span

| $\alpha$ | $X_1$ | $X_1'$ | $X_1''$ | $X_1'''$ | $X_2$ | $X_2'$ | $X_2''$ | $X_2'''$ | |
|---|---|---|---|---|---|---|---|---|---|
| 0 | 0.000 | 0.000 | 11.35 | −41.57 | 0.000 | 0.000 | 25.07 | −106.3 | 1 |
| 0.05 | 0.0133 | 0.5153 | 9.269 | −41.54 | 0.0291 | 1.121 | 19.76 | −106.1 | 0.95 |
| 0.1 | 0.0498 | 0.9269 | 7.197 | −41.30 | 0.1076 | 1.976 | 14.47 | −105.1 | 0.90 |
| 0.15 | 0.1042 | 1.235 | 5.145 | −40.69 | 0.2223 | 2.569 | 9.280 | −102.3 | 0.85 |
| 0.2 | 0.1716 | 1.442 | 3.136 | −39.59 | 0.3603 | 2.907 | 4.280 | −97.33 | 0.80 |
| 0.25 | 0.2468 | 1.550 | 1.196 | −37.92 | 0.509 | 3.003 | −0.4117 | −89.92 | 0.75 |
| 0.3 | 0.3250 | 1.563 | −0.6461 | −39.79 | 0.6568 | 2.874 | −4.670 | −79.97 | 0.70 |
| 0.35 | 0.4017 | 1.488 | −2.358 | −39.73 | 0.793 | 2.545 | −8.368 | −67.58 | 0.65 |
| 0.4 | 0.4725 | 1.330 | −3.909 | −29.23 | 0.9085 | 2.048 | −11.39 | −53.03 | 0.60 |
| 0.45 | 0.5335 | 1.100 | −5.272 | −25.20 | 0.9956 | 1.419 | −13.64 | −36.74 | 0.55 |
| 0.5 | 0.5814 | 0.8066 | −6.422 | −20.73 | 1.049 | 0.6978 | −15.05 | −19.26 | 0.50 |
| 0.55 | 0.6133 | 0.4616 | −7.340 | −15.94 | 1.065 | −0.0710 | −15.56 | −1.174 | 0.45 |
| 0.6 | 0.6269 | 0.0767 | −8.013 | −10.97 | 1.042 | −0.8428 | −15.16 | 16.85 | 0.40 |
| 0.65 | 0.6205 | −0.3355 | −8.436 | −5.968 | 0.981 | −1.573 | −13.88 | 34.15 | 0.35 |
| 0.7 | 0.5931 | −0.7627 | −8.612 | −1.101 | 0.8858 | −2.217 | −11.77 | 50.12 | 0.30 |
| 0.75 | 0.5442 | −1.193 | −8.551 | 3.461 | 0.7614 | −2.737 | −8.905 | 64.21 | 0.25 |
| 0.8 | 0.4740 | −1.614 | −8.274 | 7.546 | 0.6148 | −3.097 | −5.390 | 75.97 | 0.20 |
| 0.85 | 0.3831 | −2.017 | −7.808 | 10.99 | 0.4548 | −3.268 | −1.351 | 85.11 | 0.15 |
| 0.9 | 0.3060 | −2.392 | −7.189 | 13.62 | 0.2916 | −3.226 | 3.075 | 91.47 | 0.10 |
| 0.95 | 0.1444 | −2.734 | −6.461 | 15.30 | 0.1361 | −2.956 | 7.75 | 95.10 | 0.05 |
| 1 | 0.000 | −3.038 | −5.677 | 15.94 | 0.000 | −2.449 | 12.54 | 96.16 | 0 |

| | $X_1$ | $-X_1'$ | $X_1''$ | $-X_1'''$ | $-X_2$ | $X_2'$ | $-X_2''$ | $X_2'''$ | $\alpha$ |
|---|---|---|---|---|---|---|---|---|---|

### Third Span

Third and fourth eigenfunctions

### First Span

| $\alpha$ | $X_3$ | $X_3'$ | $X_3''$ | $X_3'''$ | $X_4$ | $X_4'$ | $X_4''$ | $X_4'''$ | |
|---|---|---|---|---|---|---|---|---|---|
| 0 | 0.000 | 0.000 | 25.83 | −120.1 | 0.000 | 0.000 | 40.99 | −274.6 | 1 |
| 0.05 | 0.0297 | 1.142 | 19.83 | −119.8 | 0.0455 | 1.706 | 27.28 | −273.0 | 0.95 |
| 0.1 | 0.1091 | 1.984 | 13.88 | −118.2 | 0.1592 | 2.732 | 13.83 | −263.1 | 0.90 |
| 0.15 | 0.2232 | 2.532 | 8.060 | −114.1 | 0.3078 | 3.103 | 1.199 | −239.6 | 0.85 |
| 0.2 | 0.3576 | 2.795 | 2.525 | −106.8 | 0.4596 | 2.878 | −9.874 | −200.7 | 0.80 |
| 0.25 | 0.4983 | 2.792 | −2.563 | −96.10 | 0.5873 | 2.154 | −18.63 | −147.4 | 0.75 |
| 0.3 | 0.6327 | 2.549 | −7.027 | −81.92 | 0.6689 | 1.064 | −24.43 | −83.38 | 0.70 |
| 0.35 | 0.7498 | 2.102 | −10.70 | −64.57 | 0.6902 | −0.2327 | −26.88 | −14.04 | 0.65 |
| 0.4 | 0.8403 | 1.494 | −13.44 | −44.61 | 0.6451 | −1.565 | −25.86 | 54.10 | 0.60 |
| 0.45 | 0.8974 | 0.7756 | −15.13 | −22.79 | 0.5359 | −2.764 | −21.60 | 114.4 | 0.55 |
| 0.5 | 0.9169 | 0.000 | −15.70 | 0.00 | 0.3734 | −3.681 | −14.65 | 160.8 | 0.50 |
| 0.55 | 0.8974 | −0.7756 | −15.13 | 22.79 | 0.1746 | −4.198 | −5.832 | 188.7 | 0.45 |
| 0.6 | 0.8403 | −1.494 | −13.44 | 44.61 | −0.0386 | −4.249 | 3.867 | 195.6 | 0.40 |
| 0.65 | 0.7498 | −2.102 | −10.70 | 64.57 | −0.2421 | −3.815 | 13.38 | 181.2 | 0.35 |
| 0.7 | 0.6327 | −2.549 | −7.027 | 81.92 | −0.4125 | −2.932 | 21.67 | 147.7 | 0.30 |
| 0.75 | 0.4983 | −2.792 | −2.563 | 96.10 | −0.5292 | −1.682 | 27.90 | 99.54 | 0.25 |
| 0.8 | 0.3576 | −2.795 | 2.525 | 106.8 | −0.5767 | −0.1858 | 31.48 | 42.95 | 0.20 |
| 0.85 | 0.2232 | −2.532 | 8.06 | 114.1 | −0.5460 | 1.418 | 32.18 | −14.54 | 0.15 |
| 0.9 | 0.1091 | −1.984 | 13.88 | 118.2 | −0.4355 | 2.987 | 30.15 | −64.88 | 0.10 |
| 0.95 | 0.0297 | −1.142 | 19.83 | 119.8 | −0.2500 | 4.396 | 25.94 | −100.2 | 0.05 |
| 1 | 0.000 | 0.000 | 25.83 | 120.1 | 0.000 | 5.560 | 20.50 | −113.5 | 0 |

| | $X_3$ | $-X_3'$ | $X_3''$ | $-X_3'''$ | $-X_4$ | $X_4'$ | $-X_4''$ | $X_4'''$ | $\alpha$ |
|---|---|---|---|---|---|---|---|---|---|

### Third Span

**B.5**  (*cont'd*) Three–span beam.
First and second eigenfunctions

**Second Span**

| $\alpha$ | $X_1$ | $X_1'$ | $X_1''$ | $X_1'''$ | $X_2$ | $X_2'$ | $X_2''$ | $X_2'''$ | |
|---|---|---|---|---|---|---|---|---|---|
| 0 | 0.000 | −3.038 | −5.677 | 57.48 | 0.000 | −2.449 | 12.54 | −10.16 | 1 |
| 0.05 | −0.1577 | −3.250 | −2.813 | 56.86 | −0.1069 | −1.834 | 12.02 | −11.12 | 0.95 |
| 0.1 | −0.3226 | −3.320 | −0.0123 | 54.94 | −0.1839 | −1.248 | 11.41 | −13.64 | 0.90 |
| 0.15 | −0.4874 | −3.253 | 2.659 | 51.70 | −0.2323 | −0.6965 | 10.64 | −17.23 | 0.85 |
| 0.2 | −0.6457 | −3.057 | 5.136 | 47.16 | −0.2542 | −0.1879 | 9.672 | −21.42 | 0.80 |
| 0.25 | −0.7912 | −2.744 | 7.355 | 41.41 | −0.2520 | 0.2671 | 8.493 | −25.77 | 0.75 |
| 0.3 | −0.9184 | −2.327 | 9.259 | 34.56 | −0.2286 | 0.6577 | 7.099 | −29.9 | 0.70 |
| 0.35 | −1.022 | −1.824 | 10.79 | 26.78 | −0.1875 | 0.9738 | 5.512 | −33.47 | 0.65 |
| 0.4 | −1.100 | −1.254 | 11.92 | 18.28 | −0.1326 | 1.206 | 3.766 | −36.22 | 0.60 |
| 0.45 | −1.147 | −0.639 | 12.61 | 9.271 | −0.0684 | 1.348 | 1.908 | −37.94 | 0.55 |
| 0.5 | −1.163 | 0.000 | 12.85 | 0.000 | 0.000 | 1.396 | 0.000 | −38.52 | 0.50 |
| | $X_1$ | $-X_1'$ | $X_1''$ | $-X_1'''$ | $-X_2$ | $X_2'$ | $-X_2''$ | $X_2'''$ | $\alpha$ |

Third and fourth eigenfunctions    **Second Span**

| $\alpha$ | $X_3$ | $X_3'$ | $X_3''$ | $X_3'''$ | $X_4$ | $X_4'$ | $X_4''$ | $X_4'''$ | |
|---|---|---|---|---|---|---|---|---|---|
| 0 | 0.000 | 0.000 | 25.84 | −120.1 | 0.000 | 5.560 | 20.56 | −388.0 | 1 |
| 0.05 | 0.0297 | 1.142 | 19.83 | −119.8 | 0.2955 | 6.104 | 1.345 | −373.3 | 0.95 |
| 0.1 | 0.1091 | 1.984 | 13.88 | −118.1 | 0.5948 | 5.720 | −16.32 | −328.1 | 0.90 |
| 0.15 | 0.2232 | 2.532 | 8.067 | −114.0 | 0.8539 | 4.522 | −30.98 | −254.2 | 0.85 |
| 0.2 | 0.3575 | 2.795 | 2.525 | −106.8 | 1.036 | 2.693 | −41.36 | 157.8 | 0.80 |
| 0.25 | 0.4982 | 2.792 | −2.562 | −96.09 | 1.117 | 0.4728 | −46.54 | −47.91 | 0.75 |
| 0.3 | 0.6327 | 2.549 | −7.026 | −81.92 | 1.082 | −1.867 | −46.11 | 64.33 | 0.70 |
| 0.35 | 0.7497 | 2.102 | −10.76 | −64.57 | 0.9327 | −4.048 | −40.26 | 167.2 | 0.65 |
| 0.4 | 0.8402 | 1.494 | −13.44 | −44.61 | 0.6839 | −5.815 | −29.73 | 249.7 | 0.60 |
| 0.45 | 0.8973 | 0.7756 | −15.13 | −22.79 | 0.3615 | −6.964 | −15.77 | 303.1 | 0.55 |
| 0.5 | 0.9168 | 0.000 | −15.70 | 0.00 | 0.000 | −7.362 | 0.00 | 321.6 | 0.50 |
| | $X_3$ | $-X_3'$ | $X_3''$ | $-X_3'''$ | $-X_4$ | $X_4'$ | $-X_4''$ | $X_4'''$ | $\alpha$ |

**B.6**  Four–span beam.
First and second eigenfunctions

**First Span**

| $\alpha$ | $X_1$ | $X_1'$ | $X_1''$ | $X_1'''$ | $X_2$ | $X_2'$ | $X_2''$ | $X_2'''$ | |
|---|---|---|---|---|---|---|---|---|---|
| 0 | 0.000 | 0.000 | 7.259 | −25.86 | 0.000 | 0.000 | 18.85 | −75.99 | 1 |
| 0.05 | 0.000 | 0.3306 | 5.966 | −25.84 | 0.0219 | 0.8477 | 15.06 | −75.89 | 0.95 |
| 0.1 | 0.0319 | 0.5966 | 4.677 | −25.71 | 0.0816 | 1.506 | 11.27 | −75.22 | 0.90 |
| 0.15 | 0.0671 | 0.7984 | 3.398 | −25.39 | 0.1694 | 1.976 | 7.550 | −73.56 | 0.85 |
| 0.2 | 0.1107 | 0.9368 | 2.142 | −24.80 | 0.2761 | 2.263 | 3.941 | −70.57 | 0.80 |
| 0.25 | 0.1597 | 1.013 | 0.923 | −23.91 | 0.3927 | 2.373 | 0.5181 | −66.07 | 0.75 |
| 0.3 | 0.2111 | 1.030 | −0.243 | −22.68 | 0.5107 | 2.319 | −2.640 | −59.98 | 0.70 |
| 0.35 | 0.2618 | 0.9901 | −1.339 | −21.11 | 0.6221 | 2.115 | −5.453 | −52.33 | 0.65 |
| 0.4 | 0.3092 | 0.8975 | −2.349 | −19.22 | 0.7200 | 1.781 | −7.849 | −43.26 | 0.60 |
| 0.45 | 0.3508 | 0.7570 | −3.256 | −17.02 | 0.7984 | 1.338 | −9.760 | −33.00 | 0.55 |
| 0.5 | 0.3842 | 0.5739 | −4.047 | −14.58 | 0.8525 | 0.8136 | −11.13 | −21.84 | 0.50 |
| 0.55 | 0.4075 | 0.3544 | −4.711 | −11.95 | 0.8788 | 0.2344 | −11.94 | −10.14 | 0.45 |
| 0.6 | 0.4191 | 0.1050 | −5.240 | −9.207 | 0.8755 | −0.370 | −12.14 | 1.718 | 0.40 |
| 0.65 | 0.4177 | −0.1673 | −5.631 | −6.426 | 0.8419 | −0.9703 | −11.77 | 13.33 | 0.35 |
| 0.7 | 0.4021 | −0.4557 | −5.884 | −3.700 | 0.7790 | −1.537 | −10.82 | 24.29 | 0.30 |
| 0.75 | 0.3719 | −0.7535 | −6.004 | −1.126 | 0.6892 | −2.044 | −9.356 | 34.21 | 0.25 |
| 0.8 | 0.3267 | −1.054 | −6.001 | 1.198 | 0.5761 | −2.465 | −7.425 | 42.76 | 0.20 |
| 0.85 | 0.2666 | −1.352 | −5.890 | 3.173 | 0.4444 | −2.780 | −5.107 | 49.66 | 0.15 |
| 0.9 | 0.1917 | −1.642 | −5.691 | 4.700 | 0.3001 | −2.971 | −2.490 | 54.69 | 0.10 |
| 0.95 | 0.1026 | −1.920 | −5.429 | 5.683 | 0.1496 | −3.026 | +0.3284 | 57.73 | 0.05 |
| 1 | 0.000 | −2.184 | −5.133 | 6.046 | 0.000 | −2.936 | 3.248 | 58.64 | 0 |
| | $-X_1$ | $X_1'$ | $-X_1''$ | $X_1'''$ | $X_2$ | $-X_2'$ | $X_2''$ | $-X_2'''$ | $\alpha$ |

**Fourth Span**

Third and fourth eigenfunctions — **First Span**

| $\alpha$ | $X_3$ | $X_3'$ | $X_3''$ | $X_3'''$ | $X_4$ | $X_4'$ | $X_4''$ | $X_4'''$ | |
|---|---|---|---|---|---|---|---|---|---|
| 0 | 0.000 | 0.000 | 25.04 | −110.0 | 0.000 | 0.000 | 22.37 | −104.0 | 1 |
| 0.05 | 0.029 | 1.114 | 19.54 | −109.8 | 0.0258 | 0.9887 | 17.18 | −103.8 | 0.95 |
| 0.1 | 0.1068 | 1.955 | 14.08 | −108.5 | 0.0945 | 1.718 | 12.02 | −102.3 | 0.90 |
| 0.15 | 0.2199 | 2.524 | 8.727 | −105.3 | 0.1933 | 2.193 | 6.981 | −98.77 | 0.85 |
| 0.2 | 0.3549 | 2.831 | 3.593 | −99.62 | 0.3097 | 2.420 | 2.186 | −92.50 | 0.80 |
| 0.25 | 0.4989 | 2.889 | −1.188 | −91.15 | 0.4315 | 2.418 | −2.219 | −83.22 | 0.75 |
| 0.3 | 0.6401 | 2.721 | −5.474 | −79.84 | 0.5480 | 2.208 | −6.086 | −70.94 | 0.70 |
| 0.35 | 0.7677 | 2.353 | −9.127 | −65.84 | 0.6493 | 1.821 | −9.268 | −55.92 | 0.65 |
| 0.4 | 0.8726 | 1.821 | −12.02 | −49.52 | 0.7277 | 1.294 | −11.64 | −38.63 | 0.60 |
| 0.45 | 0.9477 | 1.165 | −14.05 | −31.41 | 0.7771 | 0.6717 | −13.10 | −19.73 | 0.55 |
| 0.5 | 0.9878 | 0.4313 | −15.14 | −12.15 | 0.7940 | 0.000 | −13.60 | 0.00 | 0.50 |
| 0.55 | 0.9903 | −0.3327 | −15.26 | +7.537 | 0.7771 | −0.6717 | −13.10 | 19.73 | 0.45 |
| 0.6 | 0.9548 | −1.078 | −14.39 | 26.90 | 0.7277 | −1.294 | −11.64 | 38.63 | 0.40 |
| 0.65 | 0.8836 | −1.756 | −12.58 | 45.19 | 0.6493 | −1.821 | −9.268 | 55.92 | 0.35 |
| 0.7 | 0.7811 | −2.322 | −9.903 | 61.75 | 0.5480 | −2.208 | −6.086 | 70.94 | 0.30 |
| 0.75 | 0.6540 | −2.734 | −6.448 | 76.02 | 0.4315 | −2.418 | −2.219 | 83.22 | 0.25 |
| 0.8 | 0.5109 | −2.956 | −2.346 | 87.60 | 0.3097 | −2.420 | 2.186 | 92.50 | 0.20 |
| 0.85 | 0.3620 | −2.960 | 2.263 | 96.26 | 0.1933 | −2.193 | 6.981 | 98.77 | 0.15 |
| 0.9 | 0.2189 | −2.724 | 7.231 | 102.0 | 0.0945 | −1.718 | 12.02 | 102.3 | 0.10 |
| 0.95 | 0.0939 | −2.233 | 12.42 | 105.1 | 0.0258 | −0.9887 | 17.18 | 103.8 | 0.05 |
| 1 | 0.000 | −1.480 | 17.70 | 106.0 | 0.000 | 0.000 | 22.37 | 104.0 | 0 |
| | $-X_3$ | $X_3'$ | $-X_3''$ | $X_3'''$ | $X_4$ | $-X_4'$ | $X_4''$ | $-X_4'''$ | $\alpha$ |

**Fourth Span**

**B.6** (*cont'd*) Four–span beam.
First and second eigenfunctions

**Second Span**

| $\alpha$ | $X_1$ | $X_1'$ | $X_1''$ | $X_1'''$ | $X_2$ | $X_2'$ | $X_2''$ | $X_2'''$ | |
|---|---|---|---|---|---|---|---|---|---|
| 0 | 0.000 | −2.184 | −5.133 | 42.60 | 0.000 | −2.936 | 3.248 | 32.55 | 1 |
| 0.05 | −0.1147 | −2.387 | −3.009 | 42.23 | −0.142 | −2.733 | 4.859 | 31.58 | 0.95 |
| 0.1 | −0.2369 | −2.486 | −0.9235 | 41.06 | −0.272 | −2.452 | 6.375 | 28.77 | 0.90 |
| 0.15 | −0.3615 | −2.481 | 1.084 | 39.08 | −0.3861 | −2.099 | 7.709 | 24.31 | 0.85 |
| 0.2 | −0.4834 | −2.379 | 2.971 | 36.28 | −0.4809 | −1.686 | 8.783 | 18.45 | 0.80 |
| 0.25 | −0.5979 | −2.187 | 4.698 | 32.69 | −0.5538 | −1.226 | 9.535 | 11.45 | 0.75 |
| 0.3 | −0.7007 | −1.913 | 6.228 | 28.38 | −0.6030 | −0.7383 | 9.914 | 3.620 | 0.70 |
| 0.35 | −0.7880 | −1.568 | 7.526 | 23.43 | −0.6275 | −0.2415 | 9.888 | −4.703 | 0.65 |
| 0.4 | −0.8565 | −1.164 | 8.562 | 17.97 | −0.6274 | 0.2435 | 9.441 | −13.19 | 0.60 |
| 0.45 | −0.9037 | −0.7162 | 9.316 | 12.12 | −0.6037 | 0.6955 | 8.572 | −21.51 | 0.55 |
| 0.5 | −0.9277 | −0.2378 | 9.771 | 6.043 | −0.5587 | 1.0940 | 7.297 | −29.37 | 0.50 |
| 0.55 | −0.9272 | 0.2557 | 9.919 | −0.1185 | −0.4956 | 1.419 | 5.647 | −36.50 | 0.45 |
| 0.6 | −0.9021 | 0.7490 | 9.761 | −6.195 | −0.4183 | 1.653 | 3.664 | −42.67 | 0.40 |
| 0.65 | −0.8526 | 1.227 | 9.304 | −12.02 | −0.3320 | 1.781 | 1.399 | −47.74 | 0.35 |
| 0.7 | −0.7799 | 1.675 | 8.565 | −17.45 | −0.2422 | 1.789 | −1.090 | −51.61 | 0.30 |
| 0.75 | −0.6859 | 2.079 | 7.568 | −22.32 | −0.1552 | 1.669 | −3.742 | −54.28 | 0.25 |
| 0.8 | −0.5729 | 2.428 | 6.345 | −26.50 | −0.0776 | 1.413 | −6.499 | −55.84 | 0.20 |
| 0.85 | −0.4442 | 2.710 | 4.932 | −29.88 | −0.0162 | 1.018 | −9.310 | −56.45 | 0.15 |
| 0.9 | −0.3031 | 2.918 | 3.372 | −32.36 | +0.0218 | 0.4821 | −12.13 | −56.38 | 0.10 |
| 0.95 | −0.1537 | 3.046 | 1.712 | −33.88 | 0.0295 | −0.1948 | −14.94 | −56.00 | 0.05 |
| 1 | 0.000 | 3.089 | 0.000 | −34.40 | 0.000 | −1.012 | −17.73 | −87.20 | 0 |
| | $-X_1$ | $X_1'$ | $-X_1''$ | $X_1'''$ | $X_2$ | $-X_2'$ | $X_2''$ | $-X_2'''$ | $\alpha$ |

**Third Span**

Third and fourth eigenfunctions      **Second Span**

| $\alpha$ | $X_3$ | $X_3'$ | $X_3''$ | $X_3'''$ | $X_4$ | $X_4'$ | $X_4''$ | $X_4'''$ | |
|---|---|---|---|---|---|---|---|---|---|
| 0 | 0.000 | −1.480 | 17.70 | −49.54 | 0.000 | 0.000 | 22.37 | −104.0 | 1 |
| 0.05 | −0.0529 | −0.6573 | 15.21 | −50.13 | 0.0257 | 0.9885 | 17.18 | −103.7 | 0.95 |
| 0.1 | −0.0678 | 0.0401 | 12.68 | −51.38 | 0.0945 | 1.718 | 12.02 | −102.3 | 0.90 |
| 0.15 | −0.0510 | 0.6092 | 10.08 | −52.61 | 0.1933 | 2.192 | 6.981 | −98.77 | 0.85 |
| 0.2 | 0.000 | 1.047 | 7.426 | −53.24 | 0.3096 | 2.420 | 2.187 | −92.50 | 0.80 |
| 0.25 | 0.0514 | 1.352 | 4.768 | −52.85 | 0.4315 | 2.418 | −2.219 | −83.22 | 0.75 |
| 0.3 | 0.1238 | 1.525 | 2.163 | −51.12 | 0.5479 | 2.207 | −6.085 | −70.94 | 0.70 |
| 0.35 | 0.2017 | 1.570 | −0.3187 | −47.90 | 0.6493 | 1.820 | −9.267 | −55.92 | 0.65 |
| 0.4 | 0.2788 | 1.496 | −2.601 | −43.12 | 0.7276 | 1.294 | −11.64 | −38.63 | 0.60 |
| 0.45 | 0.3495 | 1.315 | −4.606 | −36.87 | 0.7771 | 0.6717 | −13.10 | −19.74 | 0.55 |
| 0.5 | 0.4088 | 1.041 | −6.266 | −29.33 | 0.7940 | 0.000 | −13.60 | 0.00 | 0.50 |
| 0.55 | 0.4524 | 0.6947 | −7.522 | −20.75 | 0.7771 | −0.6717 | −13.10 | 19.74 | 0.45 |
| 0.6 | 0.4774 | 0.2964 | −8.330 | −11.50 | 0.7276 | −1.294 | −11.64 | 38.63 | 0.40 |
| 0.65 | 0.4816 | −0.1304 | −8.667 | −1.947 | 0.6493 | −1.820 | −9.267 | 55.92 | 0.35 |
| 0.7 | 0.4643 | −0.5622 | −8.527 | 7.472 | 0.5479 | −2.207 | −6.085 | 70.94 | 0.30 |
| 0.75 | 0.4257 | −0.9755 | −7.929 | 16.34 | 0.4315 | −2.418 | −2.219 | 83.22 | 0.25 |
| 0.8 | 0.3674 | −1.348 | −6.910 | 24.23 | 0.3096 | −2.420 | 2.187 | 92.50 | 0.20 |
| 0.85 | 0.2919 | −1.660 | −5.528 | 30.80 | 0.1933 | −2.192 | 6.981 | 98.77 | 0.15 |
| 0.9 | 0.2026 | −1.896 | −3.857 | 35.73 | 0.0945 | −1.718 | 12.02 | 102.3 | 0.10 |
| 0.95 | 0.1038 | −2.043 | −1.986 | 38.78 | 0.0257 | −0.9885 | 17.18 | 103.7 | 0.05 |
| 1 | 0.000 | −2.093 | −0.0127 | 39.93 | 0.000 | 0.000 | 22.37 | 104.0 | 0 |
| | $-X_3$ | $X_3'$ | $-X_3''$ | $X_3'''$ | $X_4$ | $-X_4'$ | $X_4''$ | $-X_4'''$ | $\alpha$ |

**Third Span**

**B.7** Two–span beam.

First and second eigenfunctions

### First Span

| $\alpha$ | $X_1$ | $X_1'$ | $X_1''$ | $X_1'''$ | $X_2$ | $X_2'$ | $X_2''$ | $X_2'''$ |
|---|---|---|---|---|---|---|---|---|
| 0 | 0.000 | 4.368 | 0.000 | −48.63 | 0.000 | 2.959 | 0.000 | −56.37 |
| 0.05 | 0.2173 | 4.307 | −2.419 | −47.91 | 0.1467 | 2.889 | −2.794 | −54.90 |
| 0.1 | 0.4287 | 4.127 | −4.767 | −45.76 | 0.2866 | 2.682 | −5.443 | −50.59 |
| 0.15 | 0.6281 | 3.833 | −6.973 | −42.25 | 0.4128 | 2.349 | −7.808 | −43.62 |
| 0.2 | 0.8102 | 3.433 | −8.971 | −37.48 | 0.5197 | 1.908 | −9.766 | −34.33 |
| 0.25 | 0.9699 | 2.940 | −10.70 | −31.56 | 0.6022 | 1.381 | −11.21 | −23.16 |
| 0.3 | 1.103 | 2.368 | −12.11 | −24.68 | 0.6568 | 0.7968 | −12.06 | −10.62 |
| 0.35 | 1.206 | 1.735 | −13.16 | −17.01 | 0.6814 | 0.1861 | −12.26 | 2.706 |
| 0.4 | 1.276 | 1.059 | −13.80 | −8.765 | 0.6756 | −0.4178 | −11.78 | 16.22 |
| 0.45 | 1.311 | 0.3618 | −14.03 | 0.1719 | 0.6403 | −0.9813 | −10.64 | 29.32 |
| 0.5 | 1.312 | −0.3361 | −13.82 | 8.541 | 0.5786 | −1.471 | −8.87 | 41.45 |
| 0.55 | 1.278 | −1.013 | −13.17 | 17.14 | 0.4949 | −1.859 | −6.523 | 52.14 |
| 0.6 | 1.211 | −1.647 | −12.11 | 25.41 | 0.3949 | −2.116 | −3.687 | 60.98 |
| 0.65 | 1.114 | −2.217 | −10.64 | 33.13 | 0.2858 | −2.221 | −0.4597 | 67.75 |
| 0.7 | 0.991 | −2.705 | −8.808 | 40.13 | 0.1756 | −2.157 | 3.051 | 72.32 |
| 0.75 | 0.8456 | −3.092 | −6.645 | 46.22 | 0.0731 | −1.913 | 6.737 | 74.77 |
| 0.8 | 0.6837 | −3.364 | −4.202 | 51.30 | −0.0124 | −1.482 | 10.50 | 75.34 |
| 0.85 | 0.5113 | −3.509 | −1.534 | 55.26 | −0.0719 | −0.8634 | 14.25 | 74.45 |
| 0.9 | 0.3351 | −3.515 | −1.304 | 58.07 | −0.0957 | −0.0587 | 17.93 | 72.72 |
| 0.95 | 0.1622 | −3.376 | −4.254 | 59.71 | −0.0747 | 0.9277 | 21.52 | 70.95 |
| 1 | 0.000 | −3.089 | −7.257 | 60.20 | 0.000 | 2.092 | 25.04 | 69.93 |

Third and fourth eigenfunctions

### First Span

| $\alpha$ | $X_3$ | $X_3'$ | $X_3''$ | $X_3'''$ | $X_4$ | $X_4'$ | $X_4''$ | $X_4'''$ |
|---|---|---|---|---|---|---|---|---|
| 0 | 0.000 | 8.211 | 0.000 | −352.3 | 0.000 | 4.931 | 0.000 | −284.8 |
| 0.05 | 0.4032 | 7.775 | −17.30 | −333.6 | 0.2406 | 4.580 | −13.9 | −264.5 |
| 0.1 | 0.7636 | 6.512 | −32.77 | −279.6 | 0.4470 | 3.575 | −25.82 | −206.7 |
| 0.15 | 1.043 | 4.556 | −44.77 | −196.0 | 0.5895 | 2.059 | −34.08 | −119.6 |
| 0.2 | 1.211 | 2.114 | −52.02 | −91.61 | 0.6479 | 0.2475 | −37.50 | −15.59 |
| 0.25 | 1.250 | −0.5545 | −53.77 | 22.36 | 0.6137 | −1.602 | −35.61 | 90.45 |
| 0.3 | 1.156 | −3.168 | −49.83 | 133.8 | 0.4914 | −3.229 | −28.68 | 183.2 |
| 0.35 | 0.9392 | −5.450 | −40.63 | 230.8 | 0.2984 | −4.402 | −17.71 | 249.7 |
| 0.4 | 0.6211 | −7.159 | −27.16 | 303.1 | 0.0615 | −4.959 | −4.307 | 280.0 |
| 0.45 | 0.2358 | −8.118 | −10.86 | 342.8 | −0.1859 | −4.825 | 9.605 | 269.6 |
| 0.5 | −0.1764 | −8.228 | 6.503 | 345.6 | −0.4097 | −4.024 | 21.99 | 219.6 |
| 0.55 | −0.5726 | −7.481 | 23.07 | 310.9 | −0.5793 | −2.682 | 31.01 | 136.5 |
| 0.6 | −0.9116 | −5.965 | 37.02 | 242.2 | −0.6723 | −1.003 | 35.27 | 31.43 |
| 0.65 | −1.159 | −3.848 | 46.84 | 146.4 | −0.6783 | 0.7528 | 34.01 | −81.94 |
| 0.7 | −1.290 | −1.369 | 51.37 | 33.05 | −0.6004 | 2.305 | 27.18 | −189.2 |
| 0.75 | −1.295 | 1.191 | 50.03 | −86.55 | −0.4556 | 3.388 | 15.41 | −277.6 |
| 0.8 | −1.175 | 3.536 | 42.81 | −200.8 | −0.2731 | 3.783 | −0.1174 | −338.4 |
| 0.85 | −0.9493 | 5.382 | 30.23 | −298.9 | −0.0913 | 3.339 | −17.91 | −368.4 |
| 0.9 | −0.6491 | 6.486 | 13.32 | −372.7 | 0.0454 | 1.979 | −36.50 | −371.2 |
| 0.95 | −0.3162 | 6.665 | −6.551 | −417.1 | 0.0911 | −3.3054 | −54.77 | −358.3 |
| 1 | 0.000 | 5.807 | −27.88 | −427.0 | 0.000 | −3.4860 | −72.38 | −340.9 |

**B.7** (*cont'd*) Two–span beam.
First and second eigenfunctions

**Second Span**

| $\alpha = \dfrac{x}{l}$ | $X_1$ | $X_1'$ | $X_1''$ | $X_1'''$ | $X_2$ | $X_2'$ | $X_2''$ | $X_2'''$ |
|---|---|---|---|---|---|---|---|---|
| 0 | 0.000 | −3.089 | 7.257 | 8.538 | 0.000 | 2.092 | 25.04 | −149.8 |
| 0.05 | −0.1451 | −2.715 | 7.676 | 8.047 | 0.1327 | 3.157 | 17.57 | −148.6 |
| 0.1 | −0.2711 | −2.322 | 8.047 | 6.656 | 0.3094 | 3.851 | 10.23 | −144.2 |
| 0.15 | −0.3771 | −1.912 | 8.329 | 4.496 | 0.5118 | 4.185 | 3.206 | −136.1 |
| 0.2 | −0.4622 | −1.491 | 8.486 | 1.703 | 0.7223 | 4.180 | −3.311 | −123.9 |
| 0.25 | −0.5261 | −1.066 | 8.491 | −1.584 | 0.9246 | 3.866 | −9.113 | −107.5 |
| 0.3 | −0.5688 | −0.6449 | 8.322 | −5.225 | 1.104 | 3.284 | −14.00 | −87.34 |
| 0.35 | −0.5908 | −0.2369 | 7.965 | −9.080 | 1.249 | 2.484 | −17.79 | −63.92 |
| 0.4 | −0.5929 | 0.1482 | 7.412 | −13.01 | 1.350 | 1.525 | −20.35 | −38.06 |
| 0.45 | −0.5765 | 0.5009 | 6.664 | −16.90 | 1.400 | 0.4713 | −21.57 | −10.68 |
| 0.5 | −0.5435 | 0.8115 | 5.725 | −20.62 | 1.397 | −0.609 | −21.41 | 17.16 |
| 0.55 | −0.4962 | 1.071 | 4.606 | −24.07 | 1.340 | −1.647 | −19.87 | 44.36 |
| 0.6 | −0.4375 | 1.269 | 3.324 | −27.17 | 1.234 | −2.574 | −17.00 | 70.00 |
| 0.65 | −0.3704 | 1.400 | 1.896 | −29.86 | 1.086 | −3.326 | −12.91 | 93.08 |
| 0.7 | −0.2987 | 1.457 | 0.3458 | −32.07 | 0.9052 | −3.847 | −7.746 | 112.9 |
| 0.75 | −0.2261 | 1.433 | −1.303 | −33.81 | 0.7057 | −4.086 | −1.685 | 128.9 |
| 0.8 | −0.1567 | 1.325 | −3.028 | −35.08 | 0.5020 | −4.004 | 5.074 | 140.8 |
| 0.85 | −0.0950 | 1.130 | −4.804 | −35.91 | 0.3112 | −3.570 | 12.33 | 148.9 |
| 0.9 | −0.0453 | 0.8444 | −6.612 | −36.37 | 0.1512 | −2.765 | 19.90 | 153.4 |
| 0.95 | −0.0121 | 0.4682 | −8.436 | −36.55 | 0.041 | −1.577 | 27.63 | 155.2 |
| 1 | 0.000 | 0.000 | −10.260 | −36.57 | 0.000 | 0.000 | 35.40 | 155.5 |

Third and fourth eigenfunctions  **Second Span**

| $\alpha$ | $X_3$ | $X_3'$ | $X_3''$ | $X_3'''$ | $X_4$ | $X_4'$ | $X_4''$ | $X_4'''$ |
|---|---|---|---|---|---|---|---|---|
| 0 | 0.000 | 5.807 | −27.88 | −67.02 | 0.000 | −3.486 | −72.38 | 751.2 |
| 0.05 | 0.2541 | 4.332 | −31.02 | −54.80 | −0.2491 | −6.170 | −35.12 | 732.4 |
| 0.1 | 0.4309 | 2.723 | −33.03 | −22.75 | −0.5865 | −7.033 | 0.0131 | 663.6 |
| 0.15 | 0.5255 | 1.061 | −33.09 | 21.78 | −0.9249 | −6.250 | 30.27 | 537.5 |
| 0.2 | 0.5379 | −0.5456 | −30.77 | 71.19 | −1.189 | −4.133 | 52.91 | 360.5 |
| 0.25 | 0.4739 | −1.975 | −26.01 | 118.2 | −1.323 | −1.123 | 65.76 | 149.7 |
| 0.3 | 0.3453 | −3.111 | −19.10 | 156.2 | −1.295 | 2.260 | 67.73 | −70.04 |
| 0.35 | 0.1693 | −3.859 | −10.63 | 180.1 | −1.100 | 5.472 | 59.06 | −271.2 |
| 0.4 | −0.0331 | −4.161 | −1.387 | 186.5 | −0.7593 | 8.015 | 41.36 | −427.4 |
| 0.45 | −0.2390 | −4.001 | 7.701 | 173.9 | −0.3163 | 9.504 | 17.43 | −517.7 |
| 0.5 | −0.4259 | −3.409 | 15.70 | 143.2 | 0.1696 | 9.715 | −9.102 | −530.0 |
| 0.55 | −0.5740 | −2.463 | 21.76 | 96.93 | 0.6331 | 8.617 | −34.24 | −462.6 |
| 0.6 | −0.6682 | −1.277 | 25.21 | 39.48 | 1.012 | 6.378 | −54.18 | −324.5 |
| 0.65 | −0.7000 | 0.000 | 25.61 | −23.80 | 1.257 | 3.338 | −65.81 | −133.9 |
| 0.7 | −0.6685 | 1.231 | 22.83 | −87.06 | 1.340 | −0.0304 | −67.11 | 84.16 |
| 0.75 | −0.5805 | 2.239 | 17.00 | −144.8 | 1.258 | −3.189 | −57.40 | 302.1 |
| 0.8 | −0.4506 | 2.886 | 8.52 | −192.3 | 1.034 | −5.597 | −37.33 | 494.1 |
| 0.85 | −0.2998 | 3.056 | −2.017 | −226.8 | 0.7187 | −6.780 | −8.752 | 640.5 |
| 0.9 | −0.1544 | 2.662 | −13.93 | −247.5 | 0.3827 | −6.373 | 25.78 | 731.6 |
| 0.95 | −0.0439 | 1.651 | −26.57 | −256.2 | 0.1118 | −4.148 | 63.54 | 771.2 |
| 1 | 0.000 | 0.000 | −39.43 | −257.6 | 0.000 | 0.000 | 102.3 | 777.6 |

**B.8**   Three–span beam.
First and second eigenfunctions

**First Span**

| $\alpha$ | $X_1$ | $X_1'$ | $X_1''$ | $X_1'''$ | $X_2$ | $X_2'$ | $X_2''$ | $X_2'''$ |
|---|---|---|---|---|---|---|---|---|
| 0 | 0.000 | 3.474 | 0.000 | −36.00 | 0.000 | 3.297 | 0.000 | −48.08 |
| 0.05 | 0.1729 | 3.429 | −1.792 | −35.52 | 0.1638 | 3.237 | −2.388 | −47.10 |
| 0.1 | 0.3414 | 3.296 | −3.536 | −34.09 | 0.3217 | 3.060 | −4.678 | −44.21 |
| 0.15 | 0.5010 | 3.076 | −5.186 | −31.75 | 0.4679 | 2.772 | −6.778 | −39.50 |
| 0.2 | 0.6478 | 2.779 | −6.697 | −28.56 | 0.5973 | 2.387 | −8.601 | −33.15 |
| 0.25 | 0.7778 | 2.410 | −8.028 | −24.59 | 0.7052 | 1.918 | −10.07 | −25.38 |
| 0.3 | 0.8878 | 1.980 | −9.145 | −19.97 | 0.7881 | 1.387 | −11.12 | −16.48 |
| 0.35 | 0.9750 | 1.500 | −10.01 | −14.79 | 0.8432 | 0.8139 | −11.70 | −6.753 |
| 0.4 | 1.037 | 1.127 | −10.62 | −9.200 | 0.8692 | 0.2245 | −11.79 | 3.456 |
| 0.45 | 1.073 | 0.4430 | −10.93 | −3.337 | 0.8658 | −0.3561 | −11.36 | 13.80 |
| 0.5 | 1.081 | −0.1051 | −10.95 | 2.648 | 0.8341 | −0.9024 | −10.41 | 23.93 |
| 0.55 | 1.062 | −0.6468 | −10.67 | 8.605 | 0.7765 | −1.389 | −8.972 | 33.53 |
| 0.6 | 1.017 | −1.167 | −10.09 | 14.38 | 0.6966 | −1.792 | −7.072 | 42.31 |
| 0.65 | 0.9461 | −1.651 | −9.234 | 19.84 | 0.5991 | −2.089 | −4.759 | 50.02 |
| 0.7 | 0.8527 | −2.086 | −8.115 | 24.84 | 0.4898 | −2.262 | −2.090 | 56.51 |
| 0.75 | 0.7388 | −2.459 | −6.760 | 29.26 | 0.3753 | −2.293 | 0.8696 | 61.65 |
| 0.8 | 0.6081 | −2.759 | −5.201 | 33.00 | 0.2630 | −2.171 | 4.052 | 65.44 |
| 0.85 | 0.4643 | −2.976 | −4.193 | 35.98 | 0.1609 | −1.886 | 7.392 | 67.95 |
| 0.9 | 0.3119 | −3.104 | −1.617 | 38.13 | 0.0772 | −1.430 | 10.83 | 69.34 |
| 0.95 | 0.1555 | −3.136 | 0.3258 | 39.43 | 0.0207 | −0.802 | 14.31 | 69.89 |
| 1 | 0.000 | −3.071 | 2.312 | 39.83 | 0.000 | 0.000 | 17.81 | 69.87 |

Third and fourth eigenfunctions      **First Span**

| $\alpha$ | $X_3$ | $X_3'$ | $X_3''$ | $X_3'''$ | $X_4$ | $X_4'$ | $X_4''$ | $X_4'''$ |
|---|---|---|---|---|---|---|---|---|
| 0 | 0.000 | 1.816 | 0.000 | −36.95 | 0.000 | 6.969 | 0.00 | −286.6 |
| 0.05 | 0.0900 | 1.770 | −1.831 | −35.94 | 0.3424 | 6.613 | −14.08 | −272.0 |
| 0.1 | 0.1755 | 1.635 | −3.561 | −32.95 | 0.6500 | 5.584 | −26.74 | −229.7 |
| 0.15 | 0.2521 | 1.417 | −5.095 | −28.13 | 0.8913 | 3.985 | −36.67 | −164.1 |
| 0.2 | 0.3161 | 1.130 | −6.348 | −21.74 | 1.042 | 1.980 | −42.87 | −81.81 |
| 0.25 | 0.3642 | 0.7885 | −7.248 | −14.08 | 1.086 | −0.2284 | −44.71 | 8.771 |
| 0.3 | 0.3943 | 0.4120 | −7.741 | −5.543 | 1.019 | −2.415 | −42.01 | 98.40 |
| 0.35 | 0.4052 | 0.0217 | −7.794 | 3.455 | 0.8485 | −4.358 | −35.04 | 177.9 |
| 0.4 | 0.3966 | −0.3598 | −7.395 | 12.48 | 0.5909 | −5.860 | −24.52 | 239.2 |
| 0.45 | 0.3697 | −0.7104 | −6.553 | 21.10 | 0.2725 | −6.768 | −11.53 | 276.0 |
| 0.5 | 0.3265 | −1.008 | −2.342 | 28.93 | −0.0744 | −6.993 | 2.602 | 284.4 |
| 0.55 | 0.2701 | −1.234 | −3.679 | 35.64 | −0.4150 | −6.513 | 16.42 | 263.6 |
| 0.6 | 0.2045 | −1.371 | −1.757 | 40.98 | −0.7149 | −5.380 | 28.50 | 215.5 |
| 0.65 | 0.1347 | −1.406 | 0.3935 | 44.78 | −0.9441 | −3.713 | 37.59 | 144.9 |
| 0.7 | 0.0658 | −1.329 | 2.695 | 47.02 | −1.080 | −1.687 | 42.73 | 58.72 |
| 0.75 | 0.000 | −1.135 | 5.071 | 47.78 | −1.110 | 0.4844 | 43.35 | −34.49 |
| 0.8 | −0.0456 | −0.822 | 7.453 | 47.29 | −1.033 | 2.570 | 39.31 | −125.7 |
| 0.85 | −0.0764 | −0.3908 | 9.784 | 45.88 | −0.8586 | 4.344 | 30.96 | −206.2 |
| 0.9 | −0.0828 | 0.1549 | 12.03 | 44.04 | −0.6073 | 5.606 | 19.00 | −268.5 |
| 0.95 | −0.0591 | 0.8109 | 14.19 | 42.39 | −0.3091 | 6.201 | 4.500 | −307.4 |
| 1 | 0.000 | 1.573 | 16.29 | 41.53 | 0.000 | 6.034 | −11.30 | −317.6 |

**B.8**  (*cont'd*) Three–span beam.
First and second eigenfunctions

**Second Span**

| $\alpha$ | $X_1$ | $X_1'$ | $X_1''$ | $X_1'''$ | $X_2$ | $X_2'$ | $X_2''$ | $X_2'''$ |
|---|---|---|---|---|---|---|---|---|
| 0 | 0.000 | −3.071 | 2.312 | 23.78 | 0.000 | 0.000 | 17.81 | −70.00 |
| 0.05 | −0.1501 | −2.925 | 4.213 | 23.36 | 0.0208 | 0.8039 | 14.31 | −69.92 |
| 0.1 | −0.2915 | −2.722 | 4.634 | 22.13 | 0.0774 | 1.432 | 10.82 | −69.37 |
| 0.15 | −0.4214 | −2.463 | 5.694 | 20.14 | 0.1611 | 1.887 | 7.388 | −67.97 |
| 0.2 | −0.5370 | −2.154 | 6.637 | 17.48 | 0.2633 | 2.173 | 4.047 | −65.46 |
| 0.25 | −0.6361 | −1.802 | 7.432 | 14.22 | 0.3757 | 2.295 | 0.8635 | −61.67 |
| 0.3 | −0.7166 | −1.414 | 8.050 | 10.46 | 0.4902 | 2.263 | −2.097 | −56.52 |
| 0.35 | −0.7771 | −1.000 | 8.471 | 6.308 | 0.5996 | 2.090 | −4.766 | −50.03 |
| 0.4 | −0.8164 | −0.5707 | 8.677 | 1.881 | 0.6971 | 1.792 | −7.079 | −42.32 |
| 0.45 | −0.8341 | −0.1364 | 8.656 | −2.706 | 0.7771 | 1.389 | −8.979 | −33.52 |
| 0.5 | −0.8301 | 0.291 | 8.405 | −7.330 | 0.8346 | 0.9019 | −10.42 | −23.92 |
| 0.55 | −0.8053 | 0.7002 | 7.925 | −11.87 | 0.8663 | 0.3553 | −11.36 | −13.78 |
| 0.6 | −0.7606 | 1.080 | 7.221 | −16.22 | 0.8696 | −0.2256 | −11.79 | −3.430 |
| 0.65 | −0.6979 | 1.419 | 6.307 | −20.28 | 0.8436 | −0.8152 | −11.71 | 6.783 |
| 0.7 | −0.6196 | 1.707 | 5.200 | −23.94 | 0.7884 | −1.388 | −11.12 | 16.51 |
| 0.75 | −0.5282 | 1.936 | 3.922 | −27.12 | 0.7055 | −1.920 | −10.07 | 25.42 |
| 0.8 | −0.4271 | 2.097 | 2.497 | −29.78 | 0.5975 | −2.388 | −8.599 | 33.19 |
| 0.85 | −0.3197 | 2.184 | 0.9538 | −31.85 | 0.4680 | −2.774 | −6.774 | 39.54 |
| 0.9 | −0.2100 | 2.191 | −0.6778 | −33.32 | 0.3217 | −3.061 | −4.672 | 44.25 |
| 0.95 | −0.1020 | 2.115 | −2.368 | −34.18 | 0.1638 | −3.238 | −2.379 | 47.15 |
| 1 | 0.000 | 1.954 | −4.086 | −34.44 | 0.000 | −3.297 | 0.0106 | 48.06 |

Third and fourth eigenfunctions          **Second Span**

| $\alpha$ | $X_3$ | $X_3'$ | $X_3''$ | $X_3'''$ | $X_4$ | $X_4'$ | $X_4''$ | $X_4'''$ |
|---|---|---|---|---|---|---|---|---|
| 0 | 0.000 | 1.573 | 16.29 | −105.7 | 0.000 | 6.034 | −11.30 | −175.9 |
| 0.05 | 0.0968 | 2.256 | 11.02 | −104.6 | 0.2839 | 5.252 | −19.89 | −163.6 |
| 0.1 | 0.2211 | 2.677 | 5.865 | −101.1 | 0.5183 | 4.065 | −27.30 | −129.4 |
| 0.15 | 0.3602 | 2.846 | 0.9587 | −94.61 | 0.6850 | 2.558 | −32.54 | −78.03 |
| 0.2 | 0.5018 | 2.779 | −3.543 | −84.94 | 0.7709 | 0.8586 | −34.92 | −15.98 |
| 0.25 | 0.6347 | 2.501 | −7.484 | −72.18 | 0.7702 | −0.8800 | −34.08 | 49.68 |
| 0.3 | 0.7489 | 2.043 | −10.71 | −56.62 | 0.685 | −2.495 | −30.01 | 111.7 |
| 0.35 | 0.8366 | 1.444 | −13.11 | −38.80 | 0.5253 | −3.833 | −23.08 | 163.2 |
| 0.4 | 9.8917 | 0.7477 | −14.57 | −19.36 | 0.3084 | −4.767 | −13.96 | 198.8 |
| 0.45 | 0.9106 | 0.000 | −15.03 | 0.9114 | 0.0569 | −5.208 | −3.543 | 214.3 |
| 0.5 | 0.8921 | −0.7383 | −14.48 | 21.19 | −0.2034 | −5.117 | 7.109 | 208.1 |
| 0.55 | 0.8376 | −1.427 | −12.92 | 40.64 | −0.4462 | −4.511 | 16.91 | 180.5 |
| 0.6 | 0.7510 | −2.015 | −10.44 | 58.50 | −0.6470 | −3.458 | 24.84 | 134.0 |
| 0.65 | 0.6385 | −2.457 | −7.113 | 74.12 | −0.7864 | −2.072 | 30.06 | 72.97 |
| 0.7 | 0.5084 | −2.715 | −3.073 | 87.00 | −0.8513 | −0.5064 | 31.99 | 3.299 |
| 0.75 | 0.3707 | −2.755 | 1.537 | 96.86 | −0.8369 | 1.067 | 30.36 | −68.51 |
| 0.8 | 0.2369 | −2.554 | 6.562 | 103.6 | −0.7474 | 2.471 | 25.22 | −135.9 |
| 0.85 | 0.1196 | −2.094 | 11.85 | 107.6 | −0.5955 | 3.537 | 16.94 | −192.9 |
| 0.9 | 0.0319 | −1.366 | 17.28 | 109.2 | −0.4017 | 4.123 | 6.171 | −235.2 |
| 0.95 | −0.0124 | −0.3655 | 22.75 | 109.4 | −0.1929 | 4.126 | −6.291 | −260.3 |
| 1 | 0.000 | 0.9086 | 28.21 | 108.8 | 0.000 | 3.481 | −19.57 | −265.8 |

**B.8**  (*cont'd*) Three–span beam.
First and second eigenfunctions

**Third Span**

| $\alpha = \dfrac{x}{l}$ | $X_1$ | $X_1'$ | $X_1''$ | $X_1'''$ | $X_2$ | $X_2'$ | $X_2''$ | $X_2'''$ |
|---|---|---|---|---|---|---|---|---|
| 0 | 0.000 | 1.954 | −4.086 | −6.030 | 0.000 | −3.297 | 0.0106 | 48.04 |
| 0.05 | 0.0924 | 1.742 | −4.383 | −5.768 | −0.1638 | −3.237 | 2.396 | 47.06 |
| 0.1 | 0.1739 | 1.516 | −4.655 | −5.024 | −0.3217 | −3.059 | 4.685 | 44.17 |
| 0.15 | 0.2438 | 1.277 | −4.879 | −3.861 | −0.4678 | −2.771 | 6.783 | 39.46 |
| 0.2 | 0.3015 | 1.029 | −5.035 | −2.343 | −0.5971 | −2.385 | 8.604 | 33.11 |
| 0.25 | 0.3466 | 0.775 | −5.108 | −0.540 | −0.7050 | −1.917 | 10.07 | 25.34 |
| 0.3 | 0.3790 | 0.5198 | −5.086 | 1.477 | −0.7878 | −1.385 | 11.12 | 16.44 |
| 0.35 | 0.3986 | 0.2682 | −4.958 | 3.639 | −0.8428 | −0.8127 | 11.70 | 6.723 |
| 0.4 | 0.4059 | 0.0258 | −4.720 | 5.876 | −0.8688 | −0.2234 | 11.78 | −3.481 |
| 0.45 | 0.4015 | −0.2019 | −4.370 | 8.120 | −0.8653 | 0.3560 | 11.35 | −13.82 |
| 0.5 | 0.3861 | −0.4093 | −3.909 | 10.31 | −0.8336 | 0.9030 | 10.40 | −23.94 |
| 0.55 | 0.3610 | −0.5910 | −3.341 | 12.38 | −0.7760 | 1.389 | 8.965 | −33.54 |
| 0.6 | 0.3275 | −0.7418 | −2.674 | 14.30 | −0.6961 | 1.792 | 7.065 | −42.31 |
| 0.65 | 0.2874 | −0.8569 | −1.915 | 16.00 | −0.5986 | 2.089 | 4.752 | −50.02 |
| 0.7 | 0.2425 | −0.9320 | −1.077 | 17.47 | −0.4893 | 2.261 | 2.083 | −56.50 |
| 0.75 | 0.1949 | −0.9635 | −0.172 | 18.69 | −0.3748 | 2.292 | −0.8756 | −61.64 |
| 0.8 | 0.1469 | −0.9483 | 0.787 | 19.64 | −0.2626 | 2.170 | −4.058 | −65.42 |
| 0.85 | 0.1009 | −0.8841 | 1.787 | 20.32 | −0.1606 | 1.884 | −7.396 | −67.92 |
| 0.9 | 0.0593 | −0.7692 | 2.815 | 20.76 | −0.0770 | 1.429 | −10.83 | −69.31 |
| 0.95 | 0.0248 | −0.6023 | 3.860 | 20.99 | −0.0206 | 0.800 | −14.31 | −69.86 |
| 1 | 0.000 | 0.000 | 4.912 | 21.06 | 0.000 | 0.000 | −17.81 | −69.94 |

Third and fourth eigenfunctions     **Third Span**

| $\alpha$ | $X_3$ | $X_3'$ | $X_3''$ | $X_3'''$ | $X_4$ | $X_4'$ | $X_4''$ | $X_4'''$ |
|---|---|---|---|---|---|---|---|---|
| 0 | 0.000 | 0.9086 | 28.21 | −146.1 | 0.000 | 3.481 | −19.57 | −18.03 |
| 0.05 | 0.0776 | 2.137 | 20.92 | −145.3 | 0.1492 | 2.481 | −20.36 | −11.3 |
| 0.1 | 0.2076 | 3.002 | 13.72 | −142.2 | 0.2476 | 1.455 | −20.54 | 5.732 |
| 0.15 | 0.3719 | 3.513 | 6.756 | −135.8 | 0.2949 | 0.4445 | −19.68 | 28.99 |
| 0.2 | 0.5532 | 3.685 | 0.2098 | −125.4 | 0.2933 | −0.4929 | −17.61 | 54.14 |
| 0.25 | 0.7352 | 3.544 | −5.716 | −110.9 | 0.2479 | −1.296 | −14.30 | 77.27 |
| 0.3 | 0.9030 | 3.127 | −10.82 | −92.55 | 0.1669 | −1.906 | −9.969 | 94.99 |
| 0.35 | 1.044 | 2.479 | −14.91 | −70.66 | 0.0612 | −2.281 | −4.938 | 104.8 |
| 0.4 | 1.148 | 1.655 | −17.84 | −46.01 | −0.0567 | −2.396 | 0.3468 | 105.0 |
| 0.45 | 1.208 | 0.7165 | −19.48 | −19.52 | −0.1739 | −2.250 | 5.393 | 95.20 |
| 0.5 | 1.219 | −0.2706 | −19.78 | 7.766 | −0.2778 | −1.869 | 9.709 | 75.99 |
| 0.55 | 1.181 | −1.238 | −18.71 | 34.75 | −0.3576 | −1.299 | 12.86 | 48.97 |
| 0.6 | 1.096 | −2.120 | −16.32 | 60.36 | −0.4056 | −0.608 | 14.51 | 16.50 |
| 0.65 | 0.9714 | −2.851 | −12.71 | 83.61 | −0.4177 | 0.1243 | 14.47 | −18.51 |
| 0.7 | 0.8148 | −3.373 | −8.018 | 103.7 | −0.394 | 0.8100 | 12.67 | −53.01 |
| 0.75 | 0.6384 | −3.637 | −2.409 | 120.0 | −0.3389 | 1.364 | 9.224 | −84.14 |
| 0.8 | 0.4561 | −3.602 | 3.914 | 132.3 | −0.2611 | 1.709 | 4.353 | −109.6 |
| 0.85 | 0.2837 | −3.237 | 10.75 | 140.5 | −0.1726 | 1.781 | −1.616 | −127.9 |
| 0.9 | 0.1382 | −2.522 | 17.91 | 145.2 | −0.0883 | 1.535 | −8.315 | −138.8 |
| 0.95 | 0.0376 | −1.444 | 25.22 | 147.1 | −0.0249 | 0.943 | −15.39 | −143.4 |
| 1 | 0.000 | 0.000 | 32.58 | 147.3 | 0.000 | 0.000 | −22.59 | −144.1 |

# APPENDIX C

# SOME USEFUL DEFINITE INTEGRALS

Integrals containing the sin and cos functions

$$\int_0^{l/2} \sin^2 \frac{\pi x}{l}\, dx = \int_0^{l/2} \sin^2 \frac{3\pi x}{l}\, dx = \frac{l}{4}$$

$$\int_0^{l/2} \cos^2 \frac{\pi x}{l}\, dx = \frac{l}{4}$$

$$\int_0^{l} \sin^2 \frac{\pi x}{2l}\, dx = \frac{l}{2}$$

$$\int_0^{l} \cos^2 \frac{\pi x}{l}\, dx = \int_0^{l} \cos^2 \frac{3\pi x}{l}\, dx = \frac{l}{2}$$

$$\int_0^{l} \sin \frac{\pi x}{l} \sin \frac{2\pi x}{l}\, dx = 0$$

$$\int_0^{l/2} \cos^2 \frac{\pi x}{2l}\, dx = \frac{l}{2}\left(\frac{1}{2}+\frac{1}{\pi}\right)$$

$$\int_0^{l/2} \sin \frac{\pi x}{l} \sin \frac{3\pi x}{l}\, dx = 0$$

$$\int_{l/2}^{l} \cos^2 \frac{\pi x}{2l}\, dx = \frac{l}{2}\left(\frac{1}{2}-\frac{1}{\pi}\right)$$

$$\int_{l/3}^{2l/3} \sin^2 \frac{\pi x}{l}\, dx = l\left(\frac{1}{6}+\frac{\sqrt{3}}{4\pi}\right)$$

$$\int_0^{l} \cos^2 \frac{\pi x}{2l}\, dx = \frac{l}{2}$$

$$\int_0^{l} \sin^2 \frac{\pi x}{l}\, dx = \int_0^{l} \sin^2 \frac{2\pi x}{l}\, dx = \frac{l}{2}$$

$$\int_0^{l} \cos \frac{\pi x}{l} \cos \frac{3\pi x}{l}\, dx = 0$$

$$\int_0^{l/2} x \sin^2 \frac{\pi x}{l}\, dx = \frac{l^2}{4}\left(\frac{1}{4}+\frac{1}{\pi^2}\right)$$

$$\int_0^{l} \cos^2 \frac{\pi x}{2l}\, dx = \frac{l}{2}$$

$$\int_0^{l/2} x \sin^2 \frac{3\pi x}{l}\, dx = \frac{l^2}{4}\left(\frac{1}{4}+\frac{1}{9\pi^2}\right)$$

$$\int_0^{l} x \cos \frac{\pi x}{l}\, dx = -\frac{2l^2}{\pi^2}$$

$$\int_0^{l} x \sin \frac{\pi x}{l} \sin \frac{2\pi x}{l}\, dx = -\frac{8l^2}{9\pi^2}$$

$$\int_0^{l} x \cos^2 \frac{\pi x}{l}\, dx = \frac{l^2}{4}$$

$$\int_0^{l} x \sin^2 \frac{\pi x}{l}\, dx = \int_0^{l} x \sin^2 \frac{2\pi x}{l}\, dx = \frac{l^2}{4}$$

$$\int_0^{l} \cos^2 \frac{3\pi x}{2l}\, dx = \frac{l}{2}$$

Integrals containing the sin and cos functions

$$\int_0^l x \sin\frac{\pi x}{l}\,dx = \frac{l^2}{\pi^2} \qquad\qquad \int_0^l x \cos^2\frac{\pi x}{2l}\,dx = l^2\left(\frac{1}{4}-\frac{1}{\pi^2}\right)$$

$$\int_0^{l/2} x \sin\frac{\pi x}{l}\,dx = \frac{l^2}{\pi^2} \qquad\qquad \int_0^l x \cos\frac{\pi x}{2l}\,dx = 2l^2\left(\frac{1}{\pi}-\frac{2}{\pi^2}\right)$$

$$\int_0^{l/2} x^2 \sin^2\frac{\pi x}{l}\,dx = \frac{l^3}{8}\left(\frac{1}{6}+\frac{1}{\pi^2}\right) \qquad \int_0^{l/2} x \cos^2\frac{\pi x}{2l}\,dx = l^2\left(\frac{1}{16}+\frac{1}{4\pi}-\frac{1}{2\pi^2}\right)$$

$$\int_0^{l/2} x^2 \sin^2\frac{3\pi x}{l}\,dx = \frac{l^3}{8}\left(\frac{1}{6}+\frac{1}{9\pi^2}\right) \qquad \int_{l/2}^l x \cos^2\frac{\pi x}{2l}\,dx = l^2\left(\frac{3}{16}-\frac{1}{4\pi}-\frac{1}{2\pi^2}\right)$$

$$\int_0^{l/2} x^2 \sin\frac{\pi x}{l}\sin\frac{3\pi x}{l}\,dx = -\frac{5l^3}{32\pi^2} \qquad \int_0^l x^2 \cos^2\frac{\pi x}{l}\,dx = \frac{l^3}{2}\left(\frac{1}{3}-\frac{1}{2\pi^2}\right)$$

$$\int_0^{l/2} x \sin\frac{\pi x}{l}\sin\frac{3\pi x}{l}\,dx = -\frac{l^2}{4\pi^2} \qquad \int_0^l x^2 \cos^2\frac{\pi x}{2l}\,dx = \frac{l^3}{2}\left(\frac{1}{6}+\frac{1}{\pi^2}\right)$$

$$\int_0^l x^2 \sin^2\frac{\pi x}{l}\,dx = \frac{l^3}{2}\left(\frac{1}{3}-\frac{1}{2\pi^2}\right) \qquad \int_0^{l/2} x^2 \cos^2\frac{\pi x}{l}\,dx = \frac{l^3}{8}\left(\frac{1}{6}-\frac{1}{\pi^2}\right)$$

$$\int_0^l x^2 \sin^2\frac{2\pi x}{l}\,dx = \frac{l^3}{2}\left(\frac{1}{3}-\frac{1}{8\pi^2}\right) \qquad \int_0^l x^2 \cos\frac{2\pi x}{l}\,dx = \frac{l^3}{2\pi^2}$$

$$\int_0^l x^2 \sin\frac{\pi x}{l}\sin\frac{2\pi x}{l}\,dx = -\frac{8l^3}{9\pi^2} \qquad \int_0^{l/2} x^2 \cos^2\frac{\pi x}{2l}\,dx = l^3\left(\frac{1}{48}+\frac{1}{8\pi}-\frac{1}{\pi^3}\right)$$

$$\int_0^l x \sin^2\frac{\pi x}{2l}\,dx = \frac{l^2}{2}\left(\frac{1}{2}+\frac{2}{\pi^2}\right) \qquad \int_{l/2}^l x^2 \cos^2\frac{\pi x}{2l}\,dx = l^3\left(\frac{7}{48}-\frac{1}{8\pi}-\frac{1}{\pi^2}+\frac{1}{\pi^3}\right)$$

$$\int_0^l x^2 \sin\frac{\pi x}{l}\,dx = \frac{l^3}{\pi}\left(1-\frac{4}{\pi^2}\right) \qquad\qquad \int_0^{l/2} \cos^2\frac{3\pi x}{l}\,dx = \frac{l}{4}$$

$$\int_0^l x^2 \sin^2\frac{\pi x}{2l}\,dx = l^3\left(\frac{1}{6}+\frac{1}{\pi^2}\right) \qquad\qquad \int_0^l x^3 \sin^2\frac{\pi x}{l}\,dx = \frac{l^4}{8}\left(1-\frac{3}{\pi^2}\right)$$

$$\int_0^l x^2 \sin^2\frac{\pi x}{l}\,dx = \frac{l^3}{2}\left(\frac{1}{3}-\frac{1}{2\pi^2}\right) \qquad \int_0^l x \sin^2\frac{3\pi x}{2l}\,dx = \frac{l^2}{2}\left(\frac{1}{2}+\frac{2}{9\pi^2}\right)$$

---

Integrals containing the sin and cos functions

$$\int_0^l x^2 \sin^2 \frac{3\pi x}{2l}\, dx = l^3\left(\frac{1}{6} + \frac{2}{9\pi^2}\right) \qquad \int_0^l x^4 \sin^2 \frac{\pi x}{l}\, dx = l^5\left(\frac{1}{10} - \frac{1}{2\pi^2} + \frac{3}{4\pi^4}\right)$$

---

**Some indefinite integrals containing the sin and cos functions**

$$\int \sin^2 u\, du = -\tfrac{1}{4}\sin 2u + \tfrac{1}{2}u + c \qquad\qquad \int \cos^2 u\, du = \tfrac{1}{4}\sin 2u + \tfrac{1}{2}u + c$$

$$\int u \sin u\, du = \sin u - u\cos u + c \qquad\qquad \int_u \cos u\, du = \cos u + u\sin u + c$$

$$\int u^2 \sin u\, du = 2u \sin u - (u^2 - 2)\cos u + c \qquad \int u^2 \cos u\, du - 2u\cos u + (u^2 - 2)\sin u + c$$

$$\int u^3 \sin u\, du = (3u^2 - 6)\sin u - (u^3 - 6u)\cos u + c \qquad \int u^3 \cos u\, du = (3u^2 - 2)\cos u$$
$$+ (u^3 - 6)\sin u + c$$

# SOME ASSUMED FUNCTIONS

| Boundary condition | Assumed functions | |
| --- | --- | --- |

**Pinned–pinned**

**Origin at the end of the beam**     Cases 3, 4, 5 for symmetric vibration

1. $y = a_1 \sin\dfrac{\pi x}{l} + a_2 \sin\dfrac{2\pi x}{l} + a_3 \sin\dfrac{3\pi x}{l} + \cdots$

2. $y = a_1 x(l-x) + a_2 x^2(l-x) + a_3 x(l-x)^2 + \cdots$

3. $y = a_1 \sin\dfrac{\pi x}{l} + a_2 \sin\dfrac{3\pi x}{l} + a_3 \sin\dfrac{5\pi x}{l} + \cdots$

4. $y = a_1 x(l-x) + a_2 x^2(l-x)^2 + a_3 x^3(l-x)^3 + \cdots$

**Origin at the middle of the beam**

5. $y = a_1 \cos\dfrac{\pi x}{l} + a_2 \cos\dfrac{3\pi x}{l} + a_3 \cos\dfrac{5\pi x}{l} + \cdots$

**Clamped–free**

**Origin at the clamped end**

1. $y = a_1\left(1 - \cos\dfrac{\pi x}{l}\right) + a_2\left(1 - \cos\dfrac{3\pi x}{l}\right) + a_3\left(1 - \cos\dfrac{5\pi x}{l}\right) + \cdots$

2. $y = a_1\left(x^2 - \dfrac{1}{6l^2}x^4\right) + a^2\left(x^6 - \dfrac{15}{28l^2}x^8\right) + \cdots$

**Origin at the free end**

3. $y = \left(\dfrac{x}{l} - 1\right)^2\left(a_1 + a_2\dfrac{x}{l} + a_3\dfrac{x^2}{l^2}\cdots\right)$

**Clamped–clamped**

**Origin at the end**

$y = a_1\left(1 - \cos\dfrac{2\pi x}{l}\right) + a_2\left(1 - \cos\dfrac{6\pi x}{l}\right) + a_3\left(1 - \cos\dfrac{10\pi x}{l}\right) + \cdots$

$y = a_1 x^2(l-x)^2 + a_2 x^3(l-x)^3 + \cdots$

**Clamped–pinned**

**Origin at the clamped end**

$y = a_1\left(\cos\dfrac{\pi x}{l} - \cos\dfrac{3\pi x}{2l}\right) + a_2\left(\cos\dfrac{3\pi x}{2l} - \cos\dfrac{5\pi x}{l}\right) + \cdots$

$y = a_1 x^2(l-x) + a_2 x^3(l-x) + \cdots$

# INDEX